README

# README

**A Bookish History of Computing from
Electronic Brains to Everything Machines**

W. PATRICK McCRAY

**The MIT Press**
**Cambridge, Massachusetts**
**London, England**

The MIT Press
Massachusetts Institute of Technology
77 Massachusetts Avenue, Cambridge, MA 02139
mitpress.mit.edu

The MIT Press would like to thank the anonymous peer reviewers who provided comments on drafts of this book. The generous work of academic experts is essential for establishing the authority and quality of our publications. We acknowledge with gratitude the contributions of these otherwise uncredited readers.

This book was set in Adobe Garamond and Berthold Akzidenz Grotesk by Westchester Publishing Services. Printed and bound in the United States of America.

Library of Congress Cataloging-in-Publication Data is available.

ISBN: 978-0-262-55348-3

10  9  8  7  6  5  4  3  2  1

EU Authorised Representative: Easy Access System Europe, Mustamäe tee 50, 10621 Tallinn, Estonia | Email: gpsr.requests@easproject.com

# Contents

I am writing this book on a computer. As I look around my home office, there are two other computers within eyeshot. I can easily spot half a dozen or more, depending on how one defines "computer," silently idling on kitchen counters, tables, and nightstands.

You might be reading this book on some kind of computer. If not, it's quite likely you purchased it via a computer. Perhaps you learned that this book existed in the first place because of a computer's algorithmic recommendation, an email from a colleague, or a post you saw on a social media platform.

Books and computers are inseparable from one another. This combination would have seemed improbable to most people in the late 1940s, when the general public first began learning about the existence of "electronic brains" via books, newspapers, and magazines. But what would have seemed *much* more farfetched to the average person seventy-five years ago is the sheer omnipresence of computers today. At the end of World War II, the number of electronic digital computers in existence could be easily counted on two hands, with some fingers left to spare. British technicians exploited the aptly named Colossus machines to break German codes while their wartime opponents used a digital computer called the Z3 for aeronautical calculations until it was destroyed in an Allied bombing raid. Across the Atlantic, the Electronic Numerical Integrator and Computer (ENIAC, for short) at the University of Pennsylvania performed war-related calculations. At that time, most Americans still understood "computer" to mean a person—most likely a woman—who used paper, pencils, and maybe some type of modest-sized mechanical machine to manually solve equations.[1]

Today, billions of computers are quietly humming away in stores, offices, laboratories, and homes and, with an ease that would have surprised all but the most devoted science fiction readers, found in countless pockets and purses around the planet. These numbers don't even begin to account for the near-invisible devices that control and manage everything from appliances and automobiles to aircraft and assembly lines. Expanding our vision and sharpening our gaze reveals the even vaster digital infrastructure—the billions of routers and servers that make networking, wireless communication, streaming entertainment, and cloud computing possible. It is a constellation of technologies that, frankly, we tend to only notice when it doesn't work.

To say that computers are everywhere is an observation so obvious that it verges on the obtuse. However, this indisputable assertion carries the burden of a significant historical question: *How* did this remarkable technological transformation happen?

There were many causal factors. A political scientist might credit the national security state, its needs for a technological edge honed by the Cold War's urgency. An engineer could gesture to remarkable advances in hardware that enabled computers to become simultaneously smaller and cheaper yet far more powerful. Computer scientists, a professional community that emerged in tandem with the digital computer, might credit esoteric developments in logic and algorithms while programmers (another new, postwar profession) might point to accomplishments in coding and software. An urban planner could credit regional development that transformed places such as the San Francisco Bay Area, Boston, or Dallas, while a businessperson might give priority to pivotal corporate decisions made by such industry giants as IBM, Apple, or Google. Economists might champion the clout of consumers who brought personal computers into tens of millions of kitchens and living rooms.

However, even if we limit ourselves to developments in the United States, none of these explanations fully account for how and why people came to accept, adopt, and then surround themselves with computers and other information technologies. This question—how did computers become pervasive, popular, and personal?—can be addressed by considering another, less obvious factor. A powerful catalyst for this historical transformation was, ironically, one of the oldest information technologies we have: books.

*This* book argues that, in order for computers to become ubiquitous, people first had to learn about them, become interested in them, and take them seriously. Computers had to be both popularized and popular (the two are not the same) before they became omnipresent. Books, I contend, were an important yet overlooked part of a decades-long process that brought computing to the average person. What you have in front of you—whether you have summoned its words onto a screen or are turning crisp paper pages—is a book about books about computing. It offers a history of computing writ large as seen through the histories of a limited but well-chosen selection of books—some iconic, others obscure—and their authors, editors, publishers, and readers.

Looking at books about computing is an opportunity to recover lost or submerged visions, futures, hopes, and anxieties about this particular technology. Books provide us with a sound and steady instrument to assay past presentations of what was once a new technology. Whether they were best-selling works of nonfiction, quasi manifestos about the technological future, self-congratulatory primers by industry observers, or relatively obscure technical manuals, books afforded avenues for people to know and learn about computers. Books were essential instruments for presenting computers to the American public. Books, this book argues, helped make computers into the technology we know today.

\* \* \*

The name of this book carries a double meaning. Obviously, I'd like people who encounter this book to respond to the title's injunction by buying it and reading it. At the same time, computer programmers have, for decades, included "README" files that offered documentation about the larger software package they were writing code for. A README file might, for example, provide users with installation instructions, copyright information, and a list of known bugs. So, while this work you are gazing at uses a selection of books about computing to narrate a larger history of a particular technology, this opening chapter offers an explanation about the book you are holding.

In *README*, the books I have chosen to focus on are almost entirely nonfiction works. This was done partly out of necessity. As tempting as it was to bring classics such as Stanislaw Lem's *Cyberiad* (1965) or William Gibson's *Neuromancer* (1984) into the ensemble, doing so would have resulted in a sprawling and unwieldy project. For the same reasons, I've focused on books, authors, publishers, and the history of computing as it unfolded largely within the United States. No historical study can be unbounded; books must be finished in order to be published and read.

While it might appear as a constraint, nonfiction books offer a wide, diverse, and expansive aperture into specific historical moments as a new technology went from exotic and esoteric to essential. Each of the books examined here is situated in a particular historical moment, one which encompasses the history of computing, the history of books, and American culture. *README* offers a literary history of computers and computing between the end of World War II and the dot-com crash that marked and marred the first few years of the twenty-first century. In this period, computers transformed from mysterious "electronic brains" to the "everything machines" found in tens of millions of homes.

In this same period, the intended audiences for books about computing likewise transformed. In 1950, a typical reader would likely have been someone interested in so-called thinking machines or the new science of cybernetics. Perhaps they were an engineer or mathematician entranced by new technologies. Maybe they were a business executive anxious to remain competitive. This readership expanded throughout the 1960s and 1970s as the purview of computers expanded from number crunching to data processing and accounting. The number of books about what computers could, couldn't, and shouldn't do grew rapidly, reflecting the worries of a citizenry whose lives were increasingly influenced by the power of digital machines even if they rarely interacted with them personally.

In the 1980s, this audience grew to include technologically literate people, including teens and even children, who wanted to learn how to use computers at home or in the classroom. Likewise, many books about computing in this era reflected concerns with what the technological future might be like, especially as the United States faced competition from other advanced economies and industries. By the 1990s, computers qua computers had become so commonplace as to fade into the background of people's lives. Now readers wanted to understand the new frontiers of digital machines linked and networked to one another and used for communication, commerce, and entertainment. At the time of this writing (mid-2025), the book and the computer, twin technologies inextricably intertwined, seem poised for a new transformation amid questions about new programs with abilities that threaten the creativity and autonomy of the human writer.[2]

Focusing on books might strike some readers as an odd place from which to approach the history of a technology. Indeed, much of computing's history initially was limited to the development of iconic machines. These histories initially had an almost-Biblical cast via tales of how "ENIAC begat EDVAC begat UNIVAC." Paralleling this were complementary histories of pioneering computer designers as well as prominent companies like IBM. Eventually, historians' attention widened to include software and the people who created and used these artifacts. This scholarship expanded further to embrace the vast new worlds of networked computers, the internet, and the World Wide Web.[3] And, most recently, scholarship about computing's history, once an academic backwater, has engaged with several topics (race, gender, labor, politics, class, the environment, etc.), that are front and center in the minds of many historians.[4]

While undoubtedly tied to the activities and actions of engineers, machines, and lines of code, a surprisingly substantial part of computing's history has unfolded on the printed page. This could assume forms as diverse as a slim technical manual full of arcane instructions, an article for specialists in a trade journal, or advertising copy written to extol the virtues of a new digital product. The history of computing is ineluctably tied to the printed word, and chief among these manifestations is the book.

Books offer us a stable indicator of knowledge, attitudes, and information about computing at a particular point in time. Their very nature makes them dateable, traceable, and comparable to one another, while relative ease of preservation and access places them in contrast to the challenges of examining hardware or software from decades past. Learning about computing from a book written in 1965 is a far simpler proposition than trying to boot up a mainframe machine from that era.

Books about computing are as diverse as their authors and readers. The spectrum includes beautifully crafted works of narrative nonfiction, trade books written for a large audience, textbooks pored over by students, and anodyne instruction manuals bought by exasperated home computer users. By offering readers a shared social experience, books took what was once a technical topic familiar only to experts and translated it into something a broader audience could appreciate. Seen this way, *README* connects to a prominent theme in the histories of science and technology: the circulation of knowledge. Books were a central component in a larger circuit through which knowledge about computers and computing was created, circulated, and consumed.

To be sure, as much as authors might have loved their own words, publishing is a business and its products exist in a marketplace. *README* offers perspectives into authors' relationships with their editors, publicists, and the companies that put out their books. In this way, histories of books about computing can also tell us something about changes in the publishing business. The period this book addresses—roughly 1950 to 2000—represented a tumultuous time for publishers. One irony, of course, is that digital computers destabilized the industry many times over as typesetting was automated and books shifted from printed pages to words on electronic screens. The sprawling ecosystem of computer-focused magazines, in fact, was effectively destroyed by the very technologies they promoted and popularized. In 2023, *Maximum PC* and *MacLife*, the last "dead tree" (i.e., printed on paper) magazines devoted to computers, ceased publishing, closing out a part of the publishing business that had long been both culturally rich and exceptionally profitable.

By thinking about producing books as a business and not just a creative act, I am lightly engaging with an idea Robert Darnton first proposed in 1982, the same year that *Time* magazine named the personal computer "Machine of the Year." Darnton, a historian whose work focuses on the culture of eighteenth-century France, suggested a general "life cycle" for how books were created and subsequently spread throughout society. His innovative model included not only authors but also publishers, printers, and booksellers. What Darnton called a "communications circuit"—a pleasing electronics metaphor for a book about books about computing—was closed and completed by the reader. In other words, with *README*, I consider not just books and their authors, but also, when sources permit, the larger ensemble of editors, publishers, reviewers, and readers. This broader cast of actors helps us see books as objects not just in terms of authorship but as physical objects enmeshed in "far-flung networks" of "bibliologistics" where activities such as design, production, distribution, and consumption are just as significant as authorship.[5]

\* \* \*

People write and read books for various reasons. If we think of books as tools, albeit ones made of paper (or, more recently, electrons), we can ask what sort of work they did for their readers. Of course, what a book does need not be limited to a singular function. But, in answering this question, I can also offer a preview of the specific works on which I've chosen to focus.

First and foremost, books helped promote and popularize information technologies to a wider audience. This was often done by focusing on the people closely associated with making them, such as engineers and entrepreneurs. For example, Tracy Kidder's *The Soul of a New Machine* (1981), and Michael Moritz's *The Little Kingdom* (1984) gave readers a perspective into the activities of executives and engineers at two very different computer companies—Data General and Apple Computer respectively—as they worked furiously to bring a new product to the market. Michael S. Malone's *The Big Score* (1985) revealed a larger cast of characters, from businesspeople to factory workers, who together had transformed a sleepy agricultural area into Silicon Valley. Taking a radically different approach in presentation and style, Theodor H. Nelson's *Computer Lib/Dream Machines* (1974), a self-published cult classic, exhorted readers to seize control and spearhead a computing revolution that would be both personal and emancipatory.

Books certainly conveyed information but they also expressed authors' personal opinions and anxieties. Starting in 1948, Norbert Wiener, a polymathic professor at MIT, published a trilogy of books that presented and popularized the then-new topic of cybernetics. Wiener was especially concerned with how humans and machines interacted with each other and he eventually became a vocal critic of industrial automation made possible by computers. In 1976, another MIT scientist, Joseph Weizenbaum, published *Computer Power and Human Reason*. Weizenbaum used his book to warn readers about the limits of computing, especially the newly emerging field of artificial intelligence, and the accompanying hubris of computer scientists.

Another central task that many books about computing performed was pedagogical. Classic works, such as Donald E. Knuth's *The Art of Programming* (a series of books the computer scientist started writing in 1962) taught people how to write software for computers. In a similar vein, Carver Mead and Lynn Conway's 1980 textbook *Introduction to VLSI Design* presented an innovative new methodology for creating computer chips to thousands of engineering and computer science students. Some enthusiasts seeking instruction, however, were complete beginners. These people wanted not just advice but also assurance. With their cheery yellow covers, books in the *Dummies* series, such as *DOS for Dummies* (1991) and *Windows for Dummies* (1992) were bought by millions of computer novices.

Books, of course, have often stimulated the formation of communities. In seventeenth-century Britain, for example, literate people would gather in coffeehouses to discuss and debate newly printed volumes on business, religion, philosophy, and the natural sciences.[6] Two hundred years later, such innovations as the steam-powered printing press made books on all sorts of subjects more accessible to much wider audiences.[7] In the United States, changing practices in advertising and marketing helped the Book-of-the-Month Club, which debuted in 1926, rapidly expand until more than half a million Americans were active members.[8] More recently, talk show celebrity Oprah Winfrey gifted massive surges in sales for authors lucky enough to be featured via her book club. Winfrey's recommendations fostered vibrant discussion and debate around certain books, a pattern continued in the inclusion of suggested questions for local book clubs which are sometimes appended to the end of certain works.

Likewise, books were central to the formation of technical communities around computing. For example, Donald Knuth's *The TeXbook* (1984) helped mathematicians and computer scientists coalesce around a new digital typesetting system based on his TeX program. Today, TeX is a ubiquitous means for writing technical papers used all around the planet. Computer magazines, some long-lived and others fleeting, did something similar for non-experts. Besides entertaining and informing readers, they helped shift power away from what Ted Nelson called "the computer priesthood." Printed materials in the form of newsletters, magazines, and books helped build a community of experienced hobbyists and neophyte computer users who sought an identity based around learning and displaying their computing skills.

Some of the works I've chosen to examine performed a predictive function. Edmund C. Berkeley's 1949 book *Giant Brains, or Machines That Think* was as much about the future capabilities of electronic computers as it was a primer that explained how these devices worked. In books like these, authors described technological futures that made sense when first presented in print, but might seem somewhat strange or obvious to us today.

As computing technologies became more deeply woven into the fabric of modern society and consumers purchased tens of millions of them for home use, they still possessed an air of unfamiliarity. This strangeness became even more pronounced as personal computers were increasingly used as networked tools communicating with other machines. *The Whole Internet User's Guide and Catalog* (1992), published by O'Reilly & Associates, a company that previously put out esoteric technical manuals, became a surprise bestseller that helped newcomers navigate the exponentially growing territory of cyberspace.

Finally, books about computing also offered insights into larger economic, cultural, and political issues. George Gilder's *Microcosm* (1989) did more than explore the technological workings of computer chips. It also represented a distillation of conservative political viewpoints the author honed during the Reagan administration. Meanwhile, Gilder's musings about the intricacies of quantum physics reflected his own Christian beliefs. If we jump ahead a half decade, a political philosophy that extolled the virtues of high technology combined with an entrepreneurial zeal, free markets, and government deregulation emerged alongside the internet's commercialization. Esther Dyson, celebrated as one of the most powerful pundits in computing, wrote her book *Release 2.0* in 1997. Published as the dot-com bubble was inflating, Dyson used the most traditional of media forms to explain the legal, economic, and social implications of cyberspace to the general public.

Beyond the books themselves, there is the business of selling books. Like computing itself, the technologies and practices associated with the booksellers' business transformed markedly after 1950. The scale and scope of bookstores changed, resulting, in some cases, with a greater focus on product specialization. One sign of this was the opening of the Computer Literacy Bookshops in the 1980s. For more than a decade, computer enthusiasts, engineers, and programmers came to this small collective of stores to check out the latest books and to connect with one another. In the end, of course, brick-and-mortar bookstores faced withering, sometimes fatal, competition from online retailers, something that speaks to another powerful confluence of books and computers.

While *README* offers a bookish history of computing, it is not *the* bookish history of computing. The shelves of libraries and bookstores creak under the collective weight of thousands of books written about the subject and one could easily select a different collection of books. That said, an alternative assortment would still have addressed many of the same technological and social issues. From concerns about automation and computer-induced unemployment in the 1950s to anxieties about data privacy or software glitches that could literally cause modern civilization to crash, books about computing offer frames through which we see some of the larger concerns their authors and readers held. And, like all good histories, *README* helps reveal the hopes, desires, and anxieties people held at a particular moment in time. Such beliefs, at times, may be quite different from today's concerns and, in other cases, remained remarkably consistent.

\* \* \*

This book's narrative unfolds over roughly the half-century following the end of World War II. In this short period of time, computers morphed from mysterious machines to ubiquitous objects that were simultaneously data processing devices, interactive tools, and networked platforms for communications and entertainment. Of course, the computerization of practically everything profoundly affected both the nature of the book and the publishing industry itself.

That said, it is important not to dismiss this fifty-odd year span of time as merely a prelude to contemporary "crises of the book" marked by the emergence of phenomena such as Amazon.com, electronic books, online publishing, and, most recently, chatbot authors made possible via advances in artificial intelligence. Indeed, throughout much of the time period that *README* explores, authors, publishers, and critics alike fretted about various calamities, real or imagined. Despite recurring moments of concern, and the fact that the technologies and practices associated with book making and book selling changed significantly, the physical nature of the traditional book—paper pages bound between two covers—has remained a remarkably constant form of material culture.[9]

Engaging with this history means discounting folk fictions surrounding both books and readers. One of these myths is that the rise of television and then the internet resulted in decreased readership. But data suggests otherwise. For instance, in 2005, the average American bought ten books annually, more than double the number fifty years prior. Purchasing patterns changed also. There was a notable swing among consumers in the 1960s away from works of fiction toward nonfiction books. Part of this shift was a growing demand for books about science and technology, a change that certainly benefitted the authors and publishers of books about computing. Between 1950 and 1980, for example, the number of new titles categorized as science or technology increased by about a factor of five. If one expands this time period from 1950 to 2000, the change is an even more extraordinary factor of twenty.[10]

Several factors fueled this explosive growth. During World War II, the US government partnered with publishers, booksellers, and librarians to provide tens of millions of inexpensive, paperback Armed Services Editions to military personnel, creating a vast cohort of young adults who routinely read. When the war ended, some two million veterans entered college via the Servicemen's Readjustment Act of 1944 (i.e., the GI Bill). Many of these young people chose to pursue degrees in science and engineering, which prompted the publishing of dozens of new textbooks for growing fields such as physics, molecular biology, and electrical engineering. The upswell of new titles only increased after the Soviet Union launched its first Sputnik satellites in 1957. The Space Race generated yet more public interest in science and technology as well as the need for more textbooks, something the US government fueled by pumping millions of dollars into engineering and science education.[11]

The data on publishers is equally impressive. When World War II ended, there were about 650 publishing houses in the United States. Together, they produced some 487 million books, an output worth around $484 million. In 1947, about 9,100 new titles appeared in bookstores and catalogs. Six decades later, there were 81,000 publishing companies that issued a staggering 282,500 new titles. The three billion books that consumers purchased was worth almost $35 billion in sales. Throughout the Cold War, the book publishing business performed exceptionally well and this pattern continued into the twenty-first century.[12]

In many ways, book publishing reflected larger trends in American business. The industry grew during the boom years of the 1960s and then stalled a decade later, as did many other business sectors, due to inflation and rising costs. As was the case for other areas of corporate activity, publishing experienced waves of conglomeration spurred by Reagan-era encouragement for consolidation coupled with government indifference for antitrust cases. In the 1980s, for example, medium-size firms were gobbled up by larger publishing houses that, in turn were later consumed by new behemoths in the book business.[13] By the 1990s, books had become simply a form of *content* as publishers recast themselves as the "creator of copyrights for their exploitation in any medium or distribution system."[14]

All these books that publishers, large or small, produced still had to reach the hands and eyes of readers. In 1945, stores devoted specifically to selling books were relatively rare in the United States, especially outside of major urban centers. People turned to supermarkets, drug stores, department stores, and other similar venues to find their books. In some cases, they rented books from businesses, such as the Walden Book Company. After World War II, these for-profit rental libraries began to disappear as Waldenbooks and B. Dalton opened stores in new suburban malls, the number of which also expanded dramatically in the 1950s and 1960s. By the early 1980s, Waldenbooks was opening more than eighty new stores annually.[15]

Nonetheless, some small independent booksellers unexpectedly thrived as their owners highlighted the personal service and expert advice they could offer customers. One especially successful business opportunity was found in specialized bookstores that focused on particular clients by offering children's books, books for women, or religious books. The aforementioned Computer Literacy Bookshops were part of this larger pattern. In the 1990s, even the mall-based stores were eventually overshadowed by book superstores, a new phenomenon that brought such companies as Barnes & Noble and Borders into the picture. Customers flocked to these spaces where they could sip a cup of coffee, find a comfortable chair, and browse books and magazines (some of which they might actually buy). The sheer size and stock of book superstores started eroding the market share of independent booksellers while the appearance of online retailers—most notably Amazon—posed an existential threat to all brick-and-mortar bookstores. It's appropriate, given how computing technologies made Amazon and similar ventures possible, that the very first book Jeffrey Bezos's fledgling company sold was about computing: *Fluid Concepts and Creative Analogies: Computer Models of the Fundamental Mechanisms of Thought* by Douglas Hofstadter.[16]

Despite changes in infrastructure, business, and technology, some things in the publishing world remained relatively steady. For instance, the industry remained largely centered in and around New York City. This was, at least, the case for the publication of trade books, those works written for a broad and general readership. In terms of their provenance, however, books about computing displayed a little more geographic variability. To be sure, hundreds of books about computing were published by companies based in Manhattan. But presses in other cities were also prominent players. W. H. Freeman and Company, for example, was founded in San Francisco soon after World War II ended, while Addison-Wesley operated out of offices in a Boston suburb. O'Reilly & Associates, which published hundreds of sought-after technical manuals, was headquartered in a small town north of San Francisco. In terms of volume, few presses could match International Data Group (IDG) Books, based in San Mateo, California, which sold tens of millions of its friendly looking *Dummies* books to novice computer users in the 1990s.

While reflecting some of the changes seen in the wider publishing industry, the business of computing books had its own idiosyncrasies. These patterns were in part driven by the fact that the computing industry itself was not based in New York but rather flourished in places such as Dallas, Minneapolis, and the Bay Area. To be sure, putting out books and other printed materials about computing proved to be a profitable niche of the larger publishing industry.[17] Just as Silicon Valley capitalists sought their fortunes in the computing and software sectors, authors and publishers likewise recognized an entrepreneurial opportunity.

\* \* \*

The same period of time that witnessed this massive explosion in both computing and books about computing also saw growing enthusiasm among academics in the history of the book. In France, this scholarship had roots in the Annales school with its focus on socioeconomic history seen in works by Henri-Jean Martin and Lucien Febvre. In the United States, studies about book history were catalyzed by Elizabeth Eisenstein's *The Printing Press as an Agent of Change* and Robert Darnton's *The Business of Enlightenment: A Publishing History of the Encyclopédie 1775–1800*, both from 1979 and focusing on developments in early modern Europe.

Besides a growing roster of canonical books about books, this new academic discipline was marked by features that characterize any field of study coming into its own, such as specialized journals and a professional society (in this case, the Society for the History of Authorship, Reading, and Publishing). In 1998, the journal *Book History* appeared under the imprint of Johns Hopkins University Press. "Our field of play," its editors explained, "is the entire history of written communication," an expansive arena that included, "the social, cultural, and economic history of authorship, publishing, [and] printing" along with reading habits and readers' responses.[18] Book history, they predicted, would energize other academic disciplines (including computer science), as books were an essential part of the larger "history of information." In the years that followed, historians of the book were joined by specialists from other subfields who focused on, among other things, the histories of printing, reading, word processors, and even the anatomy of books themselves.[19]

However, compared to scholarly studies about Herman Melville's *Moby-Dick* (1851) or Jane Austen's oeuvre, prominent books in the histories of science and technology have been relatively underexamined. Much of what has been written about scientific literature has focused on early modern and Victorian-era England.[20] A good deal of this work, not surprisingly, has centered around famous books by equally famous scientists, such as Charles Darwin or Albert Einstein.[21] Other scholars have examined the history of publishing via scientific journals, both as a genre and in terms of top-shelf publications, such as *Nature*.[22]

In a sense, the history of science and the history of the book share similar concerns. Both address questions of infrastructure, power, institutions, instruments (be it a printing press or a microscope), and the reception of ideas. It would seem to be a mutually complementary pairing. However, as one moves into the second half of the twentieth century, the attention historians have given to books about science, in terms of writing, publishing, and reception, starts to dwindle. There are notable exceptions of course as influential and popular books such as Rachel Carson's *Silent Spring* (1962), James Watson's *The Double Helix* (1968) and Stephen Hawking's *A Brief History of Time* (1988) have all received consideration.[23]

While relatively understudied, books about scientific topics have received *far* more attention from historians when compared to works about technology. One explanation might be that, when compared to their scientist colleagues, engineers are papyrophobic.[24] That is to say, as members of a professional community, they are less inclined to record their recollections on paper. Many scientists have published their memoirs but few engineers have followed suit. Engineers also tend to spend their professional lives in relative anonymity by working in teams. Ask a person the name of a famous scientist and you'll get an answer. But ask them to name a famous engineer and you'll likely get a blank stare. The combined effect is that histories of books about modern engineering and technology remain scarce.[25]

It is a puzzling paucity. Engineers read and write all sorts of things every day. If one goes into an engineer's office, there are equipment catalogs, handbooks for operating various apparatuses, and weighty guides for carrying out complex technical procedures. Written materials have long been essential to how engineers learn technical skills and transmit this knowledge to others.[26] Computer programmers routinely relied on a host of handbooks when writing software. And, of course, there is the genre of software code itself, something that some scholars have recently turned a critical eye toward.[27] Technical writing is a central skill for engineers and programmers, yet one that few historians have explored.[28] By focusing on a carefully selected ensemble of books, *README* addresses this lacuna.

\* \* \*

All books have a history. They begin with an author's idea, perhaps combined with a wish to say something to one's colleagues and a broader community and maybe make some money. This particular book has its own history, one shaped by both historical context and constraints its author faced. In the summer of 2020, I had just finished another book project and was, like millions of other people, experiencing the constrictions imposed by the COVID-19 pandemic. Although I wanted something to distract me from world events and domestic politics, libraries were shut down and finding refuge among stacks of archival boxes simply wasn't possible.

My initial idea was that writing a book about books would be straightforward as I could easily access the main texts I was interested in. Computing seemed an especially felicitous subject because enthusiasts, archivists, and collectors have generously converted massive amounts of printed material into easily accessed online resources. Entire runs of magazines and newsletters as well as whole books and even personal correspondence about computing and writing about computing could be downloaded and read *on a computer*. My hard drive quickly filled up with gigabytes of publications and documents.

But I quickly realized that my original assumption was wrong. I would absolutely need to examine materials held in universities' special collections and other archives. Fortunately, I was able to work with several librarians, archivists, and graduate students who, despite pandemic restrictions, could access materials in their own institutions' collections. Their labors amidst library stacks and archival boxes helped make this book possible.

Even as my opportunity to visit archives in person was temporarily curtailed, the pandemic normalized the use of online tools such as Zoom. This proved invaluable for conducting oral history interviews. These interviews also helped me discover and dislodge new historical materials—a form of salvage history—that could not be found in official repositories. This is especially important for topics that approach the present day. It is a lamentable fact that fewer and fewer paper documents are preserved as one moves from the 1990s into the twenty-first century. Documents are born digital, people communicate via email and text message instead of paper correspondence, and digital resources (e.g., web pages and message boards) are often ephemeral. Online tools such as the Internet Archive and the Wayback Machine were absolutely essential to write this book about books about computing.

This book started as a pandemic project when the most vital scientific and technological topics were how to protect people from illness, how to test and distribute a vaccine, and how to help societies, economies, and psyches recover. It ended in a very different era. In early 2025, when I was finishing the final draft of this book, debates about artificial intelligence dominated news headlines and quarterly reports. The rapid spread of computer-based chatbot software based on large language models have raised a host of as yet unanswered questions about the very nature of creativity and authorship. It is a strange time indeed when one feels compelled to note that a book about books about computing was created by an all too human author.

# 1 THINKING MACHINES

The strange new brains were already among us. In fact, if the growing mound of books, magazine articles, and newspaper print, was any sign, they soon would be everywhere. For now, the brains could be found in shabby basements on the campuses of elite universities. Some of the brains sat in specially designed rooms, kept at frigid temperatures, behind the barbed wire fences of military installations. A "new giant brain" at a government laboratory performed, one headline blared, "wizard work."[1] An exceptionally large brain was displayed in New York at 57th and Madison Avenue in what had once been a women's shoe store. Downtown office workers on their way to lunch often stopped to gawk at the contraption and the mathematicians who tended to its needs.

The brains in question, of course, were not made of axons, synapses, or blood vessels. Rather, they were inorganic assemblages of mechanical relays, diodes, and vacuum tubes, all connected by miles of wire. As inventions of the mid-twentieth century, these brains received information from paper tapes, teletypewriters, and punch cards, not biological eyes and ears. Flashing lights and more punch cards, not words or gestures, displayed the results of whatever *thinking* these machines performed.

In 1950, electronic brains were still quite rare. This meant that nonexperts—which, at that time, meant almost everyone—learned about them not through everyday encounters but through the printed word. What started during World War II as a trickle of written descriptions, often hyperbolic in tone and sometimes technically inaccurate, had swollen into a torrent of explanatory prose by the early 1950s.[2] Newspaper features and magazines stories featuring "giant brains" appeared with a mounting frequency and books about them quickly followed. Nonetheless, these new electronic brains remained objects of puzzlement, speculation, and some trepidation. Clearly, more computers would be coming soon from a rapidly coalescing constellation of companies. But few people had any idea how they worked, what they could do, or even what they were.

Edmund C. Berkeley, an insurance actuary turned entrepreneur, was determined to change all that. As the postwar period's first prominent public evangelist for computers, Berkeley's tool of choice would be a book. Clearly composed and carefully promoted, Berkeley believed his book could inform non-expert readers about the present and future state of computing.

I have emphasized the word "brain" here quite deliberately. Throughout the 1940s and early 1950s, journalists, science writers, and editors often compared the first generation of electronic computing machines to living brains. While experts challenged such comparisons, the trope of "electronic brains" was nonetheless how the average reader was pressed to visualize computers in the years immediately following World War II. This same imagery prompted Berkeley to title his 1949 book *Giant Brains, or Machines That Think*.[3]

Today, after decades of relentless miniaturization of electronics—from the minicomputers of the 1960s to today's laptops and smart phones—the word "giant" in Berkeley's title certainly appears incongruous. Who would want a computer given to giantism? But most of the machines that Berkeley's book described *were* enormous, their size given in terms of tons and cubic feet. Despite its seemingly straightforward title, *Giant Brains* suggested a wealth of implications, many of which would figure prominently in future debates about what computers could and should do (and couldn't do). Berkeley did not choose his title hastily. In fact, Berkeley's personal papers indicate that he rarely did anything impulsively. Nestled among those tens of thousands of pages, for example, is a lengthy handwritten note meticulously laying out the pros and cons of firing an underperforming business associate.

Thinking about how people thought was something Berkeley had been doing since his undergraduate days at Harvard. As a mathematics student in the 1920s, Berkeley was drawn to the study of logic. He remained curious throughout his life about how people's reasoning processes could be understood and even optimized. Berkeley drew on this expertise when he presented readers with comparisons between human thought and the new and rapidly expanding world of electronic computers.[4] *Giant Brains* reinforced the public's image of the computer as an "awesome thinking machine" that could be analogized and anthropomorphized to the human brain.[5] This comparison was both protean and problematic.

When Berkeley wrote *Giant Brains*, computer technology was changing rapidly. In 1946, when Berkeley started outlining his book, computers were still esoteric scientific machines, akin to a giant telescope or particle accelerator. Just three years later, when his book appeared in stores, a veritable zoo of computers existed—some mechanical, some electronic, others technological hybrids—even as the machines' fundamental design principles remained unresolved. By 1952, when John Wiley & Sons was readying another print run for *Giant Brains*, the computer was already transforming from a specialized machine into widely available commercial products. *Giant Brains* offers a window both into a period of remarkable technological diversity as well as a moment when people's awareness and attitudes about computers were still developing.

Like those who wrote about nuclear weapons in the late 1940s or space travel a decade later, Berkeley wanted to explain computing technology using the most straightforward language. But he was also driven to ask what increasingly ubiquitous and capable "thinking machines" might mean for society in the decades to come. Besides featuring clear technical descriptions, *Giant Brains* gave readers some gentle warnings. But, in all cases, Berkeley believed clarity of writing translated into clear thinking. This was an ideal, in fact, Berkeley had embraced decades before he first saw a "giant brain."

### ASSEMBLING THE ATOMS OF REASONING

In June 1930, Edmund Berkeley addressed his classmates at Harvard's commencement ceremony. Graduating summa cum laude with a degree in mathematics, Berkeley called his address "Modern Methods of Thinking."[6] After progressing briskly from the logical thinking of the ancient Greeks to the "orderly set of ideas" in Einstein's theories of relativity, Berkeley lingered on the work of George Boole. He had encountered the nineteenth-century British mathematician via Boole's 1854 book, *The Laws of Thought*, and found Boole's symbolic logic, where variables can be expressed as truth values, such as true and false, tremendously appealing. Mathematicians and philosophers had extended Boole's approach to logic, which dealt with statements oriented around such operations as AND, NOT, and OR, into an "amazing method of reasoning" that suggested a more objective approach to thinking. In fact, Boole's logic would later be rendered as ones and zeros, the approach adopted by computer engineers using relays and vacuum tubes (and, later, with transistors) set in on and off states. Society, Berkeley told his classmates, was on the "threshold of even greater discoveries" as logicians continued to take the very "atoms of reasoning" and fashion them into "brilliant, sparkling crystals" that formed "orderly systems of truths."[7]

While such methods were largely concerned with *how* people would think in the future, Berkeley also noted that it mattered *what* people thought about. When he graduated from university, the Great Depression had begun and countries across the planet were already lurching toward more tragic economic and political crises. Berkeley believed the solution was "precision reasoning" applied to pressing issues of business, sociology, international relations, and even ethics. Berkeley trusted that complex social problems could be examined as a series of propositions and questions and then reduced to truths and falsehoods.

Berkeley's path from up-and-coming mathematician to the first popularizer of computers was anything but straightforward, however. Born in 1909 and educated at prestigious preparatory schools—he was the youngest student at Phillips Exeter Academy but finished first in his class—he started at Harvard intending to study rare minerals.[8] But the clarity of logic proved more appealing and Berkeley decided to pursue a career as a "creative mathematician" instead. In time, what he called the "Circe of being practical" interfered and his parents pushed him to apply his math skills to the business world. A position with a life insurance company followed but Berkeley found the tasks unchallenging and the workplace dismal.

His romantic flair remained undimmed, however. In 1934, Berkeley, after receiving a small inheritance, decided to pursue his idealistic ambitions.

With his partner Ruth Pirkle—they married later that year—he embarked on a long tour of Europe.[9] Part of their time was spent in Moscow where they took classes in economics and sociology. In the 1930s, with the Depression underway, Berkeley's interests in socialism and the Soviet Union were not that uncommon as many American citizens questioned capitalism. Two decades later, however, McCarthy-era investigators cast Berkeley's travels in a more ominous light and he was called on to defend his prewar political interests and pacificist leanings.[10]

Coming back to the United States also meant a return to the life insurance industry, this time in a position at Prudential Insurance in Newark, New Jersey. Berkeley found this new post more interesting and he eventually passed examinations to become a professional actuary. It also set him up for a new phase in his career as a computer advocate and popularizer.

Today, it is hard to imagine insurance companies at the cutting edge of information technology. The invention of modern computers is instead usually linked to machines at military laboratories performing scientific calculations or breaking codes. But, despite stereotypes of monotony and rote work, the insurance industry had long been an early adopter of information processing tools, from nineteenth-century punch card tabulators to some of the first electronic digital computers.[11] Prudential, Berkeley's new employer, in fact, had used punch card machines to handle data since 1895. By the mid-1940s, its business stirred by new federal legislation, Prudential and other insurance companies needed to process tens of millions of payments and policy changes each month, making life insurance one of the most computationally demanding business sectors.

Berkeley, always keen to apply his expertise in math and logic to practical problems, saw this as an opportunity for the insurance industry to benefit from techniques such as Boolean algebra.[12] By 1941, Berkeley had transferred from Prudential's actuarial office to the company's Methods Department where he worked as an applied mathematician. From this position, even before the debut of such machines as the iconic Electronic Numerical Integrator and Computer (more commonly known as the ENIAC), Berkeley advocated for the adoption of computing machines via memos and lengthy reports for Prudential executives.

Berkeley was also starting to meet people in the community of engineers and academics thinking about and building thinking machines. Contacts at IBM and visits to corporate labs run by General Electric and Bell Telephone allowed him to assess the state of technology and to nudge colleagues at these corporations to be mindful of the insurance industry's computing needs.[13] Later, when he sat down to write *Giant Brains*, this knowledge gave Berkeley a solid evidentiary base.

War often forces people to take stock of their lives and Berkeley was no exception. In 1942, the thirty-two-year-old went on active duty in the Naval Reserve. His personal notes suggest that he remained curious about socialism and was disenchanted with pursuing a middle-class life. He envisioned writing articles or even a book on "aspects of mathematics, science, language, economics" for a popular audience.[14] Then, in 1944, he received an exciting new assignment "on the frontiers of mathematics, symbolic logic, and calculating machines."[15] He was tasked as the liaison between a military proving ground in Virginia and a new laboratory devoted to new calculating technologies back at Harvard.

The Harvard Computation Laboratory was formally established in 1944 around an IBM-built machine called the Automatic Sequence Controlled Calculator (otherwise known as the Harvard Mark I). The project was driven forward by Howard H. Aiken, a tall and intimidating man who ran the operation like a military unit. The Mark I was equally imposing. Its five tons of gear took up some 500 cubic feet, inside of which was several hundred miles of wiring, thousands of mechanical relay switches, and three million electrical connections. The military was keen to see it enter service so it could produce mathematical tables automatically (what the Hoboken-born Aiken called "makin' numbers").[16]

Like all electronic computing machines of the era, the Mark I was custom-made and prohibitively expensive, costing IBM well over $250,000 (or roughly $5 million in today's money). It worked via a combination of electrical components and mechanical parts, such as cams and switches, making it quite different from the transistorized machines we are familiar with today. As its formal name suggested, the Mark I was sequence controlled, meaning it read instructions from paper tape, one line at a time, and executed them. Output came as punch cards or in readable numbers printed by typewriters. With so many moving parts, the Mark I was slow and noisy compared to all-electronic machines such as the ENIAC, which was being developed at the same time.[17] For example, the ENIAC could perform about 5,000 calculations a second whereas the pokey MARK I could do only about three. Nonetheless, when the Mark I was dedicated, reporters quickly made the "machine-as-brain" comparison. One weekly magazine called it "Harvard's Robot Super-Brain"—a drawing of a bespectacled sage with a machine inside their head accompanied the article—saying it could solve "centuries-old enigmas of mathematics—problems that would fatigue 20 Einsteins."[18]

Unfortunately, Berkeley's tenue at the Harvard Computation Lab was full of the personal conflicts he imagined that logic and rational thought might help avoid. He chafed at Aiken's overbearing personality—at one point, Berkeley suggested his boss read Dale Carnegie's *How to Win Friends and Influence People*—while the lab's leader ridiculed Berkeley's preoccupation with reports and documentation.[19] Despite these clashes, Berkeley got a firsthand view of how a computer was designed as well as a picture of technical developments underway in the United States.

Berkeley returned to civilian life and a desk at Prudential's newly formed Methods Research Section in the summer of 1946. His assignment was studying whether Prudential should adopt one of the new electronic computers. Within six months, he visited sixty different companies and labs, attended a series of now-famous conferences on the state of computing technology, and wrote several reports on future applications of electronic digital computers in the insurance business. Berkeley also successfully pushed Prudential to establish a modest-size electronics research laboratory with him overseeing it as Chief Research Consultant.[20] Over the next several months, Berkeley championed the general-purpose Universal Automatic Computer (UNIVAC), pointing out its advantages over rival machines from established firms (i.e., IBM, Bell Labs, and Radio Corporation of America [RCA]). Guided by Berkeley, Prudential's management agreed to buy a UNIVAC machine once one was available.

Berkeley's extensive networking brought him into contact with most of the people in the small but rapidly growing computing community. (It also added stress to his first marriage, which ended in 1948.) However, much of the activity in the "computing machine" field was being done either by private firms, which had patent concerns, or military organizations, where security and secrecy prevailed. The result was duplicated efforts and a general lack of efficient communication. At this time, unlike other engineering communities or mathematicians, the people involved in the then-esoteric craft of designing and building computers lacked a professional organization. (The first academic computer science departments wouldn't appear for several more years.) In response, Berkeley proposed an "Eastern Association for Computing Machinery."[21] The qualifier "Eastern" was soon dropped and the Association of Computing Machinery (ACM)—its stated purpose was to "advance the science, development, construction, and application of the new machinery for computing, reasoning, and other handling of information"—formed in September 1947.[22] A year later, the ACM claimed over 400 members and Berkeley was the group's first secretary.

Berkeley's lobbying for the ACM represented an idealism that valued the free circulation of ideas and information. Like many Americans, Berkeley worried about nuclear weapons. He also recognized that any future conflict where Soviet aircraft dropped atomic bombs on American cities would financially disrupt the insurance industry. This is not as callous as it might sound. In 1948, the Soviets had yet to test their first nuclear device, the size of the American atomic stockpile was relatively small, and the advent of much more devastating hydrogen bombs was still years away. Put simply, it was still possible to imagine a "small" nuclear conflict that left hundreds of thousands of people dead and several major cities wrecked but the planet largely intact.

To make the case that a future atomic war would be not only immoral but also financially ill-advised, Berkeley organized the Group on Future Catastrophe Hazards at Prudential in late 1947. He began assembling a massive archive of news clippings about nuclear weapons, international relations, and domestic politics while recruiting a group of like-minded professionals to examine what insurance companies—"the greatest losers from war"— might do to reduce the hazards posed by weapons of mass destruction.[23] In taking up this cause, Berkeley aligned himself politically with left-wing progressives such as Henry A. Wallace, who ran for president in 1948, and the short-lived One World movement started by American scientists after Hiroshima.

As might be expected, Prudential executives grew increasingly anxious as memos from Berkeley promoted a "World Federation" and cooperation between the United States and the Soviet Union. Their concern reached a tipping point when Berkeley shared an article he was writing titled "The Future Mortality Rate and Catastrophe Hazards." A Prudential vice president told Berkeley not to publish the controversial manuscript as it might be "construed as interfering with the country's preparedness efforts." Berkeley had been considering a new career path for some time and this corporate censorship convinced him he no longer wanted to be a "Prudential man." In July 1948 he submitted his resignation. "For the first time," he later wrote, "I was able to call my soul my own."[24] His first task as a free man would be to write a book that would make him nationally known as one of the foremost authorities and popularizers of modern computers.

### THINKING BRAINS

"I am," Edmund Berkeley told the *New York Times* in 1950, "descended from a long line of Frankensteins." It was an odd confession given that the forty-one-year-old entrepreneur had not produced a monster, but a book. However, for the thousands of people who purchased *Giant Brains* after its release in late 1949, the Frankenstein comparison was apt. The devices Berkeley's book described were mysterious and menacing. Like Frankenstein's monster, the machines were not alive in any biological sense but, as Berkeley wrote in his preface, they had "individuality, responsiveness, and other traits of living beings." Because computers were "closer to being a brain that thinks" than other human-made devices, Berkeley was determined to demystify them by explaining how they worked and what they did. But, as countless other writers before him had learned, producing something that was "understandable yet accurate" was "the hardest task of an author."[25]

Berkeley's original idea was to write a book called *Punch Cards and Symbolic Logic*. But even the math-minded Berkeley understood this was unlikely to appeal to a wide audience. His next iteration—*Words, Symbols, and Machines for Computing*—wasn't much more compelling but he managed to draft about four chapters by May 1946.[26] While still working full-time at Prudential, he submitted a proposal for a book, now with the more reader-friendly title of *Machines to Help Us Think* to John Wiley & Sons a few months later.[27]

In courting Wiley as a potential publisher, Berkeley was aiming high. The venerable company started in the book business in 1807 when Charles Wiley opened a printing shop in what today is lower Manhattan. The shop eventually shifted to publishing and, when Charles Wiley died in 1826, the business passed to his teenage son, John. By the mid-nineteenth century, the company's list of writers included Charles Dickens, Herman Melville, Elizabeth Barrett Browning, and Edgar Allan Poe, among others.[28] After the Civil War ended, John's son, William Wiley, decided to focus more on books about science and technology. For example, in 1880 John and William published the highly successful book *Field Engineering*. Written by William Henry Searles, it was aimed at engineers working in railroad construction—one of that era's high-tech industries—and sold tens of thousands of copies. Technical books for engineers and company managers proved very profitable. John Wiley & Sons—the company incorporated in 1904—saw sales pass the $2 million mark by 1941. Profits swelled again after the war as millions of veterans attending college on the G.I. Bill sought textbooks that John Wiley & Sons and other firms published.

J. Kenneth Maddock, an editor at Wiley, hoped Berkeley's book would ride this trend. He quickly secured an anonymous outside reader to evaluate Berkeley's proposal.[29] The response said less about the merits of the proposed book—in 1949, the community of experts able to evaluate a book proposal about computers was not large—and more about Wiley's intent to quickly get a book about the new technology on the market. "You and many other publishers can look forward to a number of manuscripts in this field," the reviewer wrote. "There undoubtedly will be prestige attached to early publication." But there was especial merit in Berkeley's book where the "principal concern is not the machines per se but what they can be made to do." This satisfied Maddock, and Berkeley soon told colleagues that he was writing a book about "new large-scale machinery for handling information."

Moreover, he hoped to sell "at least 50,000 copies" of what was now called *Giant Brains, or Machines That Think*.[30]

Berkeley signed with Wiley in the midst of some notable changes in both publishing and writing. Books that explored and explained scientific and technical topics were especially sought after by readers, forming the publishing industry's fastest growing sector.[31] Tied to this upward trend in Americans' consumption of books was the emergence of the science writer. Formed in 1934 with just a dozen members, the National Association of Science Writers grew to over sixty by 1945, an increase fostered by public interest in all things nuclear. Five years later and the group's membership had doubled, a pattern that repeated once again by 1955.[32]

Although not a professional science writer, Edmund Berkeley displayed values common to that vocation, especially his commitment to providing accurate technical facts and helping readers make informed choices about new technologies. The latter aspect was especially important for Berkeley as the tragedy of Hiroshima and Nagasaki cast a shadow over his thinking. Berkeley expressed apprehension about the social implications of computing technologies in the pages of *Giant Brains* and it concerned him throughout the rest of his career.

But why brains? Whereas his earlier iteration had been about machines that helped *people* think better, his new working title now had the *machines* doing the thinking. Perhaps following a suggestion from Wiley staff as to what would be marketable, Berkeley ultimately decided that his book would engage with the popular idea that these new machines should, in some sense, be thought of as "thinking machines."

The detailed research notes Berkeley assembled while writing *Giant Brains* have not been preserved. Fortunately, Berkeley included a list of over 250 references at his book's end, broken down by topic, which suggest what he was reading while writing his own book. Berkeley was particularly intrigued with the question of "how does a brain think?" Chief among the articles he cited were a dozen works drawn from the relatively new field of mathematical biophysics. Established at the University of Chicago in the 1930s, it combined mathematics with physics and physiology for the systematic studying of biological phenomena. Just as physicists study the natural world by reducing its complexity to fundamental units, such as atoms and electrons, mathematical biophysicists aimed to understand basic biological units such as neurons and their behavior in larger networks.[33]

In 1943, a paper in the *Bulletin of Mathematical Biophysics* became one of the most influential works in the new field. Written by Walter Pitts, a twenty-year-old autodidact, and Warren S. McCulloch, a prominent neuropsychiatrist some two decades older, their article "A Logical Calculus of Ideas Immanent in Nervous Activity" proved enormously influential (it has since been cited over 29,000 times). Pitts and McCulloch argued that the brain's neurons, with their "all-or-none" nature, could be understood using Boolean algebra.[34] Their work, it must be noted, was inspired by *another* publication, this from British mathematician Alan Turing in 1937. Turing's "On Computable Numbers" described a hypothetical "universal computing machine" that received instructions via paper tape of infinite length and used binary logic in an infinite array of "configurations" (what we might think of today as programs) to perform reading and writing operations.[35]

Such ideas naturally intrigued Berkeley. Treating neurons, for example, as electrical switches that could be toggled on or off suggested an abstracted model of how the mind worked. This went against decades of thinking by behavioral psychologists who treated the mind as a black box and focused their attention not on its workings but on its output (i.e., behavior). As McCulloch and Pitts eloquently stated it, the mind no longer needed to be treated "more ghostly than a ghost." Whereas journalists anthropomorphized early calculating machines like ENIAC to "brains," McCulloch proposed that the biological brain actually behaved like a "digital computing machine consisting of ten billion relays called neurons."[36]

The article by McCulloch and Pitts is important to our story in two ways. First, it had a catalyzing effect on the newly emerging cybernetics movement. As we'll see in the next chapter, Norbert Wiener, a mathematician at MIT and one of the field's prime movers, pitched cybernetics as the study of relationships between people and machines. A key debate for early cyberneticians was the degree to which their ideas could explain real behaviors in both "automatic machines and the central nervous system."[37]

But McCulloch and Pitts's mechanistic theory of mind also influenced the actual design of one of the earliest giant brains. In 1943, following a suggestion from Wiener, mathematician John von Neumann read McCulloch and Pitts's article. The timing was fortuitous as von Neumann, intrigued after seeing the ENIAC, imagined building a more capable machine called the Electronic Discrete Variable Arithmetic Computer (or, EDVAC). For his part, von Neumann, a consultant to Los Alamos, wanted to rapidly carry out calculations for the design of nuclear bombs. He was also one of the few Americans thinking about computers who had read Turing's prewar paper on "universal computing machines." His 1945 report on the EDVAC (commonly referred to as the "First Draft") was immensely influential on computer design concepts but also replete with biological references when describing the logical functions of a "very high speed automatic digital computing system."[38] Instead of conceptualizing the machine in terms of mechanical relays and other familiar components, von Neumann invoked the idealized neurons Pitts and McCulloch described.[39] Von Neumann, influenced also by his reading of Turing, made explicit comparisons between an electronic digital computer and a biological nervous system by analogizing electrical switches to the "neurons of the higher animals."[40]

"Computers as brains" was a common comparison circa 1950 but did the analogy mislead readers? At least one science journalist thought so. John E. Pfeiffer was among the emerging new cohort of science writers. From 1936 to 1942, he served as the science editor for *Newsweek* before joining the editorial board of *Scientific American*. "Mechanical brains," he told readers, were just a "journalistic cliché," one "dreamed up long before it had any particular meaning."[41] In the 1920s and 1930s, for example, comparisons were routinely made between human brains and automatic telephone exchanges.[42] Saying that "the brain is like a calculating machine"—which, Pfeiffer reminded his readers, is a simile—is much more tentative than saying "the brain *is* a calculating machine." That latter phrasing would be a metaphor, and metaphors, Pfeiffer judged, were simply "bad science," even when they "made good headlines."[43] (In 1962, in an act of considerable authorial irony, Pfeiffer published his own popular book about computers, titled *The Thinking Machine*.)

Anthropomorphic descriptions were an essential reference point in scores of popular articles and newspaper stories about electronic brains. Berkeley was well-acquainted with this literature while his connections to the nascent computer engineering community gave him access to technical discussions about logic operations and memory storage. Consequently, Berkeley's decision to call his book *Giant Brains* was eminently sensible. Just as the inner workings of such machines, even in their earliest occurrences, were exceptionally complex, he believed the implications of what happened when they carried out mathematical calculations had philosophical implications. As a result, Edmund Berkeley was determined that his forthcoming book be as clear and comprehensible as possible. In his quest for readability, he found an ally.

## PLAIN TALKING BRAINS

In 1947, when *Business Week* magazine profiled Rudolf Flesch, it identified him as the "Apostle of 'Plain Talk.'"[44] Born in Vienna in 1911, Flesch came to the United States after the Nazis seized power in Austria and earned a doctorate in library science at Columbia University. In 1955, his bestselling exposé *Why Johnny Can't Read* made Flesch a national authority on literacy. But, before this, Flesch worked as a writing consultant who stressed the importance of readability by using short sentences built from simple words, an idea he stressed in his first book, *The Art of Plain Talk*, which sold tens of thousands of copies.

Flesch's approach appealed to Berkeley because of its underlying methodology. Since his undergraduate days, Berkeley had believed that there were relationships between mathematics and language. The symbolic logic of Boolean algebra, for example, suggested a way to evaluate the truth or falsity of statements. In *The Art of Plain Talk*, Flesch proposed that the "readability" of sentences could be quantified based on what he called "affixes." This was his term for "gadgets added to root words," such as when "-ized" was added to "computer." Flesch claimed that fewer affixes, when combined with shorter sentences and more "personal references" such as "we" and "you," yielded a statistical approach for measuring readability. The result,

Flesch argued, was a writing style that was easier for the average person to comprehend.[45]

In the summer of 1947, Berkeley was revising *Giant Brains* before submitting chapters to Wiley for review. He and Flesch had become friends by this point and Berkeley shared the manuscript with him. Flesch, who studied law before emigrating to the United States, was not any sort of expert on computers. Not surprisingly, he advised Berkeley that the author's abundant references to symbolic logic would just "confuse the reader to no end." In fact, he suggested, it would "be better to leave mathematical logic out entirely." Flesch asked what sort of book Berkeley wanted *Giant Brains* to be. "You have concentrated 100% on the logic of the subject," he explained, but "neglected the psychology of the reader" who instead wanted "drama . . . the human element in the development of these machines."[46]

Flesch suggested that Berkeley should seek guidance in another book. *Mathematics for the Million* was published in 1937 by British zoologist and statistician Lancelot T. Hogben. Still in print today, the original version came with a laudatory blurb from Albert Einstein. Hogben's book emerged from the left-wing Social Relations of Science program. The movement's goal was to present science to the public as a tool for prompting political and social revolution. An essential part of this was communicating a scientific way of thinking to as many readers as possible.[47] Subtitled *How to Master the Magic of Numbers*, Hogben's book proved a bestseller (no mean feat for a 600-plus page tome packed with equations) and its friendly tenor was what Flesch encouraged Berkeley to emulate.[48]

When Berkeley summarized his own book's attributes for his publisher, he highlighted how he had followed Flesch's maxims by using simple words whenever possible. In the published preface to *Giant Brains*, he even noted that he had counted all the words in the book with two or more syllables (and found less than 1,800 of these). All of this was done to assuage any readers who might be put off by overly detailed or technical descriptions. For instance, Berkeley, without using any more physics than needed, explained what an electronic tube was (basically, a switch) and why it was important (it handled information by reducing it to fundamental units when toggled on or off). Berkeley's description invited comparisons to things that his readers would be familiar with, such as light switches, even as he highlighted features, such as low cost and high speed, which engineers valued.

Clarity remained a goal for Berkeley but so was social responsibility. In a note to himself, Berkeley asked, "Have you a striking word picture of the typical performance of a giant brain? Do you show the essential difference between the human brain and the machine?" Taking a broader view, he wondered, "Have you avoided ideas which are socially harmful? Communicating with spirits; psychiatric machines; mass opinion machine." As an as-yet-published author, Berkeley fretted about trying to meet conflicting goals with his book without "insulting the intelligence of those who would be interested in the technical exposition and boring those who want to enjoy it."[49]

There were legal issues as well. Most anyone who knew Berkeley as a computer advocate also knew that he worked for Prudential. To make matters more complicated, while Berkeley was writing *Giant Brains*, Prudential executives were negotiating with companies like IBM and Eckert-Mauchly Computer Corporation to acquire new computing machines. As a result, company managers insisted that Berkeley allow them to review the manuscript.

Zehman I. Mosesson was tasked with evaluating the propriety of *Giant Brains*. Mosesson, who had worked with Berkeley in the company's new electronics lab, had been his classmate at Harvard where he too studied mathematics. After getting his PhD, he decided that he probably would not find a secure job as a math professor with the Depression still underway. So, like Berkeley, he joined Prudential as an actuary.[50] Mosesson found several sections in the manuscript that might "reflect on the Prudential." In the chapter that discussed IBM's tried-and-true punch card-calculating machines, for example, Mosesson asked "Is IBM satisfied with this chapter?" They were. The author had already vetted the chapter with company representatives who found it "substantially correct."[51] Mosesson also took issue with Berkeley's frequent personification of computing machines by using the pronoun "he" instead if "it." Berkeley, on this point, balked.

The biggest issue that Mosesson highlighted is one, ironically, which *Giant Brains* is best known for today. Berkeley wanted his book's last two chapters to deal both with the future applications of new computers as well ways in which people should try to control the powerful new technology.

Not surprisingly, Berkeley's obvious point of reference was nuclear weapons. In his own carefully handwritten notes for the last chapter of *Giant Brains*, he jotted that "What sort of control do we need? Mechanical brain, by itself, just processes information. The use of information is what counts. Analogy to the atom bomb."[52] Berkeley's musings concerned Mosesson. "You run the risk," he told his colleague, "of attempting a social crusade." Even more concerning "in this day of Red-baiting," was Berkeley's reference to "our friends, the peoples of Great Britain and of the Soviet Union." Berkeley—unaware that the FBI was monitoring him—concurred and set to rewriting this and several other passages.[53] It speaks to the suspicions of the era that Berkeley ended his preface to *Giant Brains* by noting that, to his knowledge, no information in his book was "classified by the Department of Defense."

With Prudential's legal counsel giving its approval, Berkeley faced one last hurdle: the anonymous readers (there were at least five) to whom Wiley sent the manuscript. Among academic writers, "Reviewer #2" is an inside joke that refers to those petty and pedantic readers that an author must necessarily come to terms with before something is accepted for publication. In Berkeley's case, his biggest critic literally was Reviewer #2 (or, more precisely, a person identified simply as "Critic B") and they *did not* like the book. Their criticisms spanned the gamut from stylistic ("over-simplification of language, frequent incoherence, non sequiturs . . . almost condescending") to Berkeley's presentation of technical material ("the language is difficult to grasp"). The anonymous reader recommended that all nontechnical material be dropped, especially those chapters where Berkeley considered the social implications of computers. "The book," the outside reader concluded, "is not a good one on the whole."[54]

Critic B's reaction alarmed editors at Wiley enough that they did something quite unusual—they pivoted to a different anonymous referee ("Critic A") and asked *them* what they thought of the negative report. It was true, Critic A opined, that "more polishing" would improve Berkeley's book but they insisted the manuscript was still quite good. "Certainly, this book is not written for the highly trained person who wants to become an engineer of calculating machines," they noted. "On the whole I am not worried by any style the author has adopted." Instead, Critic A praised Berkeley's speculations about what computers might likely become over time as well as their social and economic implications.[55]

Fortunately, these comments complemented those from other anonymous readers. While "not great literature," said one, "Berkeley has the rare gift of being able to write as naturally as though he were chatting with a friend. . . . It has the virtues of brevity, clarity, informality, and liveliness." The praise was sufficient to satisfy any lingering concerns among the editors at Wiley & Sons. On the Tuesday after Thanksgiving in 1949, *Giant Brains, or Machines That Think*, jacketed in a striking yellow and black cover, began appearing on bookshelves around the United States.

### BRAINS, GIANT AND SIMPLE

To understand how readers and reviewers reacted to Berkeley's *Giant Brains*, one must consider another book in whose shadow it appeared. Norbert Wiener's now-classic *Cybernetics, or Control and Communication in the Animal and the Machine* (discussed in detail in the next chapter) had appeared a year earlier than *Giant Brains*. Historians of technology usually point to *Cybernetics*, and not Berkeley's book, as the work that helped popularize computing in the immediate postwar period. This is surprising given that computers themselves scarcely appear in *Cybernetics*. Wiener's goal was not to explain how actual computing machines worked but rather to present theoretical reflections on the statistical nature of information and the analogies he saw between human nervous systems and the control of machines. These new concepts, as Wiener grandly described in a memoir, "involved a new interpretation of man, of man's knowledge of the universe, and of society."[56]

To the surprise of everyone, including the bombastic Wiener, *Cybernetics* became an immediate hit. So, when it came time to promote *Giant Brains*, Wiley had a marketing plan already primed. Often overlooked is that *both* books were associated with John Wiley & Sons. Wiley was, of course, the publisher of *Giant Brains*. But Wiley also did editing and marketing duties for the Technology Press, an imprint MIT established in 1932, which served as the American publisher for Wiener's book. To Wiley's executives, *Cybernetics* and *Giant Brains* were connected products.

People were bound to draw comparisons between the two books. Wiley, in fact, highlighted this association in its marketing campaign for Berkeley's book. If one flips over an original copy of *Giant Brains*, there isn't the usual summary of the book accompanied by those authorial endorsements infelicitously known in the publishing trade as "blurbs." Instead, Wiley's marketing department turned the entire back of *Giant Brains'* dustjacket into an advertisement for *Cybernetics* and included a short statement from Wiener along with commendatory statements about that book. The advertisements Wiley placed in newspapers for *Giant Brains* linked it to the better-selling *Cybernetics* as well.[57] The press release John Wiley & Sons issued for Berkeley's book began by noting that "ever since the world was startled to learn that it had a new science, 'cybernetics,' the ordinary soul has been looking for two things: a hole to hide in before the robots take over, or a plain everyday explanation."[58] *Giant Brains*, Wiley promised, would provide the latter.

It is common practice, even today, for many presses to ask authors to respond to a set of stock questions about their book and how to best market it. Berkeley, not surprisingly, answered meticulously. The questions that publishers ask are mostly pro forma: What journals might review your book? Which professional societies would be interested in it? Can it be used as a textbook? When asked about competing books, Berkeley named two possibilities: *The Theory of Mathematical Machines*, by mathematician Francis Murray, had appeared in 1947 but its technical nature made it not "understandable to business men." The other book Berkeley noted was the much more abstract *Cybernetics*. But even though he included it in his own considerable list of references at the end of *Giant Brains*, *Cybernetics* was "not simply written nor easy to understand."[59] Given his efforts to distinguish his own book from Wiener's, there is considerable irony that the *New York Post*'s profile of Berkeley labeled him as an "apostle of cybernetics."[60]

It's also common for authors to suggest people who might offer an endorsement. For example, Berkeley recommended Vannevar Bush, a former engineering professor and dean at MIT, whose "differential analyzer"—a mechanical calculating machine built circa 1930 to solve differential equations—was featured in a chapter of *Giant Brains*. Just as telling are the people Berkeley asked Wiley's staff *not* to contact. Off-limits were John von

Neumann and Howard Aiken "because of their strong individuality." (Berkeley did thank Aiken in his acknowledgements but one might wonder if this was this genuine or an attempt to deflect any criticism from his former boss.)

Berkeley's efforts to achieve clarity in writing was met with Rudolf Flesch's approval. After receiving a signed copy of *Giant Brains*, Flesch praised his friend's "beautiful prose," which he read as a combination of "chaste scientific style and childish simplicity" used to "extraordinarily effective" ends.[61] Wiley's publicity people likewise highlighted readability in its promotional copy. The publisher's press release, for example, claimed *Giant Brains* proved that "scientific writing can be done in plain English, and with considerable humor" (two qualities definitely not found in Wiener's book). Indeed, Berkeley's publisher claimed *Giant Brains* offered readers an "explanation of processes that until now have been as unreadable to the ordinary mortal as Hittite writings."[62]

What would an ordinary person, interested in technology but without advanced training, have found between the cover of Berkeley's book after purchasing a copy? Given that the subtitle of *Giant Brains* was *Machines That Think*, Berkeley needed to demonstrate that so-called mechanical brains indeed did this, at least in some limited capacity. Brains, mechanical and human, he argued, make choices based on logic, displaying "the kind of behavior that we call thinking." For example: "When you and I add 12 and 8 and make 20, we are thinking. . . . Or, suppose that you are walking along a road and come to a fork. If you stop, read the signpost, and then choose left or right, you are thinking. . . . A machine can do this."[63] Berkeley similarly explained how a machine "remembers" information by storing it in their "memory," processes akin to learning and recollecting in the human brain. His conclusion, not surprisingly, was that "a machine, therefore, can think."

Berkeley also needed to quickly demystify the concept of information, a term that was in the process of acquiring a very specific connotation for electrical engineers and computer specialists. In this, Berkeley borrowed ideas from Claude Shannon, a mathematician and engineer at Bell Laboratories, who, along with Norbert Wiener, defined information in terms of probability and entropy. This approach might have seemed strange to people familiar with more vernacular uses of the word. So, unlike Wiener or Shannon, Berkeley defined his terms without resorting to equations or technical jargon:

By *thinking*, we mean computing, reasoning, and other handling of information. By *information*, we mean collections of ideas—physically, collections of marks that have meaning. By *handling* information, we mean proceeding logically from some ideas to other ideas—physically, changing some marks to other marks in ways that have meaning. For example, one of your hands can express an idea: it can store the number 3 for a short while by turning 3 fingers up and 2 down. In the same way, a machine can express an idea: it can store information by arranging some equipment.[64]

Berkeley was less concerned about describing theoretical concepts put forth by mathematicians. Instead, he wanted to tell ordinary readers *how* mechanical brains handled information. What was going on in all those large and intimidating metal boxes, with their miles of wire and cords? To do this, chapters of *Giant Brains* had sections such as "How Information Goes Into the Machine" and "How Information Comes Out of the Machine."

After Berkeley's gentle warmup, the next six chapters of *Giant Brains* became much more technical. Berkeley encouraged readers unacquainted with mathematical logic to choose their own reading adventure by following expository threads only so far as it "proves to be congenial."[65] Intrepid readers, however, received a clear explanation of a half-dozen different kinds of mechanical brains. After starting with traditional, punch card-calculating machines, Berkeley presented five other machines of increasing complexity, ranging from MIT's "differential analyzer" and the Mark I machine he had worked with at Harvard University to more complex machines, such as the ENIAC and Bell Laboratory's "General-Purpose Relay Calculator."

Given that most readers of *Giant Brains* had quite possibly never seen any of these machines in person, Berkeley had to explain some basic technological concepts. One of these was the "electronic tube." Fortunately, Berkeley's readers probably had radio sets in their homes that used similar devices. While readers might not know how such tubes worked, they had certainly seen them, and perhaps even replaced a tube or two when they burnt out.

With the exception of punch card machines, the "brains" Berkeley chose to write about were all custom-made machines. There was, for example, only one ENIAC in the world. His book appeared at a time when what we now call computers remained rare and bespoke. Built not to process data for business, they were instead specialized pieces of equipment built to carry out calculations for scientific and military functions (the two realms often overlapped). However, when Wiley authorized a second print run for *Giant Brains* in the early 1950s, computers were already transmuting from one-of-a-kind scientific tools to business-oriented machines that would soon be mass-produced.

This transformation also entailed a change in the nature of the machines themselves. Berkeley wrote his book when many computing machines were hybrids of one sort or another. Most of the machines he described were, at some level, still mechanical in some sense with moving gears and turning shafts. Iconic machines such as ENIAC, meanwhile, needed a slew of qualifiers to accurately describe them ("the first electronic, general-purpose, large scale, digital computer"). However, a narrative soon emerged in which digital calculating machines, whose data was represented as a discrete series of numbers, were promoted by a growing number of computer designers and science writers as *the* future of computing. Even though analog "brains" did not disappear—indeed, they were used for specific tasks well into the 1960s—"computer" soon came to be popularly understood as a digital, all-electronic machine that could perform a whole set of general functions.[66]

The machines Berkeley wrote about still used relays and vacuum tubes as their basic logic elements. Both devices could exist in a "yes" or "no" state (like McCulloch and Pitts's idealized neuron) thus providing the basic binary logic necessary for a mechanical brain to work. There is no foreshadowing in *Giant Brains* of what would eventually supplant them as the essential logic device in modern computers: the transistor. John Bardeen and Walter Brattain built their first working transistor at Bell Labs in late 1947 but the word itself didn't appear in major newspapers until mid-1948.[67] By this point, it would have been too late for Berkeley to substantially alter his manuscript and, in any case, it was far from obvious that transistors would become the essential element around which almost all computers would be built. In fact, the first transistorized computers didn't start appearing until the mid-1950s.[68] By the same token, Berkeley also didn't discuss the idea of a "stored program machine." Frequently associated with what John von Neumann theorized about in his now-iconic report for the EDVAC project, this referred to how a computer's memory could both hold instructions for operations it would perform as well as the actual numbers on which it operated. In all fairness to Berkeley, he was writing about a technology that was changing almost monthly. Given the often slow pace of book publishing (then as now), it would have been very challenging for him to offer an au courant account when the book first appeared in late 1949.

Despite the generous attention science writers and other journalists had been paying to "giant brains" since the early 1940s, Berkeley reminded readers that "computer" didn't have to always mean an expensive behemoth. "A Machine That Will Think" is one of the most curious chapters in *Giant Brains*. It lays out basic instructions for readers to build their own "very simple mechanical brain" he called "Simon" (named after the nursery rhyme, "Simple Simon"). Simon was, as its namesake suggested, quite modest in abilities, being able, in its original version, to work only with the numbers 0, 1, 2, and 3. However, Berkeley introduced Simon—referred to throughout the chapter as "him"—as a teaching tool, much like a "set of simple chemical experiments," that could illustrate basic concepts of logic and binary math.

In May 1950, about six months after *Giant Brains* first appeared, Berkeley unveiled a physical version of Simon at Columbia University to journalists and scientists. The machine had been built with the help of two engineering students, was only about sixty square inches in size, and cost about $550 (or, about $7,000 in today's money).[69] The *New York Times* was generous with its coverage of Simon's debut saying that the machine was "not so very dumb" nor was it intimidating. When pressed beyond its limits, such as being asked to add two and two via instructions fed into it with punched paper tape, a red light would switch on, something the more caustic *New York Herald Tribune* reporter referred to as a "feeble-minded brain" having a "nervous breakdown."[70]

Despite Simon's limits and some dyspeptic press coverage, Berkeley secured an impressive amount of publicity for his machine. *Scientific American* featured Simple Simon on the cover of its November 1950 issue. Inside, magazine readers learned how Berkeley's "small mechanical brain" could help explain the basics of larger and more expensive machines.[71] And, reaching beyond the audience of people with a general interest in science, Berkeley published a multipart series about Simon, including instructions for building one, in the hobbyist magazine *Radio-Electronics*. Simon—the "world's smallest electric brain"—was again featured on the cover, this time with a smartly dressed young woman feeding instructions to it via paper tape.[72]

All of these articles, of course, also helped promote *Giant Brains*. Simon was just the first of a series of small-scale electronic devices—Berkeley called them "small robots"—he offered for sale, either as sets of plans for hobbyists to build or as actual kits. In addition to Simon, for example, there was "Squee, the Robot Squirrel," a machine that moved about a room while hunting for "nuts." (The nuts in question were actually tennis balls, illuminated by flashlight, which light receptors on the small machine could sense.)[73]

Berkeley hoped machines like Squee and Simon would teach people the principles of Boolean logic, binary notation, and basic programming. When Ivan Sutherland was interviewed after receiving the Turing Award— touted today as computing's Nobel Prize—from the ACM in 1988, he recounted how Berkeley's "wonderful 'personal computer'" had influenced him as a teenager. The computer graphics pioneer experimented with a Simon machine, modifying it by adding a "conditional branch" (a programming instruction that, after making a comparison, directed the computer to another part of the program) using a long set of instructions coded onto paper tape. As a result, Sutherland's Simon could divide numbers as well as adding and subtracting them. "Berkeley," Sutherland recalled, "taught me quite a lot about computing."[74]

The hundreds of devices and blueprints that Berkeley's small consulting company sold throughout the 1950s and 1960s occupy a curious spot in the history of technology. Part entertainment and part educational tool, Berkeley's "small robots" harkened back to the equipment built by amateur radio operators starting in the early twentieth century.[75] Another parallel is Heathkits, the popular amateur electronics product, which first became available in 1947. Devices like Simon also presaged homebuilt personal computers, such as the Altair 8800, which computer hobbyists were passionate about in the 1970s (we will revisit this topic again in a later chapter). Simon was, after all, a small, accessible, digital computer that a knowledgeable user could program and play with. Berkeley sensed this future, or at least its general contours. At the end of his 1950 *Scientific American* article, Berkeley predicted that in the future, ordinary people would have "small computers in our homes . . . [just] like refrigerators or radios." And, just as he had hoped with *Giant Brains*, Berkeley wanted to encourage people to start considering the broader "philosophical and social implications of machines that handle information."[76]

### APPALLING, YET FASCINATING

It is the rare author who is fully satisfied with how a publisher promotes their book. Edmund Berkeley was no exception. A misplaced advertisement in *Harper's*, for example, prompted a gentle authorial rebuke and an apology from Wiley's advertising director about its "inconspicuous appearance."[77] To help his publisher, Berkeley drafted a typically methodical report titled "Doubling the Sale of Technical Books," which analyzed general attitudes among publishers, readers, and book sellers.[78] However, despite Berkeley's perhaps unrealistic anticipations, the large number of reviews and articles about *Giant Brains* suggests that Wiley was both diligent and successful in promoting the book.

Like many presses, Wiley prepared a detailed promotional plan for Berkeley's book.[79] Advertisements, often juxtaposing *Cybernetics* and *Giant Brains*, appeared in major publications ranging from the *New York Times* and *Scientific American* to the *Chicago Tribune* and *Astounding Science Fiction*. Wiley's advertising circulars went out to over 2,000 names on its science and engineering lists in both the United States and overseas. A carefully targeted group of science writers and engineers received free copies as well. The campaign paid off. The *Saturday Review* and *Newsweek* featured *Giant Brains* as did the Book of the Month Club (a major publicity coup) along with a host of newspapers.

The *New York Times* reviewed *Giant Brains* soon after it appeared in bookstores. Prepared by science writer John Pfeiffer, the assessment, titled "Mechanical Logicians," praised Berkeley's accomplishment. *Giant Brains*, Pfeiffer said, served as an "ideal companion volume" to Wiener's *Cybernetics* while also standing as the "first book about calculating machines intended for popular consumption."[80] Like many other reviewers, Pfeiffer was less drawn to Berkeley's descriptions about how "mechanical brains" worked and instead called attention to the book's last chapter. Titled "Social Control," this was the chapter that had so concerned Prudential executives. Berkeley argued that whatever likely threats posed by "robot machines" could only be met by securing full employment for people in a peaceful world. This might necessitate, he speculated, the creation of a public body "like an Atomic Energy Commission, Bacterial Defense Commission, Mental Health Commission, and Robot Machine Commission, all rolled into one."[81] Such predictions led the *New York Herald Tribune* to label *Giant Brains* "an appalling, yet, fascinating, little book."[82]

Several prominent engineers and scientists agreed to review *Giant Brains* for professional and academic journals. These included physicist Nicholas Metropolis, who helped build the Mathematical Analyzer Numerical Integrator and Automatic Computer (which cleverly reduced to MANIAC). Located at Los Alamos, New Mexico, its calculations underpinned the design of thermonuclear weapons. While noting that *Giant Brains* had not kept up with the rapid pace of technical developments, Metropolis branded it "the first of its kind; its account is lucid."[83] Richard Hamming, a mathematician at Bell Labs and eventual Turing Award winner, disliked Berkeley's last chapter on social control ("might well have been left out") but praised the book's accessibility and concluded that Berkeley made a convincing case for the book's subtitle of "machines that think."[84]

In fact, *Giant Brains* did well enough that *Publishers Weekly* noted that "advance sales of the book considerably exceeded" Wiley's expectations, prompting Berkeley's publisher to continue promotion and publicity.[85] By the end of 1950, Wiley reported that some 7,500 copies had been sold. *Giant Brains* didn't make Berkeley wildly rich; a royalty statement suggests he earned about $0.60 for every copy sold. Still, 7,500 copies sold translated to some $4,500 in 1950 or about $59,000 in today's money. A third print run was planned for 1952 and Wiley's head of advertising predicted that "we shall continue to sell *Giant Brains* for some time to come."[86] Press coverage and book reviews, meanwhile, continued well after the book first appeared giving *Giant Brains* what publishers would today call a "long tail."

*Giant Brains* was not Edmund Berkeley's only book aimed at a wide audience. In 1962, a year after *Giant Brains* was reissued, Doubleday published his new book, titled *The Computer Revolution*. That same year, John Pfeiffer also presented an "introduction to the world of electronic devices" with his own book, *The Thinking Machine*. Whereas Berkeley remained concerned about the rapidly escalating social and ethical consequences of computing, Pfeiffer was more interested in exploring "parallels between animate and inanimate thinking"—a theme *Giant Brains* hit hard twelve years earlier—especially given the attention artificial intelligence was receiving in the early 1960s. The title of John W. Mauchly's favorable review of Berkeley's and Pfeiffer's books for the journal *Science* captured this succinctly: "Revolution and Evolution."[87]

As the years passed, however, two trends in books about computers emerged. First of all, more people became familiar with what computers looked like and what they did. So, books explaining *what* they were became increasingly unnecessary. At the same time, the basic operations of computers were "black-boxed," meaning authors were inclined to spend less time explaining *how* they operated.

Throughout the rest of his professional life, Berkeley insisted that computer experts had a special responsibility to speak to the public about how to "prevent socially undesirable applications" of the machines and the code they knew so well, just like some physicists believed their community had a duty to speak out about the threat of nuclear war.[88] Berkeley returned to this theme over and over again in the pages of *Computers and Automation*, a journal he started in 1952, and as an organizing force behind the ACM's Committee on the Social Responsibility of Computer Scientists.[89]

It's difficult to assess a book's influence over time. One can consider the number of copies sold or how many times (and how well) it was reviewed. There are also the number of citations it receives in other people's work. And, of course, there are the reactions from readers, which are important but rarely preserved and often anecdotal. *Giant Brains* was the first popular book about computers written for a general audience. It captured the computer at a key transition point when it transformed from a bespoke instrument used mostly by scientists to a mass-produced data processing machine. Berkeley's book set a pattern that many other books followed. Beyond the book's technical information, much of it quickly out of date, *Giant Brains* raised two questions that would appear over and over in other books about computers: Could computers think? And, what would their increasing ubiquity throughout society mean?

The "brain is like a calculating machine" simile persisted for many years, if not decades, even as the strangeness of computers slowly dissipated. A highlight of the 1952 presidential contest between Dwight D. Eisenhower and Adlai Stevenson was the CBS television network's plan to use a "fabulous mechanical brain" for analyzing election returns and predicting the outcome. The UNIVAC machine that CBS hired consistently forecast such a lopsided victory for Eisenhower that, despite massive pre-election day hype, the network delayed reporting its predictions.

With *Giant Brains*, Berkeley confirmed the average reader's inclination to imagine that computers were, at some level, machines that could think like people did. This anthropomorphic perspective was reinforced in 1950 by a *Time* magazine cover. The January 23 issue featured an illustration of a computer (the Mark III) accompanied by a caption that read "Can man build a superman?" Drawn by the Russian American illustrator Boris Artzybasheff, the machine came clad in military garb, a nod to the powerful patrons who had made such technologies possible in the first place. But most intriguing were the artist's addition of human eyes, arms, and hands, which the computer used to monitor its electronic output and input. In Artzybasheff's image, computers and people shared an interdependent relationship. The rendering foreshadowed a prominent debate among engineers, business executives, labor leaders, and politicians throughout the 1950s: In the factories and offices of the future, would people be displaced and replaced by computers and other automated machines?

# 2 GOLEMS

In 1977, a trio of researchers traveled to a conference at the Massachusetts Institute of Technology. They went there to describe a computer-controlled electromechanical arm they had built. Their project's name was GOALEM (short for Goal-oriented Electrical Manipulator), a mischievous nod to their home institution: the Czech Technical University in Prague.[1]

Prague, of course, was where Rabbi Judah Löw ben Bezalel allegedly used an arrangement of three letters in the late sixteenth century to turn inanimate clay into an artificial creature—a golem—that did his bidding.[2] One might even say that Rabbi Löw's animating letters were the code that programmed and controlled (sometimes) the golem. On most days, Löw's golem performed menial tasks. In advance of the Sabbath, Löw would remove the animating letters from his creation's mouth, rendering it inert. But when the rabbi forgot to do this and the golem terrorized Prague, considerable ingenuity was required to undo Löw's incantation. Löw's three letters spelled "truth" (אמת) in Hebrew but, with the letter aleph removed, the word means "death" and the golem collapsed back into lifeless clay. Like the creature in Mary Shelley's *Frankenstein*, Rabbi Löw's golem became a synecdoche for the hubris of scientists claiming godlike powers and the uncontrollable consequences of their creations.

While GOALEM stands as a footnote in the history of computing, it relates to the book you are reading now in several ways. First, I learned about GOALEM when I read Pamela McCorduck's 1979 book *Machines Who Think*, her "personal inquiry" into the history of artificial intelligence.

Second, McCorduck's interviews with computer scientists revealed that some of them claimed Rabbi Löw as a distant relative. Her list included the MIT mathematician Norbert Wiener.[3] With regards to the latter, at least, McCorduck was mistaken. Wiener was quite forthright about his Jewish background but instead said the twelfth-century Sephardic polymath Moses ben Maimon (better known as Maimonides) was an ancestor.[4]

There is a third and stronger connection, however. In April 1964, a month after Wiener passed away at age sixty-nine, his last book, titled *God & Golem, Inc.*, was published.[5] In it, he referenced the golem fable and the many variations on it, such as Goethe's 1797 poem "*Der Zauberlehrling*" (also known as "The Sorcerer's Apprentice," and, for Mickey Mouse fans, a key part of the 1940 film *Fantasia*). *God & Golem's* subtitle, *A Comment on Certain Points where Cybernetics Impinges on Religion* appears odd at first but, as we'll see, Wiener's title makes much more sense once the book's history and its author are understood.

Wiener's 1964 book was really about *two* golems. One was the rapid adoption of automation in the United States. Some experts claimed the automatization of factories via the deployment of computer-controlled machines presented the specter of mass unemployment.[6] These technologies, some said, also endangered world peace as computers were more deeply woven into military decision making. Automation's repercussions had concerned Wiener ever since 1948 when he published *Cybernetics, or Control and Communication in the Animal and the Machine*, the work that made him internationally famous. Two years later, his next and equally successful book, *The Human Use of Human Beings: Cybernetics and Society*, amplified his unease regarding automation.

Then, with *God and Golem Inc.*, Wiener linked anxieties about automation to another golem: "machines that learn" or what today we might call artificial intelligence. Wiener predicted that future computers were the "modern counterpart of the Golem of the Rabbi of Prague," with the potential to replicate themselves or even "let loose the apocalyptic terrors of nuclear warfare."[7] This path would begin, he warned, when such machines could play classic board games or translate texts written in other languages into English.

In his books, Wiener sometimes obscured his ideas behind impenetrable mathematical formulae or distracted readers with seemingly irrelevant erudition. At other times, Wiener's thoughts soared to abstractions, which he tried to explain with platitudes and fairy tales. All three books in his cybernetic trilogy were dogged by writerly disorganization. One reviewer described *Human Use* as a "brilliant and disorderly book [but] what you might want to skip may delight someone else."[8]

A casual reader might miss the consistency found across Wiener's three books. But reading them in quick succession reveals how each one builds upon the last. Like a refrain in the classical music Wiener favored, certain elements appear and then, in another book, reappear but rephrased and refined due to the passage of time and technological change. *Cybernetics* was a dense book, full of equations and mathematical statements accompanied by ancillary material on topics like psychopathology and the relations between information, society, and language. *Human Use* returned to social issues Wiener hinted at in *Cybernetics* and explored them in more depth. By the time he composed *God and Golem*, these themes were reduced to ethical essentials such as the implications of machines that could learn and self-replicate, traits traditionally identified with humans made in God's image. Across nearly a quarter century, Wiener transitioned from fundamental research on the interactions of people and machines to articulating misgivings about what his words and formulae had helped create. Ultimately, books, not mathematics, made Norbert Wiener and his ideas famous.

### PRODIGIOUS

Norbert Wiener dedicated *The Human Use of Human Beings* to his father, Leo Wiener. He was, the epigraph said, Norbert's "closest mentor and dearest antagonist." Those few words contained considerable pride as well as pain. Born in Columbia, Missouri, in November 1894, Norbert Wiener grew up surrounded by books. At eighteen months, he was already learning the alphabet and, at age four, he greedily consumed tales such as *Alice in Wonderland* and *Arabian Nights*. His father was a master of languages who spent thirty-five years as a professor of Slavic literature at Harvard University. Norbert chronicled his early years in a memoir (the first of two he wrote), which he tellingly titled *Ex-Prodigy*. It was not always, not even often, a happy childhood. Wiener's original title for it was *The Bent Twig*.[9]

As the title suggests, Norbert Wiener was a reluctant prodigy, emotionally immature and socially awkward. A contemporary described him as a "baroque figure, short, rotund, and myopic. . . . his conversation was a curious mixture of pomposity and wantonness. He was a poor listener. . . . He spoke many languages but was not easy to understand in any of them."[10] Wiener's path to being pronounced a wunderkind was largely due to his father, who, as the epigraph of *Human Use* suggests, oscillated between supportive and abusive.

Leo Wiener treated his eldest son's education as something of a pedagogical experiment. He trained young Norbert in classical languages and literature with mathematics, botany, and classical physics added later to the home-schooling curriculum. Decades later, the scars from these educational experiences were still raw. In *Ex-Prodigy*, Wiener described his father berating him in a multitude of languages that when "combined with irony and sarcasm . . . became a knot with many lashes."[11]

Norbert graduated high school at age ten. Undergraduate study at Tufts University followed, where he took classes in chemistry, philosophy, and mathematics. Given his clumsiness in the lab, the latter two disciplines suited him better and he graduated from Tufts in 1909 with a degree in math, at age fourteen. Despite his private admonishments, Leo Wiener publicized his son's academic prowess. Norbert, he told one reporter, had a "keen analytical mind" and took in knowledge via reasoning, not rote learning. Still, he was "lazy" and "doesn't study as much as the average boy his age." (One wonders what "average" meant to Leo Wiener.) The article's headline designated Norbert as "The Most Remarkable Boy in the World" and was accompanied by a photograph of him standing on a pile of books.[12] Public interest in Leo Wiener's prodigy continued after Norbert started graduate school at Harvard in the fall of 1909. He graduated in 1913, age eighteen, after writing an obstruse PhD dissertation that straddled math and philosophy.

A postdoctoral appointment in Göttingen, Germany, helped convince Wiener to quit the rarified realm of pure abstract mathematics in search of practical problems to which he might apply his talents. The outbreak of World War I in 1914 brought Wiener back to the United States and further studies at Columbia University were followed with a temporary lectureship (secured with Leo's help) in philosophy at Harvard. His experiences in the classroom proved disastrous and a permanent post never materialized, a grudge Wiener carried for decades. Attempts to obtain an army commission also came to naught when Norbert flunked the physical exam and fell off a horse.

In desperation, Wiener signed on with General Electric as an apprentice technician but Leo, horrified at the thought of his son doing manual labor, browbeat his son until he quit. He then secured a job for Norbert as a writer at the *Encyclopedia Americana*, which Wiener, who professed encyclopedic knowledge, enjoyed.[13] Then, in the summer of 1918, Wiener, now twenty-three years old, relocated to Aberdeen, Maryland, where the army tested military equipment. Wiener was specifically recruited to be a "computer"—one of the men and women whose math talents enabled them to calculate such things as ballistic tables and firing solutions. At Aberdeen, he enjoyed the "cloistered but enthusiastic intellectual life which I had previously experienced at the English Cambridge, but at no other American university."[14]

The war soon ended, however, and with it, Wiener's military service. Once again, the mathematician turned to words to earn a living, this time as a feature writer for the *Boston Herald*. One of his first assignments was covering labor unrest in Lawrence, Massachusetts, a fading textile-producing town. The experience gave Wiener a sense of empathy regarding the challenges facing laborers—something he returned to in his writings about automation—and it boosted his confidence as a writer. "After I left the *Herald*," he boasted with characteristic verbosity, "I was reasonably confident that if it should ever be my duty to say anything in print, I could say it fairly correctly and forcefully the first time I should write it down."[15]

In 1919, Wiener's professional peregrinations finally ended. One of his father's colleagues encouraged him to apply as a mathematics instructor at the Massachusetts Institute of Technology. Despite having published little mathematics research and possessing an unpresuming teaching record, Wiener fit in well at MIT. He liked the seriousness of his students and they appreciated his liberal grading, if not his eccentric pedagogy. The undistinguished nature of MIT's math department at the time also did not provoke Wiener's feelings of insecurity, which never lay far below the surface. Wiener later dedicated his second memoir, titled *I Am a Mathematician: The Later Life of a Prodigy*, to MIT, reflecting the general satisfaction he found there.[16]

Wiener restarted his research in applied mathematics and MIT promoted him to an assistant professor position in 1924; he became an associate professor in 1929 and a full professor three years later. His reputation grew as prizes, invited lectures, and election to the National Academy of Sciences all followed in the 1930s. Despite his professional successes, old doubts remained and he regularly sought affirmation from his academic associates. "His usual words of greeting became," a colleague noted, "'Tell me, am I slipping?'"[17] Although Leo Wiener passed away in 1939, his prodigy very much wanted more opportunities to prove his worth. A new research area to call his own would do quite nicely.

### "I EAT MY ROYALTIES . . ."

Norbert Wiener could have written parts of *Cybernetics* even before World War II broke out in Europe. Several parts of the book relied on his earlier work, such as the mathematical problems of prediction, for example. Meanwhile, the real-world examples Wiener used to illustrate basic cybernetic principles—thermostats, the governors that regulated steam engines, and telephone exchanges—were well-known to the general public. For two decades, Wiener had been surrounded by MIT engineers who studied electrical systems that were dynamically monitored and controlled via the same processes he would describe in *Cybernetics*.[18] But there were three key ingredients Wiener needed that only came together for him in the 1940s.

The first was war. New military problems presented Wiener with opportunities to explore the principles of what he would later brand "cybernetics" in a laboratory setting and then to write about it, first in classified reports and, after the conflict ended, in a bestselling book. The second ingredient was collaboration. Partnerships Wiener formed during the war provided him with a vibrant community of physiologists, anthropologists, physicists, mathematicians, and neuroscientists. Finally, cybernetics needed a champion, someone who was not just a researcher but also a publicist, philosopher, and prophet. This was a role Norbert Wiener was willing, if not always contented, to play.

Wiener explained many times how he coined the word "cybernetics" from *kybernētēs*, Greek for a steersman who navigated a craft. Ever the son of a philologist, he also noted that the Latin corruption of the original Greek, *gubernātor*, provided the English word for "governor," a nineteenth-century term for a mechanism that regulates the output of a steam engine.[19] Linguistics aside, the history of *Cybernetics* as a book bought (if not always read) by tens of thousands of people is less well known. While cybernetics might have been of great interest to a small and eclectic community of researchers, it remained relatively obscure until Wiener synthesized the ideas he and his like-minded colleagues shared into a bestselling book.

Wiener's direct involvement with what became cybernetics started in 1940. Wiener and Julian Bigelow, an electrical engineer and amateur pilot, received a small contract to study how antiaircraft guns could track an airplane's flight, predict its location, and direct gunfire to that point. To what was already a complex math problem, Wiener and Bigelow added the murkier element of human neurophysiology. Where the plane and intercepting munitions might be in time and space depended on the reactions and behaviors of the gunner on the ground (who wants to shoot down the plane) and the pilot (who most assuredly does not want to be shot down).[20] Communication, control, and feedback between people and machines were central to solving the problem. These principles provided a foundation for almost all early cybernetic thinking. The general rule was that whatever work an entity did—be it a person, a factory, or a computer—involved the conversion of specific inputs into specific outputs by following certain rules and principles.[21]

Two wartime publications set the stage for Wiener's presentation of *Cybernetics* to a wide readership in 1948. In February 1942, he and Bigelow submitted a classified technical paper that summarized their research. With some 120 pages of dense mathematical theory and equations, it garnered the nickname The Yellow Peril, a moniker Wiener said came from the color that Springer-Verlag used for the math books it published.[22] A year later,

Wiener coauthored another (much shorter) paper with Bigelow and his long-time collaborator Arturo Rosenblueth, a Mexican neurophysiologist with an appointment at Harvard's medical school. The journal *Philosophy of Science* published it with the somewhat opaque title "Behavior, Purpose, and Teleology."[23] The article laid out core concepts Wiener would later expand on in *Cybernetics*. For example, they differentiated between two types of feedback. In technological systems, positive feedback (counterintuitively), is generally bad as it "adds to the input signals" while leaving them uncorrected. Such feedback can cause systems to spiral out of control. In contrast, negative feedback "restrict[s] outputs which would otherwise go beyond the goal" and keeps systems in check.

In hindsight, the article's most important observation was to draw parallels between the behavior of machines with that of living organisms. This insight stemmed from Rosenblueth's observations of medical patients whose neurological disorders resulted in uncontrolled behaviors, such as spilling a glass of water while raising it to their mouth.[24] In other words, fundamental cybernetic principles could be applied to all sorts of systems whether they were mechanical, electrical, or biological. (Attempts were later made to apply cybernetics to topics like economics, sociology, and meteorology, endeavors that Wiener saw as dubious as best.) The essence of cybernetics, in all cases, centered around "the theory of communication and control in the machine and in the living organism."[25]

After the war, a small community of researchers, including polymath John von Neumann, physiologist Warren McCulloch, and anthropologist Margaret Mead, continued to discuss the implications of what would become known as cybernetics. Wiener was a central figure in these early, often contentious gatherings.[26] Disagreements routinely broke out between people from the natural sciences and those from the social sciences while tensions between Wiener and von Neumann—two men with strong personalities and widely divergent views on the ethics of weapons-related research—were especially apparent. McCulloch recalled, "You have never heard adult human beings, of such academic stature, use such language to attack each other."[27]

By the spring of 1947, Wiener decided he needed to give a personal, perhaps even definitive, statement that might help settle debates swirling around the cybernetics community. He also had other motives for wanting to write a book. Wiener believed it could help "alert this larger public to the long series of analogies between the human nervous system and computation and control machine[s]."[28] It's certainly possible that Wiener, who saw cybernetics as a new scientific field, wanted to maintain a degree of ownership over it. For a community wracked by quarrels, a definitive book on the topic might generate some agreement, if not unity.

There was also an issue of professional priority. After his Yellow Peril manuscript appeared, Wiener learned that a prominent Russian mathematician named Andrey Nikolaevich Kolmogorov had done similar work before World War II and published articles about it in French and Soviet journals.[29] Kolmogorov was interested in prediction theory (one of Wiener's own research topics) and had also contributed to Soviet antiaircraft defenses. Discomfited, Wiener emphatically noted that they had never met or corresponded. Wiener would later credit Kolmogorov as an "independent discoverer of large parts" of cybernetics and "the first man to write on it."[30] But this was *after* Wiener had become famous for writing *Cybernetics*. In 1947, Wiener couldn't predict whose name would become most closely associated with the topic.

Money was another compelling factor. With personal expenses piling up, and "no accumulated wealth from which to pay," Wiener decided he would, as he recalled, "write myself out of this financial hole."[31] A testier side of Wiener's personality surfaced in a private letter to McCulloch shortly before they had an irreparable falling-out: "I am an author and I eat my royalties," he wrote when he told his perceived rival not to expect any free copies of *Cybernetics*.[32] But wanting to write a book, actually writing it, and getting your manuscript published are all very different things. While Wiener might have been well-known among mathematicians, in the book business he was an unknown, unpublished aspiring writer with an admittedly difficult personality.

Like many things in Wiener's life, the publishing history of *Cybernetics* was complicated. His path to becoming an internationally known author started at a café in Paris. In May 1947, Wiener traveled to Europe for a mathematics conference.[33] A colleague had encouraged him to meet with

Enrique Freymann, the operator of a "drab little bookshop opposite the Sorbonne." Freymann also happened to operate Éditions Hermann (referred to by Wiener as "Hermann et Cie."), a French publishing house that put out mathematical works including those written under the name Nicolas Bourbaki, a pseudonymous collective.[34] As Wiener recalled it, Freymann asked the mathematician if he wanted to write a short book about "communication, the automatic factory, and the nervous system." The press's link to Bourbaki intrigued Wiener and before he headed off to his conference, he enthusiastically agreed to write a book for Éditions Hermann.[35] Pierre de Latil, a French science writer who would later write his own popularization of cybernetics, was also present but remembered the meeting somewhat differently. Yes, the mathematician and the publisher met, and, yes, they shook hands in agreement but, after Wiener departed, "Freymann smiled and said, 'Of course, he'll never give it another thought.'"[36]

Wiener, however, stayed true to his commitment. Buried in MIT's archive is the result of what occupied much of Wiener's time after he left Freymann with a bemused smile: a handwritten draft of *Cybernetics* (with "Mathematics" misspelled but corrected on the title page).[37] Wiener did much of the writing in Mexico City in the summer of 1947, composing his words on legal pads. All through the process, he obsessively discussed his ideas with Arturo Rosenblueth, Wiener's "companion in science" to whom he later dedicated *Cybernetics*.[38] Before returning to Cambridge for the start of classes in the fall of 1947, Wiener mailed a finished draft off to Paris.

What Freymann received was not quite the short booklet that Wiener had promised; the surviving handwritten copy alone is eighty-five densely packed pages. Nonetheless, Freymann typeset it and mailed the page proofs back to Wiener a few weeks later. Because Wiener was having eye problems at the time, two graduate students, Walter Pitts and Oliver Selfridge, were tasked with reviewing the proofs. Somehow what was mailed back to Freymann were the *uncorrected* proofs. As a result, the book's first edition contained a multitude of errors.

Wiener soon found himself having to negotiate a new publishing arrangement. Because he had discussed his manuscript with editors at MIT's Technology Press, they were also keen to publish it (and reminded him that he was an MIT employee). However, he had already given Freymann the rights to the book. Eventually the lawyers were appeased.[39] Wiener's book would be published jointly in France and in the United States.

Unfortunately, the archival record doesn't contain granular detail such as referee reports and edits comparable to what exists for Berkeley's *Giant Brains*. Wiener's response to a questionnaire from Technology Press for the promotion of *Cybernetics* is only partially filled out. But it is telling that, in response to a question asking what set the book apart from others in the field, Wiener emphatically replied "There is no other book in the field. It is a new field."[40]

*Cybernetics* was published in October 1948 in the United States as a collaboration between the Technology Press and John Wiley & Sons in cooperation with Hermann et Cie. (i.e., Éditions Hermann) in France with all copies appearing, typos included, in the English language. A joint press release from the two publishing teams proclaimed *Cybernetics* "a classic." Wiener's prose, it promised, would be valued by "serious workers" in many fields including "anthropology, biology, mathematics, philosophy, physics, and radio and electrical engineering."[41] Nonetheless, in his memoir, Wiener recalled that "Freymann had not rated the commercial prospects of *Cybernetics* very highly. . . . [So] when it became a scientific bestseller we were all astonished, not least myself."[42]

Public reaction to the book was indeed remarkable and it marked the start of an unexpected craze for cybernetics. Scores of reviews and articles discussed Wiener's book, its prodigious author, and the implications of cybernetics. Wiener meanwhile found himself transformed from a scientist "with a good but limited reputation" into "something of a public figure." By the end of 1949, some 15,000 copies of *Cybernetics* had sold in the United States alone.[43] At $3 apiece, Wiener would have a good deal of royalties to eat.

Bookshop owners were just as startled by the book's brisk sales. In early 1949, a reporter at the *New York Times* discussed the book buying habits of Manhattanites. One bookshop worker reported that "people today go about buying a book as solemnly as if they were playing bridge." Another employee noted how nonfiction was outperforming fiction in sales. "It's like the trend in women's fashion," they said, "emphasis is shifting from bosom to headgear." Booksellers were especially surprised by the sales performance of two books in particular. One was Al Capp's *The Life and Times of the Schmoo*, a cartoon-based allegory about consumerist greed taken to Swiftian levels of satire. The other was *Cybernetics*. "Book dealers," concluded the writer, "cannot understand why either volume has the popular appeal it has."[44]

So, *why* was *Cybernetics* so popular? Timing was one reason. Wiener's book appeared when there was huge public demand for books explaining science and technology. And this book claimed to be about an entirely *new* field of science, something that would have certainly intrigued many potential readers. David Dietz, a long-time science writer, noted that *Cybernetics* was "not easy reading." But it was "a 'must' if you would be abreast in science."[45]

Wiener's stature as child prodigy turned eccentric MIT professor also boosted public interest in *Cybernetics*.[46] Photos from this era typically show him dressed in suit and tie, cigar in hand, and posed in front of a blackboard or with some machinery. Some readers may already have learned about Wiener via his outspoken stance against military-related research. In a short essay the *Atlantic Monthly* published, Wiener stated he would not undertake any work that "may do damage in the hands of irresponsible militarists." Wiener's stance, in marked opposition to people like John von Neumann, made national headlines after Albert Einstein publicly supported it.[47]

Finally, as any author knows, there is a big difference between people buying your book and people actually reading it. Historian Ronald Kline suggests that *Cybernetics* was an early example of what he calls "the Hawking effect." In 1988, the theoretical physicist's book *A Brief History of Time* flew off shelves at bookstores and airports. But whether people read it or simply bought it to appear au courant is hard to judge. Wiener's book may have fit this pattern as well.[48] Hawking's editors, however, had pushed him to eliminate technical prose and reduce his equations to just one: $E = mc^2$. Wiener made no such concessions. The entire middle section of *Cybernetics* is filled with pages of integrals and differential equations, scattered across chapters with titles such as "Groups and Statistical Mechanics" and "Time Series, Information, and Communication." These parts required a "mathematical competence beyond the capacity of most readers, including this reviewer," wrote one economist at Cambridge University.[49]

*Cybernetics'* topics ranged from the physiology of the human brain and communication methods used by ants to the reflexes of cats and the psychological techniques of the advertising industry. The book's eclecticism prompted a book classifier at the Library of Congress to appeal to Wiener for help. "We have read and reread reviews and explanations of the content of your book," she wrote, but found cybernetics "beyond our capabilities" when it came time to assign it an appropriate Dewey Decimal Classification number. Was *Cybernetics* a work of psychology or should it be shelved in the "field of electronic computation devices?"[50] The archival record, unfortunately, doesn't preserve Wiener's response. Nonplussed, the Library filed it under "General Science" before eventually creating a separate subclass years later for "cybernetics." In an era before networked library databases existed, other libraries took different approaches and placed Wiener's book under, for instance, "Science—Physiology." Today, "cybernetics" has its own main subject heading in the Library of Congress's system.

Despite the classificatory confusion, *Cybernetics* received extensive attention from specialty journals (*Electronics* and *Mechanical Engineering*, to name a few) along with mainstream publications such as *Time* magazine, the *New Yorker*, and *Business Week*.[51] Published reviews veered wildly between praise and puzzlement. Most reviewers either ignored (or were ignorant of) the mathematical mistakes littering the pages and focused instead on the book's broader implications. John von Neumann—who surely noticed the sloppy math—reviewed *Cybernetics* for *Physics Today*. While Wiener's sometimes-antagonist was "inclined to take exception to the mathematical discussion" and questioned the "style and nature of the work," he celebrated Wiener's "first, daring attack" on a "great and important subject." It would, he said, leave few readers "unmoved and unconvinced" about the technological and philosophical implications of future computing machines.[52] John Pfeiffer likewise praised the "provocative analysis of some of the most exciting developments in modern science" found in *Cybernetics*.[53]

Some private skepticism about Wiener's brash claims lurked behind the public acclaim for *Cybernetics*. Bell Labs, for example, had also studied the problem of antiaircraft fire control during World War II and its engineers dismissed Wiener's complex statistical theories as irrelevant to what was actually deployed in the field.[54] Wiener's insinuation in *Cybernetics* that he had invented some of the basic principles of electronic digital computers only increased animosity. Not surprisingly, technical experts who had actually soldered circuits and wires together to make the first "thinking machines" cried foul at the suggestion. In their estimation, cybernetics was neither an essential ingredient nor an actual predecessor to their engineering activities.[55]

Setting aside *Cybernetics*'s technical bloopers, what likely intrigued readers most was Wiener's opening chapter. It offered a lengthy discussion of how he came to discover this "no-man's land between the various established fields," essentially providing a first-person account of the subject's first five years. Here, Wiener generously gave credit to his long-time collaborator Rosenblueth as well as Kolmogorov, Alan Turing, and Warren McCulloch. Somewhat more complicated was Wiener's acknowledgment of his former student-turned-competitor Claude Shannon. Shannon achieved recognition in the scientific community for his wartime research on the mathematics of communication. Shannon developed an approach that, like Wiener's work, was grounded in concepts such as statistics and entropy. After the war, Shannon's once-classified work provided a foundation for the rapidly expanding field of information theory, a topic Wiener also claimed some ownership of.

*Cybernetics*'s opening chapter also introduced readers to one of Wiener's golems: the automated factory. The industrial revolution of the eighteenth century had seen the replacement of animal power with steam engines followed, a century later, by electric motors. In the future, automated factory equipment controlled by computers would necessarily compete with and perhaps supplant human workers.[56] As a key contributor to "the new science of cybernetics," Wiener acknowledged responsibility for the "unbounded possibilities for good and for evil" it posed. Wiener was fatalistic, saying that "we can only hand [cybernetics] over into the world that exists about us." Unfortunately, he concluded, "this is the world of Belsen and Hiroshima" and cybernetics would likely end up "in the hands of the most unscrupulousness."[57] The best way to avoid this scenario was by bringing the implications of cybernetics to an even wider audience. Enjoying his first authorial success, Wiener decided to make this the goal of his next book.

## AUTOMATIC FOR THE PEOPLE?

When Norbert Wiener died in 1964, the *New York Times'* frontpage obituary for him included the phrase "Father of Automation."[58] The moniker would have made the mathematician recoil. It is true that cybernetic principles Wiener developed with collaborators during World War II and then popularized were one foundation for the implementation of automation in factories and offices. However, besides being historically inaccurate, this phrase was an overstatement given the effort Wiener spent disowning his alleged offspring and warning of its dangers.

When it came to automation, Wiener presented himself as a conscientious objector, refusing to consult for companies seeking his expertise. He supported his cautionary position with two more bestselling books. *The Human Use of Human Beings* appeared in 1950, followed by *God & Golem, Inc.*, in 1964. In this period, Wiener did not abandon mathematics, but writing about the dangers of automated factories and "learning machines" occupied more and more of his time. "I could not get off the back of this bronco," he reasoned, "so there was nothing for me to do but to ride it."[59]

Wiener maintained this tough posturing when he talked with a reporter about his new writing projects. In April 1949, sitting in the offices of Wiley & Sons, Wiener explained how he had been "brought up in an atmosphere where it was expected that a person would write." He boasted that he had at least *five* new books in the works. Some were works written for the professional science community and he also wanted to make a stab at detective fiction. But what most excited Wiener was a new project he tentatively called *The Communication State.* "How's that for a list?" Wiener asked as he took "a fierce puff on his cigar."[60]

Financial needs lurked behind the bravado. In early 1949, he discussed his book ideas with Robert Morrison, a program officer at the Rockefeller Foundation. Morrison noted in his work diary that "[Wiener] really does need the money which he hopes a popular book will bring him."[61] In the fall of 1950, Houghton Mifflin published Wiener's new presentation of cybernetics, *The Human Use of Human Beings*, now shorn of the mathematics that jammed up his first book. However, the path to this more user-friendly book required Wiener to take an intellectual detour through the rapidly changing world of the American factory.

Given Wiener's career path, this diversion might have appeared odd. After all, he had spent the better part of three decades comfortably situated in the academic environs of MIT. But he remembered his brief stint as a General Electric apprentice with more than a dash of romanticism, recalling how he ended each day "tired but happy . . . begrimed by that grease which belongs to an engineering factory."[62] His turn as a reporter covering labor unrest also exposed him to the indignities workers faced. Wiener's views toward union leaders were a little more skeptical. The local shop stewards he contacted about the perils of automation were simply out of touch, he recalled, with larger social and political problems.[63]

So, to be heard, Wiener went straight to the top. In August 1949, he wrote Walter Reuther, the head of the United Automobile Workers (UAW). At the time, the UAW was arguably the most progressive and powerful labor organization in the United States, with a membership of some 1.5 million workers.[64] By way of introduction, Wiener explained that he had long been interested in "the problem of automatic machinery and its social consequences." Moreover, he told the labor leader how he had been approached by a "leading industrial corporation" (i.e., General Electric) who wanted him to teach cybernetic principles to its engineers. "I do not wish to contribute in any way to selling labor down the river," Wiener told Reuther, but noted others with his knowledge might not be so scrupulous.

Wiener described to Reuther how factories of the future would operate on basic cybernetic tenets. For example, electronic "sense organs" would gather "input" that could be stored in "memory." Accessed via magnetic tape or punched cards, this information could then be used to run machines without much human input.[65] Once such a system was installed, recording additional instructions (in other words, capturing worker skill as computer code) would be "extremely flexible" with the end easily being a "factory without employees." And if Cold War tensions boiled over to outright warfare, immediate military needs might mean rapidly adopting automation as a matter of national survival, producing "large scale industrial unemployment" within just a few years.[66] On this topic, Wiener was more adept at pointing out problems than suggesting practical answers, but he implored Reuther to take seriously the "very pressing menace" posed by machines operating on cybernetic principles. To help the labor leader understand the situation, Wiener offered him chapters from his forthcoming book.

Wiener's letter reached the labor leader at an exceptionally busy time. Reuther, still recovering from a recent assassination attempt, was carrying out difficult negotiations with auto manufacturers. Ford Motors planned to build what a trade paper called the "nearest approach to a fully automated factory"—the Brook Park plant, a new engine factory in Cleveland.[67] Despite his schedule, Reuther contacted Wiener—"DEEPLY INTERESTED IN YOUR LETTER," his telegram read—and proposed a meeting. They eventually met at a Boston hotel where Reuther proposed setting up a "Council of Labor and Science" that would be staffed by Wiener and other scientists.[68]

While nothing came of the plan, both men gained some new perspectives. Wiener saw Reuther as emblematic of a more "universal union statesmanship" and stepped up his efforts to warn people about what he increasingly referred to as the "automatic factory."[69] Reuther, for his part, grasped the threat that automation posed. A 1954 UAW report contained a now well-known "parable" in which Reuther was shown automated machines at the Brook Park factory. "How are you going to collect union dues from *those* guys?" a Ford official jeered to which the union leader shot back "How are you going to get them to buy Fords?" The report included a glossary of "words used most often in discussions of automation" as well as a pitch for Wiener's "good and interesting" new book, *The Human Use of Human Beings.*[70]

The history of this particular book starts in January 1949 when an editor at Alfred A. Knopf suggested that Wiener write a popular retelling of *Cybernetics* but with more social commentary (and no equations). The mathematician, overwhelmed by the response to his first book, rebuffed the offer but it nudged him to start sketching out plans for a more reader-friendly book, the aforementioned *The Communication State.* Later that year, Paul Brooks, an editor at Houghton Mifflin, also contacted him. Brooks, based in Boston where he was the editor-in-chief for the press' General Book Department, had already worked with many notable authors, including Winston

Churchill and J. R. R. Tolkien.[71] He hinted that Wiener "would have an important book" if the mathematician could explain, using "terms that the layman could understand," exactly how "factories would run without human hands." Wiener agreed to prepare a manuscript for Brooks that would serve a "warning against allowing modern technology to take over our lives." In return, Brooks offered Wiener an advance of $1,500 for what he expected would be "enormously exciting material."[72]

However, Wiener's first draft was not at all what the editor had anticipated. Instead of an erudite discussion about the social implications of cybernetics, Wiener's submission dealt more with the scientist's traumatic childhood. Publishing this material, Brooks recalled, was going to be quite impossible but he decided to gamble that Wiener would come through once he had "gotten the business with his father" out of his system.[73] Wiener's subsequent drafts were more in line with what he wanted. Brooks would occasionally have backchannel discussions with Wiener's office assistant as the two of them flattered and cajoled the sensitive mathematician to keep working. Certain passages, Brooks wrote Wiener, "remind me of Winston Churchill," he said, "whom I know you don't admire as a politician but must admire as a writer," before concluding "What a book!"[74]

Once Wiener delivered his manuscript, another stumbling block appeared. Instead of *The Communication State*, he wanted a different title drawn from classical mythology such as *Pandora* or *Cassandra*. Here, Brooks put his foot down, explaining that this would simply "kill the book dead." The editor instead highlighted a particularly felicitous phrase in Wiener's draft where the author noted his resistance to "the inhuman use of human beings." Many readers, Brooks noted, would "assume automatically that the scientist is not concerned with humanity" and yet here was a book showing that "the leader in this whole branch of science" was someone who cared deeply about human values. "I believe you have here, with slight alliteration, the title for your book: *The Human Use of Human Beings*," the editor wrote.[75] Wiener agreed and the manuscript moved into production.

Bookstores began carrying *The Human Use of Human Beings: Cybernetics and Society* in the summer of 1950, with copies selling for $3. Houghton Mifflin promoted it with newspaper advertisements that described Wiener as both a "philosopher of science" and "one of the nation's greatest scientists in the field of mathematics." The book, written "for the layman, direct and challenging, and wise," promised readers "new conceptions to reconstruct the course of life" in the "age of the thinking machines." Finally, the book would give readers "absorbingly graphic glimpses into experimental studies" at MIT as well as a "fresh evaluation of our moral standard in world thinking."[76]

*Human Use* provided a much gentler overview of cybernetics compared with his previous book. Nonetheless, many reviewers found *Human Use* a mess in authorial execution. A geneticist reviewing the book for *Scientific American* predicted readers would be "dazed, as well as dazzled" by Wiener's book, which despite only being 200 pages, managed to include references to "entropy, Mexican frescoes, the Industrial Revolution, Parkinson's disease, the patent system, the logarithmic scale in order-disorder relations, dietary habits, Duncan Phyfe furniture, the relation between law and communication, line noise, Rudyard Kipling, Marxism, secrecy in science, Buddhism, the history of language, Pavlov's dogs, *The Saturday Evening Post*, human morphology, Catholicism, the atomic bomb, Heinrich Heine, high-pressure distillation, the Thirty Years' War, the electric lamp, the prose of Theodore Roosevelt, sequential analysis, and the mediocrity of high school teachers of English."[77] The reviewer wasn't wrong. *Human Use* mentioned all of these (and other) topics. With its narrative flowing in places like a stream of consciousness, *Human Use* might be read as an acknowledgement of all the publicity Wiener had received as a know-it-all child prodigy. Unlike the reader-friendly tone that Edmund Berkeley took in *Giant Brains*, Wiener's approach was that of an erudite and eccentric professor holding forth on all subjects.

Some of Wiener's detours were indeed odd. At the book's end, for example, he labeled the Catholic Church a "totalitarian religion" akin to the Communist Party and likened both to McCarthyism in the United States.[78] Wiener's tirade looped back to his book's opening pages where he decried power held by "Fascists, Strong Men in Business, and Government" as well as "men of ambition for power . . . in scientific and educational institutions" who sought the "mechanization of man." (Take *that*, MIT administrators!)[79]

*Human Use* allowed Wiener to express both a frustration with American politics and an elitist's contempt for popular culture. For example, radio broadcasting, which millions of Americans enjoyed, was dominated by "the soap-opera and the hillbilly singer" instead of fulfilling its "great civilizing possibilities."[80]

Despite Wiener's writerly wanderings and sometimes-snobbish pronouncements, *Human Use* enjoyed even better sales than *Cybernetics*. These numbers were boosted further when a significantly revised version, published with Doubleday via its Anchor Books paperback imprint, appeared in 1954. For the second edition of *Human Use*, Wiener worked with Jason Epstein, an up-and-coming figure in the book business. Part of a new postwar generation of young editors, Epstein had helped bring quality paperbacks to the American book market.[81] (With Epstein's guidance, in 1959, Random House would publish Wiener's unremarkable novel, *The Tempter*.)[82] Working with Epstein, Wiener muted many of the first edition's polemical passages and political opinions.[83] Wiener also added some quasireligious thoughts that he would return to when he wrote *God & Golem, Inc.* By the end of 1960, enthusiastic readers purchased more than 50,000 copies of the new edition of *Human Use*, easily besting the sales figures *Cybernetics*.[84]

One of the book's readers was John T. Diebold. His professional reaction to the book appears in stark contrast to Wiener's strident but often vague admonitions about the dangers posed by automation and the automatic factory. If Wiener was akin to an Old Testament prophet, warning of impending danger lurking amidst factory machinery, Diebold was like a New Testament apostle spreading positive messages about automation. When Diebold died in 2005, such words as "visionary" and "evangelist" were amply scattered throughout his obituaries.[85]

Ironically, Diebold's route to automation was due, in part, to something Wiener helped develop.[86] Diebold served in the Merchant Marine during World War II. While on convoy duty in the Atlantic, he became intrigued by the ship's automatic mechanisms for controlling antiaircraft fire—observations that got him thinking about interactions between people and machines. After the war, he earned degrees in engineering and economics before enrolling at Harvard for an MBA.[87] At Harvard, Georges Doriot, an early venture capitalist, taught a popular year-long course simply called Manufacturing.[88] Diebold proved to be one of Doriot's most enthusiastic pupils.

For Doriot's class, Diebold led a small team that developed a workable design for automated factories. For their report, titled "Making the Automatic Factory a Reality," Diebold's team interviewed businesspeople and engineers at companies throughout the Northeast, including Norbert Wiener. Although the team cited both *Cybernetics* and *Human Use*, Diebold argued that a "much higher degree of industrial automation is both desirable and technologically possible."[89] New technologies based around computers, he insisted, could benefit both the American economy *and* its workers. This conclusion set Diebold on a professional path orthogonal to Wiener's and he spent the rest of his career persuading business executives and policy makers to adopt industrial automation and computers.

After graduating, Diebold prepared an expanded version of his report as a book written primarily for the business community. Simply titled *Automation*, it was dedicated to Georges Doriot and published in 1952 by D. Van Nostrand, which specialized in scientific and technical works. Like Berkeley's *Giant Brains* (which Diebold had clearly read), the prose was unpretentious and clear. At the outset, Diebold defined "automation," a term he helped popularize, as something distinct from the mere mechanization of existing factory processes. To him, it represented a "whole new system of automatic *feedback* control . . . a new technology, wonderful in its possibilities."[90] (Although "cybernetics" doesn't appear in the book's index, Wiener's principles were central to Diebold's argument.) This control was made possible by the postwar development of "high-speed digital computers or 'giant brains.'"[91]

The problem for business people was that most publications about computers focused on military or scientific applications, not manufacturing. Their confusion was exacerbated, Diebold said, in one of several oblique digs at Wiener, by "pedantic" analogies "between the operation of certain control systems and the operation of human and animal nervous systems." As a result, Diebold said, typical predictions of automated factories were correspondingly grim with images of "jobless and debased workers roaming the streets of a fully mechanized civilization."[92] The final chapter of Diebold's book turned more reflective by considering the social and economic effects of future automation. "Automatic factories," Diebold assured readers in a theme he repeated in scores of subsequent articles and public talks, "will not be workerless factories," a prognosis he acknowledged that "Norbert Wiener would probably not go along with."[93]

Would Diebold have written *Automation* in the absence of Wiener's books? Probably. But it is also fair to say that one gains a more nuanced understanding of Diebold's book by reading it in parallel with *The Human Use of Human Beings*. Cybernetic terms like "control" and "feedback" provided Diebold with a scientific-sounding underpinning for his own ideas and agenda. Meanwhile, Wiener's proclivity for extravagant and pessimistic assertions gave a foil that allowed Diebold to present himself as a pragmatic, business-oriented thinker, despite claims both men regularly made about an impending new industrial revolution.

Looking at the question of influence in another way, Wiener had publicly eschewed collaborations with industrial managers. This effectively marginalized him from any direct engagement with the business community. Diebold displayed no such reticence. In 1954, he founded a successful consulting group that focused on promoting the technologies of automation. This advocacy made Diebold rich and famous. His company's clients included Boeing, IBM, and DuPont while such public figures as Richard Nixon, Henry Kissinger, and Alan Greenspan became acquaintances. Diebold was relentlessly entrepreneurial as the automatic factory and the computing technologies associated with it presented a business opportunity. In contrast, Wiener's interests and actions were more defensive as he warned about automation while protecting and promoting his own books and reputation.

These differences had consequences. In October 1955, when Congress held nine days of hearings about automation and technological change, they asked both Diebold and Reuther to testify. Wiener was not invited, and his name appears exactly once in over 650 pages of densely printed statements entered into the record.[94] There's no indication Wiener was even aware of the lengthy discussions on Capitol Hill but, if he had, perhaps he would have muttered a phrase learned as a boy from Leo Wiener: *Sic transit gloria mundi* (or, so passes away worldly renown).

In his first memoir, Norbert Wiener admitted to being easily frightened as a child. Vaudeville shows left him in tears and Lewis Carroll's *Alice in Wonderland* was a source of terror.[95] Nonetheless, in his books and articles, Wiener regularly referenced a chilling work of Edwardian fiction he encountered as a boy. In 1902, William W. Jacobs, an English author of short fiction, published "The Monkey's Paw" in his collection *The Lady of the Barge*. A now-classic tale of "be careful what you wish for," it revolved around a magical artifact that grants wishes, albeit with an enormous cost attached. When the paw's new owner wishes for money, he gets an insurance payment but only after his son is fatally mangled in an industrial accident. When the young man's grieving mother wishes for her child's return, he appears at their door, dead and disfigured. A final wish from the boy's father makes the specter disappear. A classic horror story where the ghost is implied but never revealed, it is also a tale of unintended consequences.

Based on the number of times Wiener noted "The Monkey's Paw" in his work, the moral of Jacobs' story resonated deeply with him. An unpublished essay, from mid-1949, mentions it.[96] So do both editions of *Human Use*, a 1960 article in *Science* on the consequences of industrial automation, and a new edition of *Cybernetics*, which appeared in 1961. There are several possible explanations for the story's reoccurrence. One is that as he became more in demand as a speaker and author, Wiener returned to material he had written before. It is certainly possible to find themes, if not entire passages, in *Human Use* that resemble those in *Cybernetics*.

Another explanation is that Wiener increasingly turned to familiar stories he had encountered as a child—the myth of Prometheus, Satan from Milton's *Paradise Lost*, Goethe's *The Sorcerer's Apprentice*, the genie from *One Thousand and One Nights*, and *Alice in Wonderland*—to better communicate his dark messages. It could be quite effective. Toward the end of *Human Use*, Wiener evocatively described the "Monkey's Paw of skin and bone . . . as deadly as anything cast out of steel and iron," and, when it came to the future of automated machines, "the hour is very late, and the choice of good and evil knocks at our door."[97]

*God & Golem, Inc.*, the final book of Wiener's cybernetics trilogy, introduced a new hazard with a new fable. As the title indicates, Wiener included Rabbi Löw of Prague and his clay golem as yet another way of thinking about technologies that are programmed to do certain tasks but then escape human control. Like "The Monkey's Paw," the tale of the golem gave Wiener an expedient way to convey apprehension about unintended consequences. But, in *God & Golem, Inc.*, Wiener shifted his concerns to a new technological threat. Whereas *Cybernetics* and *Human Use*, published only two years apart, addressed the dangers of automated factories, much of *God & Golem, Inc.*, was devoted to warning about what Wiener called "machines that learn."

This idea had origins in chess, one of Wiener's favorite games. According to MIT's Marvin Minsky (about whom more in the next chapter), Wiener regularly haunted MIT's faculty club with "ambitious schemes" to best would-be challengers. But he "would always make a mistake and lose his queen or something early in the game so he couldn't carry out [his] plans."[98] In spite of, or perhaps because of, his deficiencies at chess, Wiener was captivated by computing machines that could play the game. At the end of the first edition of *Cybernetics*, he asked "whether it is possible to construct a chess-playing machine, and whether this sort of ability represents an essential difference between the potentialities of the machine and the mind." Perhaps thinking of his own limitations, he concluded that it should be possible to make a machine that would play "a chess not manifestly bad as to be ridiculous."[99]

Wiener returned to the idea of a "chess-playing machine that learns" in *Human Use*. By pitting it against "a wide variety of good chess players," it could likely learn "a great range of performance" and improve over time. Wiener also noted that one must distinguish between what a machine can learn and what it cannot. For example, engineers might be able to construct a device so that "certain features of its behavior may be rigidly and unalterably determined." However, one might also build a computer with a "statistical preference for a certain type of behavior." In this case, a chess playing machine, given the power to "learn," might morph into a "machine doing a totally different task," an idea that alarmed Wiener. Instead of being just a toy, a chess-playing machine that could also learn might just be "the first step in the construction of a machine to evaluate military situations." Drawing on his reading of the influential 1944 book *The Theory of Games and Economic Behavior*, written by John von Neumann and economist Oskar Morgenstern, Wiener mused on how two-person, zero-sum games were routinely used for "military and quasi-military aggressive and defensive purposes," including the evaluation of possible nuclear war scenarios.[100]

In none of his books did Wiener use the term "artificial intelligence" nor does one learn any technical details about the emerging field. As was the case for earlier discussions about "automatic factories," Wiener's writings about "machines that learn" tended toward the abstract. Still, one can't read Wiener's discussions about these devices today without artificial intelligence coming to mind. In describing such hypothetical machines, Wiener was connecting his ideas, consciously or not, to two other historical strands.

The first is the history of chess-playing machines. Wiener was well aware of fraudulent devices such as the Mechanical Turk and Edgar Allan Poe's exposure of it in an 1836 essay. He also could not have missed more recent linkages that emerged after World War II between artificial intelligence research and chess. Just as the fruit fly provided geneticists with an ideal model organism, chess was an "excellent model environment" in that it was accessible, familiar, and easily experimented with.[101] As an example of abstract human reasoning, chess offered scientists a way to explore whether and how computers functioned (or didn't) like human minds. In 1946, Alan Turing had speculated that one could compel a computing machine to "display intelligence," one sign of which would be "to play very good chess."[102] Twelve years later, Herbert Simon and his colleague Allan Newell, two leading artificial intelligence researchers, made a series of extrapolations as what computers would soon be able to do. Topping their list was the prediction that a digital computer "will be the world's chess champion, unless rules bar it from competition."[103] Wiener's association of "machines that learn" with chess signaled his awareness of scientists studying artificial intelligence and their broader research agenda.

Second, there is the history of artificial intelligence itself. In the mid-1950s, Wiener's institutional home was arguably *the* world's leading center of artificial intelligence research. People such as Marvin Minsky, John McCarthy, and Wiener's own advisee, Oliver Selfridge, built prominent research programs. The possibility of "artificial intelligence" and the phrase itself first appeared in technical discourse around 1955. It gradually gained traction as a research program while cybernetics became marginalized in the United States. In mentioning "machines that learn," Wiener was referring to computer research activities happening around him. To be clear, this was research he personally refused to participate in due to its links to military funding and because he had also fallen out personally with some of the people involved with it.[104]

By the early 1960s, Wiener had largely left research behind for a role as a public intellectual, a position built on the foundation of his books. As demand for Wiener's public lectures increased, his pace of writing slowed. An invitation from Yale University to give that school's prestigious Dwight H. Terry Lecture series presented an opportunity to reach a new audience and prepare a new book. Started in 1905, the Terry Lectures asked speakers to address how science, philosophy, and religion informed and complemented each other. Wiener, to be sure, had little interest in organized religion. Long conflicted about his Jewish background, his immediate family attended Unitarian services that left him indifferent. Toward the end of his life, motivated by travels to India, Wiener became interested in Hinduism, especially the idea of reincarnation.[105]

Yale University Press typically published a speaker's Terry Lectures as a book, an arrangement they extended to Wiener along with an offer of a $1,000 advance (about $10,000 today) plus royalties.[106] Wiener accepted the deal and set to writing his lectures, which he planned to present in October 1961 until an accident put him in the hospital with a broken hip. While recovering, Wiener submitted a summary of his proposed lectures to Yale so they could be publicized. Humans, Wiener said, were developing new technologies that were "beginning to encroach on certain fields of activity which in the past have been delegated exclusively to God." This included "making machines which can create other machines in their own image" as well as a "creator playing a game with his own creature as God does with the devil in the Book of Job or Paradise Lost."[107] Wiener gave his public lectures in January 1962 and Yale expressed enthusiasm for Wiener's newest book, which the author and the press agreed would be titled *Ethics in a Machine World*.[108]

As with Wiener's other books, the path from proposal to publication was not straightforward. While visiting Europe, Wiener presented his lectures to different audiences and began to think about centering cybernetics more firmly within a religious framework. Wiener also spent a month with the Danish writer, scientist, and polymath Piet Hein. After training as a theoretical physicist, Hein had used poetry during World War II to send political messages that escaped Nazi censorship. After the war, Hein continued to write poems—he called them *grooks*—while working at the intersection of math and design. With Hein's encouragement, Wiener looked at his manuscript with a fresh eye and he decided the text was "pretty scrappy and not consistent enough." He rewrote it and told his editor at Yale that it was now going to be called *God & Golem, Inc.*[109] To those puzzled by the new title, Wiener explained that the book explored how "organized systems should be regarded not from the point of view of a supposed antagonism and opposition between man and machine, or between God and machine, but from a point of view of a total organized system containing all these elements" (hence the "Inc." in the title).[110]

Once he returned to the United States, Wiener started to discuss the new project with editors at Yale as well as Jason Epstein who, in 1958, had decamped to Random House. Both publishers were somewhat baffled by Wiener's new direction. Yale had already advised Wiener that his original manuscript was "not entirely suitable for publication" but, now that he had revised it, perhaps it could be redeemed. Having already compensated him for the Terry Lectures, Yale suggested they might have some legal claim to the new version.[111] Epstein, meanwhile, was used to dealing with the temperamental author and responded that "there is a lot in this manuscript that's good and more than meets the eye" but the "theological framework imposed on the material finally defeats it." Maybe, Epstein suggested, Wiener could rewrite the entire manuscript without the overt references to theology?[112] Piet Hein complicated matters further by showing the manuscript to interested Scandinavian publishers. To make things even more convoluted, Irving Kristol, a vice president at Basic Books, contacted the mathematician to say he had heard that Wiener's new book manuscript "has yet to find an American-language publisher" and that he had the "strongest possible interest" in it.[113]

In the end, Wiener took the path of least resistance. The MIT Press—its forerunner had published *Cybernetics*—agreed to take the manuscript, rewritten and reorganized but without the major edits Epstein wanted.[114] In the winter of 1964, Wiener, who had just received the 1963 National Medal of Science from President Lyndon Johnson, was back in Europe again for a lengthy visit. The MIT Press airmailed him a dummy copy of the book's dustjacket, with a publication date set for April. Wiener, however, would never get to see the book on store shelves. He died of a pulmonary embolism in Stockholm on March 18, 1964.

*God and Golem, Inc.* appeared a month later. His widow, Margaret, decided to forego any additional revisions to her late husband's last work. As she saw it, *God and Golem* would be "his benediction."[115] The final version was quite modest in length and its hundred pages are easily read in a single sitting. *God and Golem* stands as a distillation of humanistic musings that Wiener had been writing about since *Cybernetics*, albeit with an additional layer of theological ideas. If one were to consider the sprawling erudition of *The Human Use of Human Beings* as akin to a Mahler symphony, *God & Golem, Inc.*, might be seen as a piano sonata built around a few simple themes.

In keeping with the premise of the Terry Lectures, *God & Golem, Inc.*, addressed "questions concerning knowledge, power, and worship." Knowledge, Wiener wrote, was "inextricably intertwined with communication" while power revolved around "control." The latter category also implied a consideration of how human purposes meshed with ethics and the "whole normative side of religion." Put another way, Wiener wanted to discuss theological topics which "possess a close analogy to other situations" particular to "the new science of cybernetics."[116] To do this, Wiener oriented his book around three main topics: machines that learn; machines that reproduce themselves; and the relations between people and computing machines.

For his first theme, Wiener returned to familiar territory and posed a series of abstract questions about game-playing machines. Wiener drew some inspiration from religion, asking whether a creator (be it a person or an omnipotent deity) could play a meaningful game with something they created, be it a computer or the Devil. What would it mean if the creator lost such a game? And what about machines that learned from past experience and improved their performance? Wiener, for instance, wondered whether such machines might become all-powerful, exceeding the capacity of their creators.

Wiener next considered machines that had some capacity to replicate themselves. The idea of self-reproducing automata was something his long-time colleague and sometimes antagonist John von Neumann had already considered before his death in 1957. Wiener, via a short digression through Darwinian evolution, took this in a different direction, noting that there were already machine tools in factories that "make other machines in their own image." Given that God supposedly made humans in his own image, and that living things obviously reproduced, Wiener submitted that similar behavior by machines was disturbing from a religious and ethical perspective.[117]

This led readers to Wiener's third set of points, which, more than others in his slight book, directly addressed how "cybernetics impinges on religion." (Wiener's conception of religion was capacious as his text flitted between Christianity, Judaism, Zoroastrianism, and Manicheanism.) Wiener imagined that any sane person two centuries ago who suggested a mere machine could learn to play games, translate languages, or make copies of itself would surely have been charged by the Spanish Inquisition. To Wiener, such claims represented sorcery and therefore might be condemned as sin. But what exactly was the sin? "As long as automata can be made, whether in metal or merely in principle," studying them was a "legitimate phase of human curiosity." But when a scientist's motives moved beyond innocent scholarly inquisitiveness, then beware. Wiener claimed such a person was guilty of being a "*gadget worshiper*," someone bending a knee before a profane new kind of Golden Calf. As an MIT professor, Wiener noted he was quite "familiar with gadget worshipers in my own world, with its slogans of free enterprise and the profit-motive economy."[118]

*God & Golem, Inc.*, added a sheen of religious commentary to what was a distillation of many things Wiener had already said. When it came to the relation between people and computing machines, each had place and purpose or, as Wiener phrased it, "Render unto man the things which are man's and unto the computer the things which are the computer's."[119] But any machine that learns and thinks was transgressing into territory best left to humans, perhaps becoming the "modern counterpart of the Golem of the Rabbi of Prague."[120] Edmund Berkeley had written his 1949 book around the idea that computers were akin to brains, something that Wiener would have once agreed with. While Berkeley had made a compelling case that, in limited ways, computers could think, Wiener took this a step further and warned that computers could also learn. The question remained whether this was scientific sorcery, a sinful act of hubris, or simply the next stage of people's relationship with machines.

Compared to the seismic booms that *Cybernetics* and *The Human Use of Human Beings* caused, *God & Golem, Inc.*, produced only a tremor. The *New York Times* ignored it and scholarly appraisals appeared long after Wiener's ashes had been scattered at his small farm in New Hampshire. Back in Cambridge, where Wiener had spent so much of his professional career, an undergraduate evaluated *God & Golem, Inc.*, for the *Harvard Crimson*. In retrospect, the review was more notable for its young author. Michael Crichton would go on to write celebrated screenplays and best-selling techno-thrillers, such as *Westworld* (1973), *Jurassic Park* (1990), and *Prey* (2002), all with plots based around human creations that escape the control of their makers. Crichton found Wiener's book "both fascinating and frustrating," akin to "watching a mountain goat bound from peak to peak—the display is impressive, but hard to follow."[121]

Despite the relative neglect, the last book in Wiener's cybernetics trilogy generated one final ripple after the author's passing. *God & Golem, Inc.*, won a National Book Award in the category of Science, Religion, and Philosophy. The citation said that *God & Golem, Inc.*, gave readers "a glimpse of the religious implications of this confrontation of man and his most sophisticated machines."[122] Jerome Wiesner, a professor of electrical engineering at MIT, accepted Wiener's posthumous award. Over the years, Wiener had collaborated with Wiesner but also rebuked his colleague's coziness with military patrons. At the awards ceremony, Wiesner, who later became MIT's president, explained that Wiener's last book stated that "God can in no way be threatened by his creation, so man need not be threatened by the machine." When a reporter wondered if future machines might become writers, Wiesner noted that some computers could already "write better novels than some novelists."[123] For Norbert Wiener, who prided himself on being an author as much as a mathematician, it's hard to imagine a worse insult.

# 3 APOSTLES AND APOSTATES

To some of the readers who encountered Mortimer Taube's book, *Computers and Common Sense*, the author appeared as an infuriated man.[1] Despite his book's slim size (the paperback was just 136 pages), Taube took sizable swipes at computer scientists and their promotion of machines that could purportedly think. The subtitle of his 1961 book was, in fact, *The Myth of Thinking Machines*. Taube's provocative writing style and violent refutations prompted some in the computing community to dismiss him as an "angry man" peddling ad hominem "allegations presented as fact."[2] These were tough words to describe a professional librarian.

Taube learned about computers via an information management technique he developed in the early 1950s called "coordinate indexing." This method enabled Taube to address the postwar boom in scientific and technical publications by assigning descriptive terms to documents via computer punch cards. Boolean logic (for example, searching for "red" AND "tasty" but NOT "tomato") allowed users to better locate, search, and retrieve documents.[3] For Taube, computers were tools to work with, not some transcendent technology displaying hints of human intelligence.

Taube especially objected to a particular form of verb usage and the mindset that accompanied it. He deplored how discourse around computers had shifted from what computers actually did to what they *might possibly do* one day. "All the great mechanical brains, translating machines, learning-machines, chess-playing machines," Taube fumed, "owe their 'reality' to a failure to use the subjunctive mood."[4] For readers who don't recall their grammar lessons, the subjunctive expresses a wish or imagined situation. Instead of "I am eating tacos," (the indicative mood), the subjunctive gives "I wish I were eating tacos" or "We should be eating tacos."

Taube directed especial ire toward the "intelligent-machine field" (the term "artificial intelligence" had yet to be fully adopted among computer professionals). "The game," he wrote, "is played as follows: First it is asserted that, except for trivial engineering details, a program for a machine is equivalent to a machine. The flow chart for a program is equated to a program. And finally, the statement that a flowchart could be written for a nonexistent program for a nonexistent machine establishes the existence of the machine."[5] As a result, Taube argued, the scientific literature gave the illusory impression that artificial intelligence was more capable than it actually was. "When reputable scientists begin to accept explanations merely on the basis that they *could* be true and that nothing *forbids* their being true," Taube wrote, "science becomes indistinguishable from superstition."[6]

Taube thought the problem was even more pernicious given that taxpayers generously supported scientists' research even as ill-informed journalists and grant-giving agencies aided and abetted the entire enterprise with more publicity and resources. On this latter point, Taube was absolutely right. In the wake of the launch of Sputnik in 1957, federal patrons doled out generous funding for computer science. By 1970, this annual largesse, much of it earmarked for military applications, topped $30 million (almost a quarter billion dollars in today's currency). This figure is even more munificent if one factors in the monies lavished on new facilities, equipment, and education.[7] Given such generosity, Cold War scientists, to some critics, were no longer detached truth seekers but instead just another political interest group.[8]

While there was some merit in Taube's critique, colleagues recoiled at how he "allowed his anger to carry him to excesses."[9] Taube intimated widespread intellectual fraud was happening. Even more shocking, Taube named names, calling out experts such as Herbert Simon and Marvin Minsky for talking about unrealized devices in the scientific literature "as though they existed." In 1958, for example, Simon had coauthored a much-cited article that predicted computers would soon be able to play games and exhibit artistic creativity.[10] "Electrical engineers and computer enthusiasts," Taube demanded, "should stop talking this way or face the serious charge that they are writing science fiction to titillate the public and to make an easy dollar."[11]

The tempest Taube stirred proved temporary (in part, due to his death in 1965) but the "subjunctive speak" he critiqued remained a dominant mode of discourse whenever academics, business leaders, and journalists talked about computers and, especially, artificial intelligence. Taube was not alone in his indignation, however. In 1976, Joseph Weizenbaum, a computer scientist at MIT, published *Computer Power and Human Reason: From Judgment to Calculation*.[12] Weizenbaum's book examined the question of what computers (and computer scientists) could do and, more importantly, what they *ought* to do and not do. The book's humanistic perspective was all the more surprising given the author's employment at one of the world's leading engineering schools. Weizenbaum's surprise best seller ignited a vigorous debate among public intellectuals and computer scientists the author branded as the "artificial intelligentsia."

Like Norbert Wiener, Weizenbaum drew on religious imagery—priests, gods, and idols—to convey and amplify ideas in his narrative. His book gave readers a pessimistic assessment of the computer science community and its reductionist worldview. Less the machines themselves, Weizenbaum claimed it was computer experts that citizens should fear. Today, for better or worse, computer scientists have developed, if not (yet) perfected, forms of artificial intelligence for corporate and government patrons. Weizenbaum's book takes us back to the first wave of enthusiasm for artificial intelligence, when it was still grounded in the subjunctives of what might happen and what should happen. But, to understand the consternation, we must first look at what had already happened.

### THE LAST DREAM

Weizenbaum sometimes spoke of artificial intelligence as a culmination of "man's eternal dream," perhaps, making reference to Rabbi Löw's Golem of Prague, our "last dream."[13] And, like the lingering memories of a dream, existing narratives of artificial intelligence often appear fuzzy, incomplete, and indistinct. Even if one limits the perspective to just the United States, its history still sprawls across many academic disciplines: computer science, mathematics, and electrical engineering, to be sure, but also psychology, linguistics, and philosophy. The vastness of this territory, along with the fact that its history has often been written either by journalists or the actors themselves, has produced narratives that sometimes veer toward unreliability or simplification. Even Pamela McCorduck, the first author to seriously attempt to document the field, acknowledged that her book represented a "secret history" reliant on what its practitioners—"a group of poets, dreamers, holy men, rascals, and assorted eccentrics"—had revealed to her.[14]

Presenting any history of artificial intelligence is complicated by the fact that the term today encompasses many new areas—from machine learning and large language models to autonomous vehicles—compared to what scientists and journalists spoke about during the 1960s. Moreover, contemporary artificial intelligence is firmly implicated in larger debates about environmental costs, social control, national prestige, and political power.[15] With these caveats in mind, it's practical to consider a selection of debates and developments that helped catalyze the depictions of computing and artificial intelligence found in Weizenbaum's book.

As everyone knows, it is difficult to sense exactly when a dream begins. For artificial intelligence, one place (out of many possibilities), is to situate the field's origins with the shift from "brain-speak" to "mind-speak." In *Giant Brains*, Edmund Berkeley depicted the computer as akin to the human brain but with devices like vacuum tubes serving as proxies for neurons. In this discourse, some practitioners likened computers to biological brains and based their research around the manipulation of a "large number of very simple information processing elements," sometimes building up actual physical models from scratch.[16]

By the mid-1950s, however, the analogy of neurons with electronic logic elements was displaced by another school of thought. The cognitive model school, which eventually became dominant, viewed computers as symbol-processing devices more akin to the human *mind*. Whereas the physical workings of the brain were the bailiwick of physiologists, how the mind handled symbols like shapes, numbers, and letters provided an opening for researchers from such fields as linguistics and cognitive psychology. For example, the discipline of behavioral psychology, established in the 1910s, treated the brain as a black box. What could be observed in the laboratory was how people responded to external stimuli and therefore what went on in the gray stuff between a person's ears was deemed unknowable and went unstudied. But after World War II, cognitive psychologists sought to open up that box, seeing the mind as the interface between stimulus and response, and therefore claiming the mind as an essential object to study.

More and more scientists likewise abandoned physical models of the brain for cognitive models and a symbolic approach to understanding intelligence.[17] For instance, MIT's Marvin Minsky, one of the pioneers in postwar artificial intelligence research, started his scientific career by examining networks of neurons (real or artificial), and what he called the "brain-model problem."[18] Minsky soon retreated from physical analogs of the brain and opted to focus on ways to theorize "how any kind of thinking machine might work." Older neurophysiological approaches to artificial intelligence gave way to new research programs that treated minds and computers as symbol-processing machines.[19] It was Minsky, in fact, who claimed that "the brain is merely a meat machine," a statement that infuriated Joseph Weizenbaum.[20]

In 1963, Edward A. Feigenbaum and Julian Feldman, two artificial intelligence researchers from the new symbolic school, coedited a collection of essays titled *Computers and Thought*. At the outset, they addressed the same question Edmund Berkeley had considered years earlier—what does it mean to say that a computer can think? As was the case for Berkeley, the answer depended on one's definition. If *thinking* referred to an activity that was "peculiarly and exclusively human" or if it meant something "inscrutable, mysterious, and mystical" (criteria that Feigenbaum and Feldman rejected as "unscientifically dogmatic"), then computers did not think.

However, if one compared the "behavior of the computer with that behavior of human beings to which the term 'thinking' is generally applied," Feigenbaum and Feldman concluded (as had Berkeley) that computers were indeed performing a certain type of thinking. Following this line of reasoning, Feigenbaum and Feldman stated that the basic goal of artificial intelligence researchers was to "construct computer programs which exhibit behavior that we call 'intelligent behavior' when we observe it in human beings."[21] Computers could display intelligence, in other words, when they could do things that people did.

One activity people are generally good at is solving problems. As a result, artificial intelligence researchers in the 1950s and 1960s devoted considerable time to the theory and practice of "heuristic methods." For a computer scientist, a heuristic refers to a technique that suggests solutions to a specific problem that will work *most* of the time. In a program designed to play chess, for example, a strictly algorithmic approach would use brute force to examine every possible move to find the very best solution, regardless of how long it took. Heuristic methods, in comparison, look for "good enough" results (an outcome sometimes called "satisficing," from "satisfy" and "suffice") in a manner that is more practical, whether done by machine or the brain.[22]

The essays Feigenbaum and Feldman included in *Computers and Thought* reflected growing interest among artificial intelligence researchers in how people and machines solved problems. Of the twenty papers they believed represented the state of the field, three were based on work carried out by Herbert A. Simon. Born in 1916, Simon was a polymath social scientist who spent much of his career at what was then known as the Carnegie Institute of Technology in Pittsburgh. Initially, Simon studied decision making in corporations and other institutions. He was, for instance, intrigued by the idea that people do not always make perfectly logical decisions due to inherent limits in their cognitive capacity and available information, something he called "bounded rationality." This work, which was the basis for his 1978 Nobel Prize in economics, stood in contrast to experts who assumed that economic actors always make choices that optimize profits and savings.[23]

In his 1991 autobiography, Simon credited Edmund Berkeley's *Giant Brains* for stimulating his "awareness and knowledge" about computers.[24] By the 1950s, Simon started shifting his research into more abstract realms by considering the "psychology of human problem-solving . . . [and] the symbolic processes that people use in thinking." Computer programs became an ideal tool for Simon's research and he started spending time at the Systems Research Laboratory operated by the RAND Corporation, a defense-oriented think tank in Santa Monica, California. Besides getting access to the lab's IBM 701 machine, Simon met Allen Newell, a former Princeton mathematics student who was conversant with computer technology. Simon eventually recruited Newell to join him at Carnegie Tech. Besides becoming life-long friends, the duo were instrumental in developing a powerhouse program for artificial intelligence research at their school.

Simon and Newell believed that both computers and human minds were systems that functioned by manipulating symbols and processing information. This separated them from older neurophysiological approaches to thinking about thinking. "We wish to emphasize that we are not using the computer as a crude analogy to human behavior," they wrote in 1958, "we are not comparing computer structures with brains, nor electrical relays with synapses."[25] In collaboration with RAND programmer John Clifford Shaw, they developed software that solved problems like playing chess and constructing mathematical proofs.

In 1955, Simon and Newell claimed their first major result. Called the "Logic Theory Machine," it was described as a "complex information processing system . . . capable of discovering proofs for theorems in symbolic logic."[26] At first, their "program" was just instructions written on paper notecards that were "executed" by Simon's students and family members at the dinner table. But once these instructions were converted into punch cards, Shaw, Simon, and Newell used RAND's JOHNNIAC machine (named after John von Neumann) to run their creation as an actual computer program. One result was proving dozens of the theorems in Alfred North Whitehead and Bertrand Russell's classic work *Principia Mathematica*, a book which had inspired young Norbert Wiener decades earlier.

Simon was so excited by the possibilities of the Logic Theory Machine that he announced to his Carnegie Tech students that "Allen Newell and I invented a thinking machine." A critical caveat was in order, however. The "machine" in question was not an actual device nor was it an all-purpose problem solver. Rather, it was a set of instructions which proceeded, by heuristic trial and error, to solve a narrow selection of mathematical proofs. One of the people in Simon's classroom was Edward Feigenbaum who was finishing his degree in electrical engineering. "We all looked blankly," Feigenbaum told writer Pamela McCorduck, "the words 'thinking machine' didn't quite fit together." When pressed for details, Simon offered his students a stack of

IBM 701 manuals. Feigenbaum grabbed one and "read it straight through, like a good novel."[27]

In the summer of 1956, Simon and Newell had the opportunity to show the Logic Theory Machine to their peers. The occasion was a workshop at Dartmouth College funded by the Rockefeller Foundation and organized by Marvin Minsky, Claude Shannon from Bell Labs, and John McCarthy, an assistant professor of mathematics at Dartmouth. The workshop was memorable for several reasons. McCarthy began promoting the phrase "artificial intelligence" at the gathering, partly to "escape association with cybernetics."[28] McCarthy's colleagues accepted the term and it gradually entered the vernacular. The 1956 summer study also appears notable, in retrospect, because historians have often labeled it as the "birthplace of AI."[29]

Whether or not the 1956 meeting was *the* pivotal point for the history of artificial intelligence research, there's no disputing that Simon and Newell wanted to display what the Logic Theory Machine could do.[30] Unlike other attendees at the Dartmouth gathering, whose work was still in speculative stages, Newell and Simon could actually show off a working program. Moreover, it was developed at RAND, a hardheaded place where one was expected to show proof of concept in the form of results and code, not handwaving.

Simon and Newell also wanted to reach beyond the nascent community of computer scientists interested in whether and how computers could think. So, in November 1957, Simon presented their work to members of the Operations Research Society of America. Operations research was an interdisciplinary field of study created out of the exigencies of World War II. It analyzed practical ways for business managers and military logisticians to optimize decision-making processes. A computer that could make decisions would be of great potential value to many patrons with deep pockets.

Simon's capacious after-dinner address ranged from the efficiency seeking methods of Frederick Winslow Taylor to the protocomputer inventions of Charles Babbage before settling in to describe computer programs as tools for "heuristic problem solving." Computers, Simon noted, could already distinguish between printed letters and play "a pretty fair game of checkers." Now, with the Logic Theory Machine, "intuition, insight, and learning are no longer exclusive possessions of humans" but now something "any large high-speed computer can be programmed to exhibit." Just as the steam-powered industrial revolution forced humans to reconsider their place in the world when their physical power was bested by machines, the "revolution in heuristic problem solving" would make people reassess their place in the world once their intellectual capacity was "outstripped by the intelligence of machines."[31]

Not all of Simon's and Newell's colleagues appreciated such speculations. Richard Bellman, an applied mathematician at RAND, branded Simon's address as "magnificent" but "not scientific writing." Simon's "careless usage" of phrases like "machines that think," he charged, "merely adds to the mysticism that surrounds" the existing uses of computers.[32] This was exactly the sort of "subjunctive speak" that Mortimer Taube found so objectionable.

Despite these critiques, funding for research began to pour into the artificial intelligence community, creating a boomtown mentality. Solving problems and making decisions were activities that interested the armed forces. The Advanced Research Projects Agency (ARPA), formed hastily in 1958 after the Soviets' Sputniks, became the primary patron for computer scientists in the United States by supporting work on such topics as robotics, speech recognition, and computer time-sharing. Artificial intelligence researchers, funded via ARPA's Information Processing Techniques Office, were especially well supported, creating a halcyon era for the field that lasted until the early 1970s.[33]

ARPA did not distribute this funding evenly however. Three schools—MIT, Carnegie Mellon, and Stanford (where McCarthy and Feigenbaum ultimately landed)—benefitted the most. This inequality was fostered in part by researchers who rotated between university positions and stints as program officers at grant agencies. Years later, Marvin Minsky recalled having "more freedom of research under ARPA than any other government agency." When asked why project managers were so supportive, Minsky smiled, as if explaining something to a child, "Because they were us."[34]

## SOMEWHERE BETWEEN ZERO AND ONE

The first boom for artificial intelligence also came with some busts. After World War II, a massive increase in scientific publications worldwide had occurred. This glut—the very same problem that Mortimer Taube's library-related work addressed—created a deep desire on the part of scientists and military leaders to convert materials published in other languages (especially Russian) into English.[35] For years, a concerted effort to "machine translate" documents from Russian into English via computer took place. Funding approached several million dollars a year, CIA staff monitored the efforts, and, ironically, a new academic journal sprang up to report results. The Soviets, sensing a potential "translation gap" followed suit with their own projects.

Unfortunately, reconciling the two languages' vocabularies and syntaxes in a manner legible to computers proved harder than advocates originally promised. Skeptics like Norbert Wiener (whose father was a linguist and translator) and Mortimer Taube criticized machine translation programs as wasted efforts that could be better done by simply training more humans. Taube, for instance, stated that machine translation was not "a genuine scientific investigation, but a romantic quest," a well-funded folly camouflaged by overly optimistic statements of what *might* be possible.[36] In 1966, a pessimistic assessment by the National Academy of Sciences effectively killed, for the time being, computer-based translation efforts. Machine translation represented just one attempt to exploit the imagined potential of "thinking machines." But what if the entire field of artificial intelligence was a delusion? Starting in the early 1960s, Hubert L. Dreyfus, a philosopher at MIT, started to express exactly these sentiments.

Dreyfus's skepticism started when he and his brother Stuart attended a series of lectures organized at MIT around the theme of "Management and the Computer of the Future." A talk by John R. Pierce, a renowned electrical engineer at Bell Labs, caught Dreyfus's attention. Pierce, who later led the study that torpedoed machine translation efforts, titled his lecture "What Computers Should Be Doing" (note the subjunctive). "Just because a computer *can* do something a man might do," Pierce said, "does not mean that it should." After several computer experts publicly attacked Pierce, the Dreyfus brothers felt compelled to respond. They agreed with Pierce but went even further, stating that work on language translation, game playing, and pattern recognition had "contributed nothing to the understanding of the nature of intelligence."[37]

In 1964, Hubert Dreyfus spent several months at RAND, which only deepened his suspicion about artificial intelligence. Dreyfus singled out research by Simon and Newell for scrutiny, especially their newly revealed General Problem Solver (GPS). An extension of the Logic Theory Machine, the GPS was pitched as a computer program "capable of simulating, in first approximation, human behavior in a narrow but significant problem domain." Able to do more than solve math proofs, they boldly billed it as a "program which simulates human thought."[38]

In response to such extravagant claims, Dreyfus wrote a lengthy report, published under the auspices of the RAND Corporation, provocatively titled "Alchemy and Artificial Intelligence."[39] After evaluating research efforts in language translation and problem solving, Dreyfus concluded that several questionable assumptions underpinned much of the work in artificial intelligence. For example, he challenged the original idea, drawn from neurophysiology, that the on-off logic switches of computers were like the neurons of the brain (an assumption, to be fair, that most computer scientists had already abandoned). Dreyfus also noted—correctly—that digital computers, which were binary-based, yes-or-no machines, displayed no tolerance for ambiguity whereas human brains often function quite well with it.

Like ancient alchemists who did interesting chemical experiments but, distracted by their "retorts and pentagrams," never transmuted lead into gold, Dreyfus concluded that artificial intelligence researchers were following a similar path. "Blinded by their early success," they still were nowhere near their ultimate objectives which, in fact, might be unattainable.[40] Dreyfus's report circulated outside the wood paneled rooms of RAND's headquarters in Santa Monica, garnered attention from the *New Yorker*, and was republished in the journal *Review of Metaphysics*. He eventually assembled his criticisms about the "limits of artificial intelligence" into a book, published in 1972 as *What Computers Can't Do*.[41]

Not surprisingly, the artificial intelligence community repudiated Dreyfus's opinions. Seymour Papert, a computer scientist at MIT, wrote a rebuttal that crossed from the professional into the personal (he called this piece "The Artificial Intelligence of Hubert L. Dreyfus"). Papert's attack was not without humor. For a section titled "Computers Can't Play Chess," Papert slyly provided no words—why argue with something one disagrees with? He followed this with a section titled "Nor Can Dreyfus."[42] For his part, Herbert Simon wrote a somewhat conciliatory open letter to Dreyfus after the philosopher lost a chess match to a computer program written by an MIT student.[43] Rather than crowing about Dreyfus's ignominious loss, Simon suggested that when the program bested the philosopher, it had behaved not as an omniscient machine but like "a frail and desperate humanoid . . . as you and I."[44]

As the 1970s began, the artificial intelligence community existed somewhere between zero and one, figuratively speaking. On one hand, the field claimed enormous potential to make contributions to academic disciplines like psychology, linguistics, and, of course, computer science, while potentially generating a host of products for business and military patrons. A decade and a half of work, supported by millions of dollars in funding, had produced notable results, such as John McCarthy's programming language LISP, which was widely adopted by artificial intelligence researchers. Hyper-optimists like Marvin Minsky predicted that artificial intelligence would eventually be a watershed moment in human history as people would be surrounded by the "intellectually superior beings" they had built.[45] On the other hand, some of the gloss that accompanied early enthusiasm had worn off and interest from funders waned in the early 1970s. Skeptics like Taube and Dreyfus looked at the community's brief history and found it replete with rampant boosterism, exaggerated claims, underwhelming research results, and even suggestions of a swindle. Such was artificial intelligence's liminal dream state when Joseph Weizenbaum started writing his own book.

### "ONE SHOULDN'T LIE"

Joseph Weizenbaum didn't begin his transformation from successful computer scientist to outspoken critic of the "artificial intelligentsia" until the early 1960s.[46] Yet the seeds for his moralizing can be traced back to his childhood in Germany. Born in 1923, Weizenbaum grew up in Berlin in an assimilated, upper-middle class Jewish family. By his own account, Weizenbaum was a sensitive and lonely child who never received attention and affection from his autocratic father while his mother provided too much.[47] The Nazis' takeover of Germany added to his already considerable sense of anxiety and insecurity. What he observed as an impressionable young boy, provided Weizenbaum with a visceral sense of what happens when people abdicated moral responsibilities or were "victimized by irrationality."[48] His family left for the United States in 1936, eventually settling in Detroit.

Weizenbaum was already interested in mathematics before coming to the United States and his early unfamiliarity with English only deepened his interest. "I could understand algebra," he recalled, "it didn't matter if it was in German or English." He enrolled at Wayne State University in Detroit and started to study math before the war intervened. After five years of military service, Weizenbaum finished his degrees, eventually earning a BS and an MS (but no PhD). It's worth noting, given what followed a few decades later, that Weizenbaum's interest in computing emerged around the time of his first marriage's collapse and his initial exposure to psychoanalysis.

In 1962, Weizenbaum, now remarried, poked his first needle in the inflated expectations surrounding artificial intelligence. In an article titled "How to Make a Computer Appear Intelligent," he assumed a combative tone by challenging Marvin Minsky's claim that an activity which gives results not immediately understandable to an observer will appear "to be somehow intelligent."[49] Weizenbaum countered that any person who claims to have written an "'artificially intelligent' program" is "clearly setting out to fool" people. To support his argument, Weizenbaum described a program that played the game Gomoku (or "Five-in-a-Row"). Weizenbaum observed that the computer could be programmed to play the game with a "wonderful illusion of spontaneity" so as to "appear intelligent." However, the computer's "simple algorithm" essentially fooled observers into believing something more complex was happening. What looked like "intelligence" was merely a computer performing evaluations, following rules, and optimizing its actions based on mathematical instructions. It was not intelligent, in other words.

Soon after his article appeared, MIT offered Weizenbaum a position as a visiting professor in its Department of Electrical Engineering. The appointment became permanent and Weizenbaum spent the rest of his career at the school. His initial assignment was to work on Project MAC, a flexible acronym that could mean "Multi-Access Computers," "Machine-Aided Cognition," or (to some wags), "Man against Computer."[50] One goal of Project MAC was time-sharing by having a user type at a terminal and get the impression they were the only person using the computer system. Project MAC was generously supported by defense funds channeled to MIT via ARPA. Unlike Norbert Wiener, who had rejected military monies almost two decades ago, Weizenbaum did not express similar misgivings until the late 1960s when he began speaking out against the Vietnam War.

Just before he arrived at MIT, Weizenbaum wrote a new computer language, a competitor to McCarthy's LISP, that he called Symmetric List Processor, or SLIP.[51] In an article describing SLIP, however, Weizenbaum broke with his field's tradition and insisted that the journal publish not just a description but also the code itself. "It was an attempt to be as lucid as possible and make it possible for other people to use it," he recalled. "There's no mysteries."[52] Weizenbaum's interest in transparency continued after his family settled in Massachusetts. MIT provided him with a home terminal that could connect to one of the school's large IBM machines over phone lines. Weizenbaum's children were intrigued by the equipment their father spent so much time with so he wrote short computer routines for them. The "toy programs" could answer basic questions ("Is it Tuesday?") with a yes or no. But his kids' various conjectures as to how the computer "knew" things discomfited Weizenbaum. "I worried about my children getting the idea that the computer is a magic machine and daddy is a magician," he recalled.[53]

In 1966, Weizenbaum wrote an innovative computer program he called ELIZA. It became the technical accomplishment he would best be known for and its implications were a touchstone he returned to over and over. Weizenbaum named his new program after Eliza Doolittle, the fictional character who turns to Professor Henry Higgins for elocution lessons (the Oscar-winning film version of *My Fair Lady* had recently appeared). ELIZA enabled people to interact with a computer using what experts call "natural language." Instead of having to master a specialized language such as FORTRAN, a person could *converse* with the computer using normal words, sentence structure, and punctuation as in a regular conversation.[54] Central to

ELIZA's workings were data Weizenbaum called a "script," which referenced a "particular class of conversation."

What made ELIZA (and Weizenbaum) famous was one of the scripts he used to test his program. He imagined a dialogue between a human and the computer with the former standing as a patient seeking counseling and the computer acting as a therapist. More specifically, Weizenbaum programmed ELIZA using a script called DOCTOR so it would imitate a Rogerian therapist. In the 1940s and 1950s, psychologist Carl Rogers developed a form of "person-centered therapy" in which the analyst listened to and affirmed a patient's statements. For example, if a patient said "I went for a long boat ride," the therapist might respond with "Tell me about boats." The technique provided the patient with "a sense of being heard and understood" while, for the sake of programming, it eliminated the need for the computer to store "*explicit* information about the real world" in its memory.[55] In fact, the patient often unconsciously filled in gaps with "generous assumptions" about their "conversational partner" and therefore could "clothe ELIZA's responses in vestments of plausibility." For example, a conversation between a person and a computer running Weizenbaum's DOCTOR script might unfold as follows:

Person: Men are all alike.
Computer: In what way?
Person: They're always bugging us about something or other.
Computer: Can you think of a specific example?
Person: Well, my boyfriend made me come here.
Computer: Your boyfriend made you come here?
Person: He says I'm depressed much of the time.
Computer: I'm sorry to hear you are depressed . . . [56]

Weizenbaum noted that it was possible to set up experiments "in which the subjects find it credible to believe that the responses which appear on [the] typewriter are generated by a human sitting at a similar instrument in another room." In this sense, Weizenbaum's creation seemed to fulfill part of what Alan Turing had famously called the "imitation game." In this scenario, the British computer scientist imagined a person monitoring a dialogue between a human and a machine—both unseen—who are both trying to convince the evaluator they are human. If the evaluator could not reliably delineate between the human and the computer, then the machine was judged to have won.[57] But, as Weizenbaum strenuously clarified for a *New York Times* writer, "the computer does not understand what is being said." Still—much to the discomfort of Weizenbaum—people imbued ELIZA with more talent than it actually had. The *Times*, for instance, described how Weizenbaum's assistant asked him to leave the room so she could *talk* with ELIZA in private.[58] For Weizenbaum, who had worked assiduously in therapy to recover from his own personal difficulties, the idea that a machine could replicate a human therapist was ludicrous.

While his technical publications described how ELIZA worked ("Few programs ever needed it more," he noted), Weizenbaum was bothered that his creation still deceived people into thinking they were conversing with a thing that understood their feelings. "It is said that to explain is to explain away," he wrote. "Once a particular program is unmasked, once its inner workings are explained in language sufficiently plain to induce understanding, its magic crumbles away."[59] But what really troubled Weizenbaum was that, even after he explained what ELIZA was (and wasn't), people still *wanted* to be fooled by it. For Weizenbaum, this raised a personal sense of culpability. If he was writing programs that fooled people, what did that make him? "My job should be demystification," he said, "one shouldn't lie. One shouldn't pull con jobs on the world."[60] Weizenbaum's success with ELIZA made him realize that, in addition to things computers *couldn't* actually do, there were also things that computer scientists *shouldn't* do.

This theme would become important in *Computer Power and Human Reason* but it was something he had talked about it with colleagues for years. In 1965, for example, the RAND Corporation invited Weizenbaum to its annual computing symposium. One topic his panel addressed was identifying problems facing the computer profession. Weizenbaum referenced J. Robert Oppenheimer's famous statement that "physicists have known sin" for building nuclear weapons.[61] "There is some probability that computer people will have that feeling," Weizenbaum said. "There are some things we ought not to do even though people will pay us to do them." The key difference, he explained, was in pursuing research problems that "were not worthwhile" versus "problems that are *wrong*." What computer scientists needed, Weizenbaum said, was to be brave enough to say to their patrons, "I will not do this."[62]

Later in his life, Joseph Weizenbaum claimed that he was not a computer critic but rather a social critic.[63] The problem wasn't the machines but rather the motives and mindsets of the people who built, programmed, and used them. His conversion to criticism coincided with a growing realization among layfolk and experts alike that, whatever the future might hold, it would be built around computers. These conjectures were presented quite visibly to general readers when Alvin Toffler's bestseller *Future Shock* appeared in 1970.[64] With his wife Heidi making essential research contributions, Toffler described a fast-moving society where transience, novelty, and diversity would be de rigueur. The book's title referred to a psychological state brought about the accelerating pace of social and technological change.

Norbert Wiener might have boasted that he ate his royalties but *Future Shock* produced a genuine feast of fame and riches for its author. By the end of 1970, hardback sales of *Future Shock* in the United States alone passed 100,000.[65] The Tofflers sold the paperback rights for more than $250,000, close to $2 million today, and its sales eventually crested the seven million mark. Meanwhile, Toffler's titular phrase soon became a cliché as well as a musical, an entry in the *Oxford English Dictionary*, and the title of a hit single by soul singer Curtis Mayfield.

*Future Shock* wasn't about computers per se, but rather about a future society that, to Weizenbaum's discomfort, would be fully saturated with computers and computing. Words such as "technology," "computer," and "information" appeared in its pages scores of times and stood as the prime mover for the broad changes Toffler described. When it came to the future of work, the futurist predicted employees would be less concerned with the manufacture of traditional commodities and would instead spend their days immersed in an economy based on services and information.[66] Global society would transition from an older industrial model, based on the "brute force technology" of huge, capital-intensive factories, to a new phase that would be "technological but no longer industrial."[67] The tomorrow of *Future Shock* was one where information technologies would remake work, politics, and personal lives. It was a future Weizenbaum viewed with skepticism.

In 1972, Weizenbaum broadcast his emerging opposition in the *New York Times* and *Science*—arguably the world's preeminent general journal for scientists—meaning that Weizenbaum's message reached a wide audience of both laypeople and specialists.[68] It was a controversial move. Here was a noted researcher, writing from the heart of the military-industrial-academic complex, saying that the computers were not to blame; they were just machines that did as instructed. Instead, it was the computer programmers who had "abdicated their responsibility" to society.[69] "The fundamental question the computer scientist must ask is not what shall I do?" but rather, Weizenbaum wrote, "what shall I be?"[70] When Weizenbaum sat down to compose his book, alerting the public not about what computers *might* do but rather what computers scientists *ought not* do, became his mission.

## THE COMPUTER METAPHOR

Weizenbaum's path from socially conscious programmer to successful author has to be understood in a wider context marked by two events. One was the growing mobilization against the Vietnam War that was coalescing on college campuses in the United States and overseas. By the end of the 1960s, MIT was the largest academic recipient of funding for military research, hoovering up some $118 million in 1968 alone.[71] Fields like computer science owed their existence to support from defense agencies. In order to provoke a public discussion about this state of affairs, in January 1969 several MIT faculty members called for a strike. Forty-eight professors, including Weizenbaum, signed what became known as the "March 4 Manifesto."[72] Weizenbaum believed computer scientists were as culpable, if not more so, than other researchers. Nuclear weapons relied on complex computer programs that he claimed their creators did not fully understand and which might reveal their systemic fallibilities with apocalyptic consequences. Meanwhile, computer-based "target selection systems" helped prosecute a high-tech war in Vietnam while their makers remained insulated and ignorant of the carnage they caused.

The abhorrence Weizenbaum and his colleagues expressed about the use of computers and other sophisticated military technologies in the ongoing war in Southeast Asia was related to another issue. Scholars, public intellectuals, and citizens were increasingly concerned about the deleterious influence technology as a whole had on society. While the 1960s stood as a golden era of largely optimistic statements from futurists like Alvin Toffler, the flipside was a drumbeat of gloomy proclamations about technology and social change. Technological risks, whether from chemical pollutants, badly designed automobiles, or malfunctioning nuclear plants, moved to the forefront of public awareness. Coincident with these concerns was a growing number of books, from authors like the French academic Jacques Ellul and the Marxist philosopher Herbert Marcuse. Here, the targets were not specific technological objects per se but rather the seeming self-directed nature and interconnected totality of the entire technological enterprise.[73]

Naysayers could dismiss Ellul and Marcuse for stylistic clunkiness and philosophical impenetrability but the same could not be done for works by Lewis Mumford. Born in 1895, Mumford enjoyed an extraordinarily long run as a public intellectual, writing books on American literature and urban planning (his book *The City in History* won a National Book Award) before turning to the history of technology. After Mumford's son was killed in action in 1944, his writings assumed a more pessimistic cast.[74] When one historian praised Mumford's 1970 book *The Pentagon of Power* as "fresh and passionate," he also castigated the author for strident and ultimately ineffectual denunciations of the modern "megamachine" run amok which helped foster an "anti-intellectual wave" among university students.[75] As with assessments by Ellul and Marcuse, Mumford depicted technology not as a central part of modern humanistic culture but rather as a monolith standing in opposition to it.

Weizenbaum began corresponding with Mumford around the time that *Pentagon* appeared and its ideas strongly influenced him. Some twenty-five years Weizenbaum's senior and an accomplished author, Mumford assumed the role of friend, confidant, and manuscript reader.[76] Weizenbaum's letters to Mumford oscillated between prosaic updates about health issues, university matters, and family news with anguished musings about contemporary society and Nixon-era politics. In these letters, one can also track Weizenbaum's steady progress toward completing *Computer Power and Human Reason* as well as his insecurities about how it would be received.[77] In late 1973, Weizenbaum shared a draft of his opening chapter with Mumford, noting that a career spent at MIT, where everything was "polarized around science and technology," had left him questioning whether he could even write a meaningful book critical of such subjects. "I am as frightened of the task ahead of me," Weizenbaum confessed, "as a laymen might be were he called to do surgery."[78]

Setting fears aside, Weizenbaum finished a full draft by September 1974, taking advantage of a grant from the National Science Foundation and two sabbatical fellowships. In his application to Stanford's Center for Advanced Study in the Behavioral Sciences, Weizenbaum noted his wish to engage with scholars "who know more about society than I do" and who could help him "survey humanistic and radical critiques of technology."[79] While at Stanford, he learned from Steven Marcus, an English professor from Columbia, who "struggled mightily to educate a primitive engineer." A second fellowship, taken at Harvard, brought Weizenbaum into contact with Hilary Putnam and Daniel C. Dennett, two Boston-area philosophers whose "patience with my philosophical naivety was unlimited."[80] Weizenbaum also corresponded intermittently with Mumford ("that grand old man") who read the entire manuscript for his friend.

When it came time to select a publisher, Weizenbaum opted for the familiar. He had recently written an introductory essay for a collection of articles about computers released by W. H. Freeman and Company.[81] The company, founded in San Francisco in 1946, primarily published textbooks. (One of their bestsellers was Linus Pauling's now-classic textbook, *General Chemistry*.) Once committed to W. H. Freeman, Weizenbaum worked closely with Peter Renz, an editor with a PhD in mathematics, and Aidan Kelly. Weizenbaum displayed an especial fondness for Kelly, who had a side career leading a Wiccan coven, saying that "perhaps readers will understand if I say simply that Aidan Kelly is a poet." Kelly returned the affection, at one point telling Weizenbaum how he had been "waving your manuscript at people and saying 'You know what this is? It's Nobel prize juice!'"[82]

Juicy or not, W. H. Freeman's staff found Weizenbaum's project a marked departure from the company's usual fare of math and science textbooks. Renz noted that, if they were lucky, what the press initially planned to title *Not So Human a Machine* might be adopted for the new "computer and society" courses that colleges were starting to offer.[83] Aidan Kelly imagined an even larger audience and "looked forward to the screams of anguish from those whose oxen will be gored by your forthright stand." The trick would be finding the best combination of marketing and title. For the latter, author and publisher agreed on *Computer Power and Human Reason*. The choice revealed Weizenbaum's foremost concern: by embracing the power of what computers could do, humans' ability to make informed and ethical judgments had been subordinated.[84] The book's subtitle, *From Judgment to Calculation*, meanwhile referred to the ethical compass he found lacking among computer scientists who became, in essence, amoral calculating machines.

For a book cover, Weizenbaum was particularly drawn to art made in the 1920s by the German Dadaist artist George Grosz, which showed the dehumanizing effects of machines and war. His editors at W. H. Freeman were unconvinced, thinking that a poorly chosen image might dilute Weizenbaum's argument.[85] In the end, they agreed to include a watercolor by Grosz called *Republican Automatons* as a frontispiece. Made in 1920, the image depicts two people, possibly war veterans. One figure, missing both their right leg and hand, waves a small German flag while the other figure, with gears and belts attached to their side, wears the Iron Cross, a military medal of the German Empire and later the Nazis. The image, quite fitting for Weizenbaum's book, encapsulated a critique of both militarism and nationalism while illustrating how people, as their lives became more intertwined with machines, could become less than human.

W. H. Freeman's editors expressed uncertainty about how (and how much) to promote Weizenbaum's book. As Peter Renz joked to Weizenbaum, marketing strategies were a "split second matter decided months in advance." But W. H. Freeman hoped that, once the book was out, a trade press would see that Weizenbaum had not written "just another computer book" and acquire the rights for it. They encouraged Weizenbaum to ask Noam Chomsky and Lewis Mumford for blurbs (both complied) but the publisher didn't initially advertise in newspapers, a decision they reversed once *Computer Power* started to do better than expected.[86]

In anticipation of its appearance, the Book of the Month Club named *Computer Power and Human Reason* an "alternate selection," calling the book "an eloquent and moving defense of the sanctity of the human spirit."[87] Now sensing they might have a stronger-than-expected performer, W. H. Freeman issued a press release describing Weizenbaum's rebuttal of "extravagant predictions that computers will someday replace uniquely human functions," claims the author branded "obscene" and a "fundamental failing" of modern society.[88]

Assessments of Weizenbaum's book from experts and laypeople were aided by his clear and accessible prose. Two somewhat pro forma chapters he included about how computers and programs worked welcomed readers "not comfortable with technical details."[89] (Compared to early drafts, Weizenbaum had tempered the technical material substantially.) Weizenbaum signaled his arguments quite clearly at the very outset of the book. Indeed, his very first sentence ("This book is only nominally about computers"), marked his intent to not write about computing machines per se, but rather the technocratic mindset that accompanied their use and misuse. Too few people, he explained, "have any idea where the power of a computer comes from," a situation he set out to remedy.[90]

To Weizenbaum, the influence of computers was not located in its logic elements or memory banks but in the misleading language that accompanied them. Terms like "computer" and "artificial intelligence" had become dangerous representations for "what we have done and are doing" in uncritically accepting technologies. Weizenbaum said the root of the problem was linguistic, found in what he called "the technological metaphor." Scientists such as Newton, Darwin, and Einstein had introduced scientific theories—the clock-like universe, evolution, and relativity, respectively—which eventually became larger cultural metaphors that ordinary people used. Computer scientists, he argued, were introducing new metaphors that the larger culture was accepting and internalizing even if they didn't fully understand their implications.

Dozens of reviews of *Computer Power and Human Reason* appeared in newspapers and the specialty journals read by mathematicians, psychologists, and, of course, computer scientists. With the exception of the latter group (more on that in the next section), the reactions were very favorable. One psychologist who reviewed *Computer Power* for *Nature* concluded that "Weizenbaum deserves our gratitude" for producing a clear and accessible book that "should force us to think more carefully about what we set our computers to do."[91] Another reader said the book sounded the "call of a respected teacher to other teachers of computer science" to control this "powerful new metaphor."[92] Meanwhile, *Byte*, a magazine read by computer hobbyists, praised what it saw as an "exciting, challenging book," that "is as much about life as it is about computers."[93]

Weizenbaum started receiving letters from readers in the United States and overseas, which he preserved in folders labeled "Fan Mail." Some readers wanted to argue with him, sending rebuttals to points his book raised. A few sent him poems. Others simply expressed thanks for how he addressed difficult moral questions. A professor of information science called *Computer Power* a "beautifully written book" that his graduate students would read for a seminar on the personal implications of computers.[94] And a sociologist at the University of California shared some high praise with Weizenbaum, giving the book a "place of honor" alongside Wiener's *God & Golem, Inc.*, on their bookshelf.[95]

Publisher and author alike benefitted from positive reviews and reader enthusiasm. Monthly sales hovered around 1,000 copies and the book quickly came out as a paperback. A German translation also appeared via Suhrkamp Verlag, a company that had a strong roster of books in literature and the humanities, including works by Theodor Adorno and Bertolt Brecht. *Computer Power and Human Reason* appeared as *Die Macht der Computer und die Ohnmacht der Vernunft*, which can be translated as *The Rise of the Computer and the Collapse of Reason* (the German version dropped the subtitle). The book eventually appeared in several other languages, including Japanese, Dutch, Swedish, and Portuguese. Adoptions of Weizenbaum's book for such university courses as The Computer Threat to Society (Grinnell College) and Computers and Society (University of California at Santa Barbara) boosted sales too.[96] One professor called it a "veritable Rosetta Stone . . . for a multiplicity of subjects." The overall reaction pleased Weizenbaum's editors. Aidan Kelly gushed that Weizenbaum's book was "the finest book this company has ever published" while the more restrained Peter Renz wrote, "We are most pleased with your book, intellectually and financially, and we expect to continue to be pleased."[97]

Weizenbaum was delighted to see his book getting traction with younger readers. "More and more students," Weizenbaum told Mumford, were showing up at his office to discuss "the problems of technology and society." To Weizenbaum, this suggested a power greater than what any machine could offer. As he wrote in his book, there were most certainly "tasks which computers *ought* not to be made to do, independent of whether computers *can* be made to do them."[98] Here, he drew comparisons between artificial intelligence research and the emerging field of genetic engineering based around using recombinant DNA. Both were ethically fraught fields where scientists, as well as students preparing for careers in science, should exercise self-restraint. For Weizenbaum, the maxim "Can does not imply 'ought'" was the crux. A scientist's actions should ideally be based on moral, not technical, grounds as they struggled against the "imperialism of instrumental reason."[99] "How absurd," Weizenbaum mused to Mumford, "do the megalomaniacal fantasies of technological madmen look by comparison."[100] Of course, those "madmen" had their own thoughts about what computers (and computer scientists writing about them) ought to do and not do.

## DEBATING THE ARTIFICIAL INTELLIGENTSIA

It had been easy for computer scientists to dismiss Hubert Dreyfus. He was an outsider—an infidel—to their community. Even Weizenbaum paid Dreyfus little attention, noting that "he just doesn't know much about computers."[101] But Weizenbaum, whether he liked it or not, was a member of the "computer priesthood" and could not be similarly dismissed.[102] Practically every review of Weizenbaum's book mentioned his ELIZA program, an accomplishment that, in the eyes of many readers, granted him the authority to critique the computer science community. As a result, Weizenbaum assumed the role of an apostate, a former believer who now throws stones through the windows of his own church. "I have pronounced heresy," he told the *New York Times*, "and I am a heretic."[103]

Computer scientists who read Weizenbaum's book typically focused on two central critiques it made. The first was personal. In what was probably the most oft quoted passage of his book, Weizenbaum derided the social shortcomings of the "compulsive programmer."[104] His indelible description is worth quoting at length:

> Wherever computer centers have become established, that is to say, in countless places in the United States, as well as in virtually all other industrial regions of the world, bright young men of disheveled appearance, often with sunken glowing eyes, can be seen sitting at computer consoles, their arms tensed and waiting to fire their fingers, already poised to strike, at the buttons and keys on which their attention seems to be as riveted as a gambler's on the rolling dice. When not so transfixed, they often sit at tables strewn with computer printouts over which they pore like possessed students of a cabalistic text. They work until they nearly drop, twenty, thirty hours at a time. Their food, if they arrange it, is brought to them: coffee, Cokes, sandwiches. If possible, they sleep on cots near the computer. But only for a few hours—then back to the console or the printouts. Their rumpled clothes, their unwashed and unshaven faces, and their uncombed hair all testify that they are oblivious to their bodies and to the world in which they move. They exist, at least when so engaged, only through and for the computers. These are computer bums, compulsive programmers.[105]

Looking beyond their pathological behavior and poor hygiene, Weizenbaum savaged "computer bums" for actually not being good programmers. These young men (here, we get a sense of the resolutely masculine world of computer coders at places like MIT) failed to provide sufficient documentation about their programs thus leaving future users uninformed about how they worked. While perhaps superb technicians, compulsive programmers saw their activity not as the means to an end (solving a problem) but as an opportunity to interact with computers, and not people, as much as possible. This was the goal and the grail. Unlike the engineer, who functioned in the real world of budgets and schedules, the compulsive programmer created his own digital realm inside the machine where he could revel in "illusions of grandeur" (as long as the system administrator allowed it).

The term "computer bum" was not Weizenbaum's invention. A few years earlier, the countercultural publisher Stewart Brand had given readers of *Rolling Stone* a compelling depiction of the "computer bums" (his term) obsessed with the "fanatic life and symbolic death" they experienced while playing games on their school's digital machines.[106] But, instead of doing real programming, what these "bum" programmers mostly engaged in was "hacking," a decidedly lesser activity in Weizenbaum's judgment. The questionable skills of these hackers—a term that Weizenbaum helped bring to the vernacular—resembled those of the "monastic copyist who, though illiterate, is a first-rate calligrapher." Indeed, Weizenbaum diagnosed computer bums as having a form of mental illness, akin to a compulsive gambler. Weizenbaum's book referenced Dostoevsky's 1866 novella *The Gambler* where the denizens of the casino remain glued to roulette tables at all costs. The Russian writer, Weizenbaum wryly observed, "might as well have been describing a computer room."[107] With depictions like these, it is easy to imagine how Weizenbaum's prose peeved some programmers.

The second main critique Weizenbaum made—one that drew even more ire from his colleagues—was both professional and philosophical. Several responses to Weizenbaum's book came from scientists who had been called out by name in the book. Since he figured as a "principal hero" in *Computer Power*, Herbert Simon puckishly asked the author to be gracious enough to mail him a signed copy (Weizenbaum complied).[108] A continent away, two Stanford researchers were discomfited by how Weizenbaum had portrayed them. Joshua Lederberg was a Nobel-winning molecular biologist who, with Edward Feigenbaum, had written an artificial intelligence program in the 1960s called DENDRAL. The code helped chemists identify unknown organic molecules. In a review commissioned by the *New York Times*, Lederberg expressed misgivings with Weizenbaum's book, saying the author had shrouded his argument in "lyrical anti-technology slogans" and "out-and-out-obscurantism." While acknowledging that some researchers had regrettably made "extravagant prophecies" in the "early adolescence of computer science," Lederberg disputed Weizenbaum's claim that there were things a computer could never "know." By focusing too much attention on marking "the bounds of AI"—that is, what might or might not be possible—Lederberg charged Weizenbaum with disparaging investigators whose day-to-day work focused on "more concrete, modest, and achievable goals."[109]

Lederberg's comments fueled a heated debate between Weizenbaum and his critics, which occurred in print throughout the mid-1970s. Stanford's John McCarthy launched the most sustained objection with a review that started as an internal laboratory memo. McCarthy's lengthy comments were lightly edited and reprinted in *SIGART Newsletter*, a specialty publication read by the artificial intelligence community. McCarthy also wrote a third version for *Physics Today*, a publication that reached over 60,000 scientists monthly.[110] Attacking Weizenbaum's arguments in the pages of science journals was akin to trying to defrock him in his own church.

Setting aside the propriety of a single reviewer critiquing the same book multiple times, McCarthy savaged both Weizenbaum's mindset and motives. *Computer Power* was "unreasonable," he fumed, "a moralistic and incoherent book" that supported the "view promoted by Lewis Mumford, Theodore Roszak, and Jacques Ellul." "Obviously," McCarthy said, "one shouldn't program computers to do things that shouldn't be done." But McCarthy objected to how Weizenbaum had labeled certain areas of research as "obscene," lumping him in with other supposedly irrational critics, such as Taube and Dreyfus. What might happen, McCarthy asked, if "some bureaucracy e.g., an Office of Technology Assessment, that acquired power over the computing activities of a university, state, or country" and made such work illegal?[111] McCarthy also took issue with Weizenbaum's epistemological claim that science was "not the sole or even main source of reliable general knowledge."[112] Without science and scientists, McCarthy asked, how does one come to understand the world?

Weizenbaum stood fast in the face of such criticism. He objected to McCarthy's insinuation that somehow his alliance with such writers as Mumford originated with the "New Left political ideology." While he and McCarthy agreed that there were "tasks computers should not be programmed to do," the salient issue was determining how this demarcation might be enforced, especially if computer scientists themselves didn't mark or respect the boundaries. When it came to science as the best means to reveal the workings of the natural world, including how the human mind functioned, he and McCarthy clearly were at odds. Weizenbaum asked if there could really be some "authentic model of man that does not include and ultimately rest on philosophical and moral thinking," instead of McCarthy's science-based rationality. Computers offered programmers, compulsive or otherwise, a host of seductive powers but these were no substitute for striving to become "a whole person." Given the fundamentally different places he and McCarthy started from, Weizenbaum rued, "No wonder we talk past one another."[113]

Despite professing a philosophical attitude (in both senses) toward his critics, Weizenbaum admitted in private that his colleagues' attacks were wounding. When Pamela McCorduck was researching her history of artificial intelligence (her book *Machines Who Think* first appeared in 1979 with W. H. Freeman) some of his colleagues dismissed Weizenbaum's turn toward social issues. "He fell out of the field ten years ago, and hasn't done a damn thing since ELIZA," one scientist (likely Edward Feigenbaum) told McCorduck.[114] Such negative reactions prompted Weizenbaum to reflect that "once one goes public, one is subject to the interpretation of both the clever and stupid, both honest and dishonest . . . still, this sort of thing hurts."[115]

Heartened by readers' letters and cheered by students who regularly visited his MIT office, Weizenbaum accepted his role as gadfly and social critic. He produced scores of public lectures and writings on the potential of computers to reduce people to the level of machines. When Penguin Books published a new edition of *Computer Power and Human Reason* in 1984, Weizenbaum supplied a fresh preface. He observed, with some satisfaction, that he encountered much less "euphoric claiming and predicting from the artificial intelligence community" compared to a decade earlier. Perhaps, he speculated, his book helped dampen the enthusiasm. (The first so-called "AI winter," when military and corporate funding began dropping in the mid-1980s, was undoubtedly a more important factor.[116]) Nonetheless, Weizenbaum linked the increased militarism he saw on his own campus and in global politics to computer games that allow enthusiastic players—a new type of compulsive programmer perhaps—to maintain "clinical distance" from the electronic killing they engaged in for entertainment. The technological metaphor of the computer, he believed, still fostered a lack of empathy and moral responsibility.[117]

Late in his life, Weizenbaum recalled his daughter Naomi's reaction when she read his book for a university class. Playing psychologist, she surmised that he had not written the book as a screed against the "artificial intelligentsia" as he so often claimed. "You have to read what you wrote," Naomi said, "what you wrote *against yourself*, against your own belief that the world is just ones and zeros." Perhaps her father's real sparring partner, she said, had been his own fears about "those who appear to have power over us." It was, Weizenbaum conceded, a "clever observation."[118]

When travelers cross the Delaware River on the Lower Trenton Bridge, they can look out their train windows and see the sign. Erected in 1935 and more than 300 feet long, its bright red neon letters proclaim, "TRENTON MAKES—THE WORLD TAKES." Originally the phrase referred to Trenton, New Jersey's secure place in the global economy, as its factories churned out cigars, ceramics, and auto parts. By the 1970s, however, Trenton had fallen on hard times as factory closures and rising unemployment exacerbated entrenched social ills.

To some cynics, the city's motto was a passive aggressive poke at commuters who quickly changed trains at the Trenton Transit Center and departed for less desperate destinations, and the iconic sign itself fell into disrepair. This depredation occurred even as the Garden State was enjoying a modest boom of interest in computing, which, if people like Alvin Toffler were right, would be the industry and activity of the future.

In May 1976, some 1,500 people, many of them teens and children, traveled to Trenton State College to attend the Trenton Computer Festival. Organized by two engineering professors, it featured lectures, commercial exhibits, and a marketplace where enthusiasts and hobbyists could swap gear with one another. The upsurge of amateur computing was driven in part by the plummeting prices of electronics components.[1] New specialty stores such as the Hoboken Computer Works sold computer kits as well as all-in-one machines that came with a keyboard, display terminal, and a host of other peripheral devices. A community of aficionados coalesced around the new machines, including the Amateur Computer Group of New Jersey, which formed in 1975. A multitude of new computer-related books and magazines helped fuel the boom by offering readers essential information as well as a sense of community. Children and teens seemed especially drawn to the new machines. At the festival, one reporter observed twelve-year-olds "as familiar with microprocessors as their predecessors were with model trains."[2]

A few months after the Trenton festival, John Dilks, an engineer and computer hobbyist, organized an even larger event. Personal Computing '76 (better known as PC '76) was held at the end of August and it enticed even more enthusiasts and technophiles to Atlantic City, New Jersey.[3] Dilks had initially intended it as a trade show for companies to show off their products but the idea quickly morphed into a full-fledged techno-happening.

Some 5,000 people crowded into the venerable Shelburne Hotel for sessions titled "Software for Speech Synthesis" and "Making Big Bucks for the Computer Hobbyist." Some PC '76 attendees were computer experts who had trained on mainframe machines but were interested in what smaller, inexpensive computers could do. Others were novices including one housewife from New York City who used a computer for the first time at PC '76 to play the hit game *Star Trek*. Scores of teenagers, eager to buy their own machines, showed up wearing shirts proclaiming "DO IT WITH A COMPUTER" and "FORTRAN FREAK." Rumors circulated that a writer from *Playboy* was in the crowd collecting material for an article about the new technology.[4]

The exhibition floor of PC '76 was crowded with scores of businesses promoting their wares. Some companies, like MITS and IMSAI, were relatively new players. Others had been in the computer business for years, if not decades. Relegated to a miniscule booth, IBM simply hung a sign that directed curious consumers to a lounge where public relations staff passed out complimentary drinks. In addition to free booze, people were handed information about the company's new 5100 Portable Computer. One could order the fifty-five-pound machine for around $9,000 (about $50,000 today).[5] Dilks, meanwhile, persuaded the conference hotel to extend credit to two cash-strapped young entrepreneurs from California. Steve Jobs and Steve Wozniak had traveled to New Jersey to show off their new hand-built computer, which they called the Apple I.[6]

One of the speakers Dilks invited to PC '76 was Ted Nelson. A boyish-looking computer evangelist just shy of forty years old, Nelson was already celebrated among computer hobbyists because of his 1974 book *Computer Lib/Dream Machines*.[7] Nelson's fervent belief that "anyone can understand computers if they want to" was central to his book's message. Rather than being "implacable, dictatorial, and unapproachable" corporate tools that only produced "bleakness and oppression," Nelson maintained that computers could and should be "fun, friendly, helpful, exciting, and cheap."[8] Access to computers, Nelson believed, was a person's right, while its converse—computer illiteracy—was unnecessary and avoidable.

At Atlantic City, Nelson proclaimed a similar message in a talk he called "The Meaning of Computer Liberation." Nelson envisaged a shift in power that would occur once people became more familiar with computers. By running user-friendly programs and being accessible to ordinary people via easy-to-understand programming languages, home computers would be the solution to both corporate dominance as well as computer phobia. "COMPUTERS," as Nelson wrote in his 1974 book, "BELONG TO ALL MANKIND," a view he expressed with a vibrant slogan appropriate to the era: "COMPUTER POWER TO THE PEOPLE!"[9]

Nelson's opinions about computers were, like the machines he championed, deeply personal. They reflected years of thought, writing, and engagement with a diverse range of computer-focused communities. He predicted more than ten million home computers, functioning like "home appliances," would be in use by 1985. Nelson's forecast, while radical at the time, actually proved conservative. Scores of companies—familiar names like IBM and Apple as well as forgotten ventures like Polymorphic Systems and Parasitic Engineering—sold some 20 million home computers between 1976 and 1985. Meanwhile, the market for hardware and software products expanded to over $5 billion annually.[10] "Revolutionary" is an overused adjective in the history of technology but does seem appropriate when describing consumers' enthusiastic acquisition of home computers along with the feverish amount of media hype and business activity which accompanied it.

Narratives about the personal computer abound. Historians, museum curators, and journalists have filled up bookshelves with their accounts. In addition, by the early 1980s, the actual actors themselves—engineers and entrepreneurs—had already started to put their own recollections into print. The result is a vast and sprawling assortment of source material. There are over 600 books on Apple Computer alone.[11] What initially appears merely daunting quickly becomes utterly overwhelming once popular magazines, industry publications, catalogs, and newsletters are taken into account. Sampling, selection, and separating signals from noise becomes paramount.

This chapter assays an eclectic but necessarily limited array of publications that emerged alongside the personal computer. Besides books by Ted Nelson, it branches out to consider publications that computer experts and enthusiasts relied on to stay informed, promote their products, and argue with one another. Whereas much (but not all) of the action in hardware and software happened in and around the Bay Area, these publishing activities were more diverse. Some were short-lived business ventures while others proved both lucrative and durable.

All of these efforts embraced the idea that the personal computer represented a sudden shift in access and availability of computers. It *was* a shift in control. But before there was something called a personal computer, there was the activity of personal *computing*.[12] And part of that story unfolded in a big red barn just a short train ride north from Trenton.

### BARNSTORMING

In July 1968, the trade publication *Datamation* reported on the recent "Spring Joint Computer Conference," held that year in Atlantic City. One news item described how an ill-timed strike by telephone operators had prevented exhibitors from linking their terminals to off-site computers as union-sympathetic workers refused to wire up the necessary connections. Although companies' displays were effectively dead, a small cohort of teenaged computer enthusiasts from the Princeton area flaunted a clever work-around. They borrowed an acoustic coupler, a forerunner to computer modems, and connected it to a nearby pay phone. With this hardware in place the youngsters could dial in to an off-site minicomputer.

The teenagers called themselves the RESISTORS, a retronym (they picked the moniker first and then matched words to the letters) for "Radically Emphatic Students Interested in Science, Technology, Or Research Studies."[13] *Computerworld*, another industry-oriented publication, gave the RESISTORS front page billing—"Students Steal Show as Conference Opens"—and noted how the group drew a "fascinated crowd" of computer professionals.[14] A reporter even suggested that the RESISTORS represented the vanguard of a small-scale social movement as the teens wanted to engage with their counterparts from "underprivileged areas of Trenton" and introduce them to personal computing.[15]

In a book about books about computing, a small cohort of teens "playing" with computers might seem tangential. But the previously untold history of the RESISTORS highlights the fact that, years before there were machines called personal computers, some people regularly accessed computers for activities unrelated to their professional lives. Motives varied but entertainment as well as the display of technical prowess mattered. Just as importantly, the story of the RESISTORS expands our sense of the hobbyist community's membership and geography beyond better known groups like the Bay Area-based Homebrew Computer Club.

Fewer than seventy kids claimed membership in the RESISTORS over the group's roughly decade-long existence. Nonetheless, a surprisingly large number of them went on to have careers in technology and science. Two members wrote books about computing that would sell millions of copies (more on this in a later chapter). Another member cofounded Cisco Systems, a Bay Area company that made a fortune manufacturing internet routers and other hardware. Others became college professors, professional programmers, or started their own companies.[16] And then there is the fact that, starting around 1969, Ted Nelson's evangelism for more personal forms of computing became linked with the RESISTORS.

Engineer Claude Kagan and his farm, located a short drive from the town of Princeton, was the nucleus around which the RESISTORS organized. Born in 1924 in Orval, France, as Claude Ancelme Roichel Kagan, he moved to the United States as a teen, served in the army, and earned an MS from Cornell University in 1950. Kagan took a position with Western Electric, the manufacturing arm of AT&T, and then moved to Hopewell, about ten miles from Princeton, in 1958. Kagan's specialty was high-level computer languages, such as FORTRAN and BASIC, in which programmers write code that is independent of the particular type of computer. Kagan was also an inveterate collector of old computers and other electronics, which he stored in a large red barn on his property along with an ensemble of donkeys and large malamutes.[17]

Chuck Ehrlich, one of the original RESISTORS and, later, an entrepreneur and venture capitalist, recalls that in late 1966, he and a small group of "brainy social outcasts" were looking for some sort of clubhouse.[18] Less interested in smoking pot and social protests than they were in electronics, the kids were also disenchanted with the science classes their local schools offered. Kagan knew one of the teens' fathers and offered to let the group use his barn. They soon discovered Kagan's collection of artifacts including a surplus IBM paper tape punch, analog telephone equipment, and a Friden Flexowriter (a kind of heavy-duty typewriter that could be linked to a computer). The main attraction for the teens was Kagan's collection of computers.

The most imposing of these was a Burroughs Datatron 205, a computer first manufactured in the mid-1950s and based on vacuum tube technology. The enormous machine weighed several tons and stories circulated about how Kagan had borrowed a tractor trailer and heroically transported the behemoth from Michigan to New Jersey. Only slightly less imposing was an inoperable Packard Bell 250, a refrigerator-size computer of more recent vintage that the teens managed to get working. Kagan also allowed the teens to connect to his employer's DEC PDP-8 machine via teletype and phone lines so they could run programs written in TRAC. Originally developed in the late 1950s (and later trademarked) by Calvin Mooers, a computer scientist based in Massachusetts, TRAC was an efficient language amenable to being implemented on machines with relatively little memory.[19] The teens were fond, one member recalled, of connecting to the offsite computer and accessing a version of Joseph Weizenbaum's ELIZA program.[20] The opportunity the group had to work with computers interactively and in real time was something generally unavailable for nonprofessional computer users at the time. Kagan eventually persuaded DEC to donate a PDP-8 machine—no trivial gift as new models sold for $15,000 or more—which the RESISTORS worked with in the barn.

The bargain Claude Kagan struck with the RESISTORS was unusual for many reasons. First, Kagan was gay, a fact that the teens (and their parents) were aware of but which, by all accounts, bothered no one. When the Hopewell Valley Jaycees held a house tour in the spring of 1966, the brochure encouraged people to visit Kagan's "unique bachelor setting" that he shared with artist George Furnish.[21] Furnish passed away around the time the RESISTORS were forming and the grieving Kagan assumed the role of "guru, mentor, preceptor, fund-raiser, publicity agent, and landlord" for the RESISTORS. Kagan provided the space but the teens were responsible for maintaining both it and the equipment as well as covering the cost of electricity.

In contrast with most amateur computer clubs of the era, which were predominantly masculine spaces, photographs of the RESISTORS almost always show one or more young women working at a terminal or solving a programming problem.[22] When it came to deciding whose turn it was to use a teletype or printer, Jean Hunter—later a biology professor at Cornell—likened it to social time-sharing that required "beating people over the head to make them give you a turn." John R. Levine, who was a RESISTOR before studying computer science at Yale, recalled, "We were so nerdy that it didn't occur to us that girls were any different in terms of what they could do."[23] There were also efforts to recruit African American teens from schools in towns like Trenton. One of these kids, Joseph Tulloch, provided dozens of quirky, Dr. Seuss-like illustrations for a programming manual that Kagan and the teens assembled and published.[24] Tulloch later became a programmer for the state of New Jersey.

New members were initiated into the group by having an omega sign, the engineer's symbol for electrical resistance, drawn on their face with a magic marker (these were teenagers after all). One of the first things a new member would learn was how to use TRAC to write programs. For his part, Kagan held a dim view toward traditional learning as practiced in local classrooms. He instead insisted that the RESISTORS learn by doing. The group's pedagogical approach was adopted from the African American motto "Each one, teach one." As one member recalled, "if you want to teach someone how to do something, you *had* to let *them* sit at the key board."[25]

The RESISTORS's location in the Princeton area was another factor in their success. Several members had parents employed at one of the nearby technology companies, such as AT&T or RCA while others, such as Nat Kuhn, had parents who worked at Princeton University. Kuhn's father was Thomas Kuhn, a historian and author of *The Structure of Scientific Revolutions* (1962), a landmark book that introduced "paradigm shift" into the vernacular. As a kid, Nat Kuhn built devices from hobbyist electronics kits with his father, a former physicist. He joined the RESISTORS after attending an open house the group sponsored in February 1968 at the Princeton Junior Museum. Kuhn was just ten years old at the time. "I was super geeky," he recalled, "and the computer became my hobby and obsession. You could understand things through it and make things happen."[26] Then, soon after Nat Kuhn had his face inked with an omega sign, another person, much older but just as passionate about personal computing, started showing up at Claude Kagan's barn in New Jersey.

### INTERTWINGLED

Theodor Holm Nelson had not one, but two famous parents. His mother was Celeste Holm, an Academy Award-winning actress, and his father, Ralph Nelson, directed three films that won Oscars. The couple was also famously fractious and Nelson's maternal grandparents largely took responsibility for Ted's upbringing after he was born in 1937.[27] Books, reading, and wordplay were central features of a comfortable childhood spent in Chicago and New York. "We had a home of wonderful words," Nelson recalled. "A new word was a gift, a lens, a construction piece." Bedtime reading, radio shows, and listening to his grandparents compose correspondence together taught him the "delicacy of the writing process." "I was not a prodigy," Nelson later wrote, "I was just very clever, high-strung, interested in a lot of things . . . a lover of reading, movies, and ideas."[28] Nelson's childhood heroes included Walt Disney, Buckminster Fuller, and P. T. Barnum, a list that makes sense given his predilection for technological predictions and talent for showmanship.

For college, Nelson opted for Swarthmore and eventually majored in philosophy. He also published a modest-size magazine, designed to fit in the palm of one's hand, which he called *Nothing*. Nelson later recalled that the college environment was a disappointment in that ideas were not valued as much as professors seemed to suggest.[29] Despite these reservations, he graduated from Swarthmore in 1959 and went on to study sociology at the University of Chicago and then at Harvard where Nelson took his first computer course. Nelson's autobiography includes a whole chapter, titled "The Epiphany of Ted Nelson," about this revelatory experience. According to Nelson, when he realized the computer, instead of a dreary number-crunching device, "could be whatever it was programmed to be" his "world exploded."[30]

An even bigger revelation for Nelson, given his penchant for writing, was that computers could handle text by manipulating, storing, printing, and, above all, displaying it on screens. And, if this could be done with text, it could probably also be done with images and sound. "The future of mankind was at the computer screen," he decided, as the "interactive computer would become the workplace of the future."[31] As early as 1960, Nelson was already imagining a company—he christened it General Creative—which would offer interactive computers to artists and writers.[32]

Equally profound for Nelson was recognizing that once a person had manipulable text on a computer screen, they could use it to construct parallel, nonsequential textual passages. These word assemblages could then be linked to one another or branch off in entirely new directions. These were farsighted ideas at a time when access to actual computers remained limited to scientists, engineers, and other technical experts. Personal computing, let alone personal computers, still sat well over the horizon.

In 1964, Nelson accepted a teaching position at Vassar College where his new colleagues invited him to describe how the future of work as well as artistic creativity would happen on computer screens. In the promotional flyer for the talk, titled "Computers, Creativity, and the Nature of the Written Word," he introduced a new word. "Hypertext" would eventually become a term indelibly associated with him.[33]

As Nelson defined it in 1965, hypertext meant "a body of written or pictorial material interconnected in such a complex way that it could not conveniently be presented or represented on paper." Almost any topic could, in principle, be represented on a computer screen with "links" connecting one entry to another along with annotation, footnotes, and summaries while also including "every feature a novelist or absent-minded professor could want."[34] Ultimately, hypertext represented a system for nonsequential writing, something Nelson claimed had already been demonstrated on the printed page via novels like Laurence Sterne's *Tristram Shandy* (1759–1767) or Vladimir Nabokov's *Pale Fire* (1962). Nelson imagined his system of information storage, retrieval, and documentation could "grow indefinitely," containing more and more of the world's knowledge while revealing important connections between all of the entries. As Nelson said, showing his fondness for inventing new words, "EVERYTHING IS DEEPLY INTERTWINGLED."[35]

Nelson soon quit Vassar and started raising money and his professional profile. In 1969, *Electronics* favorably assessed Nelson's ideas for reading and writing using digital tools. Describing him as "lean, well-educated, and fast-talking [with] a real flair for showmanship," the trade magazine imagined using Nelson's proposed machine to navigate a "multilevel mélange of characters, diagrams, images and movies." In the future, an author might be able to jot down writings "whenever and wherever thoughts occur on a small pocket keyboard." This data could then be stored on magnetic tapes before it was transferred to a larger machine, something Nelson cleverly called, referencing a magician's sleight of hand, "prestidigitative publishing."[36]

All of this hardware would, in theory, become part of an even larger and more ambitious plan that Nelson started in 1966. Nelson's ultimate goal was to design and implement a universal text handling, publishing, and globally-connected electronic library system.[37] Nelson named his project Xanadu, after the imaginary home of Kubla Kahn in Samuel Taylor Coleridge's famous poem "Kubla Khan: or, A Vision in a Dream" (it was also the name of Charles Foster Kane's mansion in Orson Welles's 1941 film, *Citizen Kane*). Xanadu would become the "fundamental text system of the future, the magic carpet of the mind." It would also grow into Nelson's lifelong obsession.[38]

Around the time that the *Electronics* article appeared, Nelson met Claude Kagan and the RESISTORS. Ironically, the catalyst that brought them together wasn't some new computer hardware but an avant-garde art show. In the fall of 1970, a lavish new exhibition titled *Software*, inspired by Norbert Wiener's cybernetic concepts, opened at the Jewish Museum in New York. The show's curator, Jack Burnham, wanted to explore how conceptual artists might experiment with new computing technologies, such as "real-time computing" and "interactivity," in a gallery setting. *Software* gave thousands of museum visitors an opportunity to see, and in some cases use, an array of modern information technologies, including minicomputers, teletype equipment, high-speed copy machines, and closed-circuit television.

A contributor to the show, Ted Nelson recruited the RESISTORS to help him and some of the artists. As he wrote in *Computer Lib*, "Some people are too proud to ask children for information. This is dumb. Information is where you find it."[39] For Agnes Denes, a Hungarian-born conceptual artist, the teens coded a minicomputer to graphically animate triangles on a screen for a piece called *Trigonal Ballet* (1970).[40] For another conceptual artist, Carl Fernbach-Flarsheim, the teens used the I Ching to program a piece called *Conceptual Typewriter* (1970). A visitor could select one of several labeled buttons, such as "the silent" (represented by a circle) or "the providing" (illustrated by sheaves of wheat). A light pen allowed users to interactively alter the image that was generated.[41] In both cases, the artists provided the initial ideas but let the teenaged RESISTORS execute them via software code.

Nelson, working with programmer Ned Woodman, also contributed a piece that they titled *Labyrinth* (1970). Based around a PDP-8 machine that DEC provided, *Labyrinth* was explained as "the first public demonstration of a hypertext system."[42] To use it, a visitor would sit at a terminal and begin reading the displayed text. Keystrokes (e.g., "F" for "Forward") allowed them to navigate text as if in a maze. For example, a visitor might read the statement: "The exhibition you are attending is called *Software*. It was organized by Jack Burnham." You could then decide if you wanted a definition of "software" or get biographical details about "Jack Burnham." For many museum goers, *Labyrinth* suggested a technological future where people easily navigated the information-rich realm of what would become known as cyberspace. Unfortunately, this turned out to be largely a missed opportunity as the PDP-8's "failure to function was a mystery to everyone and a source of embarrassment to D.E.C."[43] Despite the company's efforts to debug it, more than a month passed before *Labyrinth* worked properly.

The RESISTORS gradually faded throughout the 1970s as its members went off to college and the supply of new recruits dwindled. Nonetheless, members like Nat Kuhn and John Levine recalled that ideas bantered about in bull sessions with Nelson in Kagan's barn materialized later in the pages of *Computer Lib/Dream Machines*. As Levine recalled, "There was certainly very little in that book that we hadn't already heard about before it appeared."[44] As had been the case with their personal computing experiments, the New Jersey teens were already ahead of the curve.

### CONFRONTING CYBERCRUD

Nelson started composing *Computer Lib/Dream Machines* a few years after the *Software* show. One motivation was his experience using the Programmed Logic for Automatic Teaching Operations (PLATO) system at the University of Illinois. Even more than Claude Kagan, Nelson had strong opinions about teaching (and teachers) and, not surprisingly, this extended to the use of computers in the classroom.[45] Engineers had built PLATO to do "computer-assisted instruction" but Nelson found the results unimpressive. In his view, the system gave students "an orange screen that connected to a mainframe computer" but offered "no way for students to create or store content." Nelson saw it as yet another example of the "computer priesthood" inflicting programs on users.[46] In response, Nelson decided to write a book expressing his ideas for how all people, especially children and teens, could and should think about and use computers. (Making money to break free of peripatetic part-time work and generate support for Xanadu supplied additional motivation.)

Unlike other books discussed thus far, there isn't a robust archival record that documents the writing of what became *Computer Lib/Dream Machines*. This absence is exacerbated by the fact that it was self-published so there is no recourse to the publisher's records either. When Nelson began writing it, he expected to have it in bookstores by the summer of 1973. Instead, the process took closer to eighteen months, in part, because Nelson "had so much to say."[47] A roommate suggested that Nelson relax his mind with some controlled substances but this wasn't helpful either. In the end, he churned out a flood of verbiage on a dizzying array of topics. *Computer Lib/Dream Machines* came in at around a staggering 300,000 words, most of it composed, ironically, on an electric typewriter. Nelson laid out the many, many pieces of the book along with hand-drawn images and assembled the entire thing on his kitchen table by hand. Decades later, when asked about the labor involved, Nelson responded, "I can no longer imagine how I created that book."[48]

Nelson originally wanted his book to provide a general and gentle introduction to computers. But he soon realized he was more interested in presenting an "*exposé* of computers, revealing all the things they've [the computing establishment] been hiding from the public."[49] With that provocation in mind, he considered calling his book something like *The Counterculture Computer Handbook* but, after a girlfriend discussed the women's liberation movement with him, Nelson had a new revelation. In keeping with his belief that computers should be used by ordinary people, he likened *Computer Lib/ Dream Machines* to then-controversial feminist works, such as *Our Bodies, Ourselves* (1970) and Ellen Frankfort's *Vaginal Politics* (1972). The book would fundamentally be about emancipation as "the reader and I would be setting the computer free—and ourselves." *Computer Lib* captured everything Nelson wanted for his book—"its slant, its direction, its tone, its agenda."[50]

He hammered this point home in book's opening pages with a "chant you can take to the streets":

COMPUTER POWER TO THE PEOPLE!
DOWN WITH CYBERCRUD!

That latter term, which Nelson coined, referred to the tendency of computer companies to "confuse, intimidate, or pressure" consumers so as to keep them in "controlled ignorance."[51] The widespread promulgation of "cybercrud" erected barriers that prevented ordinary people from engaging with computers in a creative and personal manner. Nelson went so far as to compare big computer companies with how the Catholic Church used to conduct liturgical services in Latin, a practice that alienated aspiring believers. "You shall not crucify mankind on a cross of expertise!" he declared.[52]

Nelson finally had *Computer Lib* (with *Dream Machines* available, as explained presently, on the flip side) ready for the public in mid-1974. "The joy and exhilaration of publishing my own book filled my soul," he recalled, "laying it out, explaining it to the printer, loading the boxes in the car, selling it to stores."[53] Within a year, *Computer Lib/Dream Machines* was in its third printing and Nelson was pleasantly surprised to hear the book was popular on college campuses. However, its circulation was handled by a group literally called "the distributors" (all lower case) with whom Nelson had struck a handshake deal. "I didn't like the guys in suits, so I trusted hippies instead, and they turned out to be swindlers," he recalled, "I never saw my share for the 50,000 copies they said they sold."[54]

The format and look of *Computer Lib/Dream Machines* was as much responsible for the book's success as its content. Nelson's creation was an example of what bookbinders call a tête-bêche ("head to toe") volume. Rather than having a back cover, the book had two front covers with a single spine. When a reader reached the end of one book, they flipped it over and started reading the other one.

Nelson's cover artwork combined an aspirational sensibility with a quirky "do-it-yourself" look. The *Dream Machines* side featured a "super-student," dressed like Superman in flight, reaching out to touch a computer screen, itself a noteworthy image in an era before touchscreens were commonplace. Underneath the picture, Nelson handwrote "New Freedoms Through Computer Screens." The cover of *Computer Lib* was even more remarkable. Nelson drew a raised fist against the backdrop of a black computer punch card and had the title prominently displayed as if issuing a challenge. At the top, scrawled in a way that implied urgency, he wrote, "You can and must understand computers NOW." Nelson maintained a sense of whimsy, having the "super-student's" foot from the *Dream Machines* cover stretch across the book's spine so that it peaked out on the cover of *Computer Lib*, with the words "Something is afoot" barely visible. This quirkiness continued inside with an eclectic array of artwork (over 500 images in toto), some of them borrowed from illustrations done by John R. Neill for L. Frank Baum's Land of Oz books, which Nelson enjoyed as a child.

Nelson's prolific use of images, photos, and cartoons combined with various fonts and text—some typed, some handwritten, some of it small enough to require a magnifying glass—invited comparisons to other books. Chief among these was *The Whole Earth Catalog*, which had won a National Book Award in 1972. First published by Stewart Brand in 1968, its many editions served a generous helping of how-to articles and product reviews. *Whole Earth* provided readers with "access to tools," a category that included not only self-composting toilets, hiking boots, and geodesic domes but also informational gear like maps and books. Besides appearing in a similar out-sized format (eleven inches by fifteen inches) as *The Whole Earth Catalog*, Nelson's book also presented an eclectic array of opinions, essays, and product reviews. A promotional poster even likened *Computer Lib/Dream Machines* to "that holy earthy catalog."[55] Years later, Nelson recalled other influences. These included *Domebook 1* and *Domebook 2* (1971), books that described how to build alternative shelters for aspiring communalists. He credited some of his authorial tone to "Pete Seeger's wonderful banjo book [*How to Play the 5-String Banjo* from 1962] and Tom McCahill's automobile reviews in *Mechanix Illustrated*." Finally, Nelson, eager to take his argument directly to the public, cited inspiration from Alexander P. de Seversky's 1942 book *Victory Through Air Power*, a controversial bestseller that propagandized military aviation.[56] Books, in other words, shaped Nelson's vision for his own book.

The complementary halves of Nelson's creation reflected different facets of his personal and professional identities. In previous biographical statements accompanying his talks and articles, Nelson described himself not as a programmer or computer scientists but as a "technology critic" and "computer consultant." Now, in the opening pages of *Computer Lib*, Nelson disclosed, "I am no longer calling myself a computer professional. I am a computer fan." For this, Nelson listed his qualifications: university education, a listing in the *New York Times*' "Who's Who in Computing," teaching positions, and membership in the Association for Computing Machinery. Nelson included a hand-drawn map showing the primary computing institutions in the United States (he included the RESISTORS for good measure) and provided copious information about mainstream computer publications, professional conferences, and the "main computer organizations" that readers could turn to for guidance and community.

But flip the book over to *Dream Machines* and one encounters a different version of Nelson. Here, the author highlighted his "counterculture credentials" as a "writer, showman, generalist" and a "computerman by accidestiny." Nelson noted his collaborations with dolphin scientist John Lilly and his experience at Woodstock. *Dream Machines* included a short essay called "The Greening of Computerdom" (a riff on Charles A. Reich's best-selling 1970 homage to the counterculture, *The Greening of America*) and offered a list of "Underground Computer Mags" that might be found on newsstands. A variation on the era's sexual liberation movement can be seen in an essay on computer sex in which Nelson created the memorable portmanteau "dildonics" from "dildo" and "electronics." When asked years later about the book's bifurcated character, Nelson explained that he was "a legitimate computer guy" but also a "counterculture sympathizer."[57]

Nelson's two-in-one-package is hard to easily summarize given the plenitude of topics and themes packed between its Janus-facing covers. *Computer Lib/Dream Machines* touched on practically every possible subject related to computing and then some more. Nelson's comments veered from the proliferation of junk mail and the doomsday-saturated *Limits to Growth* (1972) report to Buckminster Fuller's wristwatches and jokes about Richard Nixon. But Nelson consistently used *Computer Lib* to puncture the myth that computers were mysterious, all-powerful machines.

When a much-revised version of *Computer Lib/Dream Machines* appeared in 1987, Stewart Brand contributed a foreword that likened Nelson and his book to Thomas Paine and his 1776 pamphlet *Common Sense*. *Computer Lib*, Brand said, had rallied readers "around a common cause many of them hadn't realized was so worthy." Nelson's vision, Brand noted, was not about personal computers per se. Instead, with *Computer Lib*, Nelson offered a compelling vision of personal computing as an activity. For the uninitiated, Nelson's book was an ideological statement, setting up those who wanted to "understand computers NOW" in opposition to the repressive enemy of "Central Processing, in all its commercial, philosophical, political, and socio-economic implications. Big Nurse."[58]

*Dream Machines* took readers on a different adventure, focusing not on the technical details of computers (what Nelson called "changeable devices for twiddling symbols") but on the capabilities computer users would soon enjoy. People would be able to interact directly with them to generate, manipulate, and integrate animation, graphics, and sound into new creations that Nelson called "hypermedia." "Presentation by computer," Nelson said, "is a branch of show biz and writing, not of psychology, engineering or pedagogy."[59] It was an optimistic vision of what all people might be able to do once they had access to computers as tools to write, show, and think with.

Starting with a description of cathode ray tubes (which he called "lightning in a bottle"), *Dream Machines* described how scientists already used computers to produce images. He gave special attention, for example, to Ivan Sutherland, a computer scientist at MIT who had invented a program called Sketchpad in 1962. A person could use Sutherland's innovative "graphical user interface" and a light pen to draw directly on a computer screen. "If computers are the wave of the future," Nelson predicted, "displays are the surfboards." Nelson ended *Dream Machines* with a lengthy description of his own screen dream. He predicted that Xanadu would soon offer consumers "high performance computer graphics and text services" at an affordable price, perhaps via franchised venues people would visit like "hamburger stands." Opposing the myth that computers were mechanistic and incomprehensible machines, Nelson described a future where people would become "freer through computers."[60]

*Computer Lib/Dream Machines* didn't receive the media attention that accompanied books like *Cybernetics* and Nelson had no public relations department or editorial team to boost his book's profile. Despite this, thousands of people bought it or read the illicit photocopied versions that circulated widely, samizdat-style. By and large, most readers of Nelson's book were either computer professionals of various stripes, academics considering it for their courses, or people curious about the technology and intrigued by the book's appearance.

Despite the obstacles that came with self-publication, a few reviews of Nelson's book helped it get wider attention. Software entrepreneur Dan Fylstra pronounced it in a relatively new magazine for the computer hobbyist community called *Byte* as "a marvelous, delightful, one-of-a-kind book." The computer world, Fylstra said, "desperately needed" Nelson's "grandiose plans" and encouraged readers to read the book and "share its visions."[61] Former RESISTOR John Levine, now a computer science student at Yale, appraised Nelson's book for another magazine, *Creative Computing*. He observed that "university computer people" had thoughts about Nelson's work, ranging from "the best book I've ever seen" to dismissing it as a mere "book of gossip." Levine concluded that, while the "author is no technical computer whiz," this was a selling point since "he doesn't assume you are one either."[62]

Much of the recognition Nelson's book received came long after it first appeared. In 1984, Lee Felsenstein, a founding member of the Homebrew Computer Club, recalled how hobbyists like himself had "pored over" *Computer Lib/Dream Machines* to extract as "many nuggets of computer science as we could handle" while they navigated its eccentric layout and tiny text. A few years later, the *New York Times* noted that the "wonderful, infuriating" *Computer Lib/Dream Machines* had "defined many of the computer issues that are still with us."[63] Mitch Kapor, another pioneer in the software business, was working as a junior programmer in the Boston area in the 1970s when he saw Nelson's book at a store in Harvard Square. "It inspired me as no other book has before or since," Kapor said, and "pointed me in the direction of a career in the as-yet-then-uninvented field of personal computers."[64]

Today, *Computer Lib/Dream Machines* appears prescient in its claims about how computers would become available to almost everyone and how people would interact with screens on a near-constant basis. But an unfortunate thing about being ahead of your time is that when people finally catch up, they often claim your ideas were obvious all along. When Nelson first started assembling *Computer Lib/Dream Machines* in his kitchen, small and inexpensive computers didn't exist in any meaningful way. But their availability was something he longed for. "I was totally impatient," he recalled decades later, "why had personal computing taken so fucking long?"[65] Nelson's vision for a multimedia machine was already capacious. "Imagine," he suggested in 1970, "a device with a red oval 2-inch TV screen, a set of chimes in the natural key of C, a smell generator capable of giving off most smells, and a foghorn."[66] While the first personal computers didn't emit smells (although some frustrated users would say their performance stunk), the machines of Nelson's dreams had started to become a reality.

## "THE HOME COMPUTER IS HERE!"

By the early 1980s, journalists were already observing that Steve Jobs projected a "reality-distortion field." By this, they meant that the cofounder of Apple Computer could make people acquiesce to a certain perception of how things were. Consider the title of a short essay that Jobs published in Edmund Berkeley's journal, *Computers and People*. In "When We Invented the Personal Computer . . ." Jobs predicted that the machines would become like bicycles: inexpensive and accessible devices that would enhance the capabilities of people worldwide. Jobs would later refine his analogy by calling the personal computer the "bicycle of the mind."[67]

While catchy, Jobs's essay distorted history in several ways. First, in no way did Steve Jobs or his collaborator Steve Wozniak "invent" the personal computer. That credit would be spread among an entire cohort of people, from the builders of the first microprocessors to visionaries like Ted Nelson who popularized the idea of accessible and affordable computers. Invention is a singular act while the development of personal computers occurred over many years and in many places. Jobs however implied that creative credit rested on one specific machine—"*the* personal computer"—by which, of course, he meant his own company's products. In reality, Apple's machines accounted for a small fraction (about 15 percent) of all personal computer sales in 1981. The best-selling brand that year was actually the Atari 400/800, with the TRS-80, sold by the Tandy Corporation via their nationwide network of Radio Shack stores, coming in as a close second. Industry data, meanwhile, attributed almost half of all sales that year to "other companies," a catchall category indicating how much product diversity there was at the time.[68]

One milestone in what came to be seen as the "personal computer revolution" (it was hard to recognize it as such at the time, of course), was announced in the January 1975 issue of *Popular Electronics*.[69] This monthly magazine, largely read by hobbyists, spoke to a community whose members liked to tinker with electronics and build things like home stereo equipment and digital calculators. The magazine's cover blared "Project Breakthrough! World's First Minicomputer Kit to Rival Commercial Models" and featured a color photograph of a new machine called the Altair 8800.[70]

Inside that issue, Arthur P. Salsberg, the magazine's new editorial director, proclaimed that "The Home Computer Is Here!" For years, Salsberg and other electronics enthusiasts had been "hearing about how computers will one day be a household item." And, for years, those hopes had been quashed by "demonstrator" machines that were "fun to build" but suffered from "limited usefulness." Now, things were changing. The Altair 8800 was a "minicomputer that will grow with your needs" and not just an "expensive toy." Salsberg noted that industry experts expected to see a 50 percent annual growth rate for the "minicomputer market" (terminology for the devices at this point was still inconsistent). For now, Salsberg was pleased to say that the "home computer age is here—finally."[71]

Built around an Intel microprocessor, the Altair was made and marketed by Micro Instrumentation Telemetry Systems (MITS), a small electronics company started in 1975 and based in Albuquerque, New Mexico. The initial price for a basic machine was less than $500, an astonishingly low figure given that traditional computers that companies bought still cost at least twenty times as much. The Altair had obvious limitations. MITS initially sold Altairs as kits, which owners assembled by carefully soldering together parts and connections. Once they had built *their* machine, users entered programs by toggling switches on the machine's front panel, with on or off translated into ones and zeros. At the end, the reward might simply be a series of blinking lights indicating that the program had run correctly.

The quality of the products MITS hurriedly shipped was uneven at best, further complicating the user experience. Nonetheless, thousands of orders poured into the company's modest headquarters. While barely recognizable as a "computer" today, machines such as the Altair, the IMSAI 8080, and the Sol-20 prompted entrepreneurs to start their own companies. By 1976, scores of brick-and-mortar businesses with names like The Computer Store and Byte Shop had opened up. Two years later, the National Computer Conference, held in Anaheim, California, featured a separate "Personal

Computing Festival," which, over the course of three days, attracted several thousand industry and hobbyist participants.

Ted Nelson briefly explored this enticing financial frontier himself via a short-lived project called the Itty Bitty Machine Company (Nelson, who regularly castigated IBM as a corporate villain, gleefully wrote his own company's name as "ibm"). It was based in Evanston, Illinois, and had a phone number that ended in 6800, for the Motorola 6800 processor. Itty Bitty's catalog featured dozens of products, from kit machines and circuit boards to books such as *My Computer Likes Me When I Speak BASIC*. To help promote the company, Nelson designed a logo: a friendly looking computer with big, cartoonish eyes, shaking off its shackles and asserting "Computers Arise!"[72] Nelson composed a letter to accompany the drawing, which announced "We are the Itty Bitty Machine Co., Inc. We are about to change the world. For years people have talked about the home computer, but nobody's been building it. We are. We know how it ought to be. And it will."[73]

Nelson's tenure with Itty Bitty proved short-lived. While his soon-to-be-former colleagues wanted to sell goods to technologically savvy consumers who could build computers out of kits and parts, Nelson derided their focus on the hobbyist ("hobby, hobbier, hobbiest," he scoffed). Instead, he insisted that the "real market" would be the "private consumer" who wanted to lead "a new kind of life around the computer screen."[74] In 1976, when he spoke at the first (and last) World Altair Computer Convention, Nelson offended many in the room when he predicted that hobby computing would "remain a cult" until machines could be bought and used right off the shelf. What was needed, he claimed, were "canned systems or black boxes," not finicky machines that required considerable technical knowledge to build and operate.[75] Nelson, of course, would be proven right, but not before he quit Itty Bitty.

Once computer enthusiasts had bought or built their own personal minicomputers, they wanted software to run on it. This opened up another space for entrepreneurs who would write programs that they could sell, share, or give away. In the grand narrative of computing's history, the inventive exploits of Paul Allen and Bill Gates stand as an obvious landmark. Allen and Gates, two young men from well-off Seattle families, had been fascinated with computers and programming since their teens. *Popular Electronics* brought the Altair 8800 to their attention. In the early months of 1975, their fledgling company Micro-Soft (Gates and Allen dropped the hyphen a year later) developed a version of BASIC, a language familiar to and favored by many hobbyists, for the Altair 8080.

Within the next few years, hundreds of new companies started selling software to the owners of new personal computers.[76] Operating systems let users control the eclectic range of peripheral devices they hooked up to their machines while versions of high-level languages such as BASIC let them write their own programs. Digital Research, Personal Software, and Atari, as well as many other such companies soon began selling programs to users who could now do rudimentary word processing, handle spreadsheets, and play a rapidly expanding universe of fantasy and science fiction games.

The proliferation of both hardware and software in the mid-1970s were two important elements in consumers' acceptance of personal/home/mini/ micro-computers (the terms were used interchangeably while iconoclast Ted Nelson simply called them all "dinky computers"). Despite the expanding diversity of both machines as well as programs, a vital third ingredient was needed to help nudge the personal computer into people's living rooms, offices, and kitchens.

## PUBLISH AND PROFIT

When it came to connecting accomplished enthusiasts and neophytes alike into a vast *community* of informed computer users, the printed word proved essential. As David Bunnell, who led marketing efforts for the Altair before becoming a successful publisher, remarked, computer magazines allowed someone to "automatically identify" with "thousands of other personal computer users who have a similar orientation."[77] The role of magazines was especially significant between 1975 and 1980, a brief but vibrant era when advocates for personal computing struggled to define, explain, and promote the technology to consumers. Besides providing some of the first marketplaces for computers via copious advertisements for mail order purchases, magazines offered forums for expressing opinions, revealed and reviewed new products, and shared valuable technical knowledge.

Compared to books, which take longer to write, produce, and publish, magazines could respond more nimbly to the vicissitudes of rapidly changing technologies and consumer tastes. But magazines did have a synergistic relationship with books. Magazines helped guide readers to books about computing. They often included regular columns devoted to book reviews and also promoted these publications via curated lists of recommended readings. In less than a decade, what started as a modest-size ecosystem of specialized magazines and books had blossomed into a "hyperactive field" in the publishing world that was worth close to $100 million annually. The most fortunate magazine publishers (what the *New York Times* called "computing's lusty offspring") could claim a subscriber base of 100,000 readers or more.[78]

Before 1974, computer magazines could be roughly divided into two categories. For instance, there were established trade monthlies like *Datamation* and *Computers and Automation* that reported on the computing industry. Advertisements tended to be bought by large computer manufacturers and other sizable companies while the articles were written with the computer professional or businessperson in mind. At the other end of the spectrum were magazines like *Radio-Electronics* and *Popular Electronics*. These were read by knowledgeable hobbyists keen to keep up with the technology and use published schematics to build their own gear.

In this latter category, *Radio-Electronics*, for example, had been founded in 1929 by Hugo Gernsback, a technology enthusiast (he popularized the term "science fiction") and inveterate publisher with dozens of titles under his watch. The magazine, whose original title was *Radio-Craft*, published articles for decades about amateur radio gear as well as television and hi-fidelity audio equipment. Its subtitle—"For Men With Ideas In Electronics"—indicated the assumed masculine nature of these interests. Personal computing offered a new topic to draw in some more readers. *Popular Electronics*, which broke the news about the Altair, was a more recent magazine but also aimed at the electronics hobbyist market. Started in 1954 by the Ziff-Davis Publishing Company, it presented news and information to readers who wanted to tinker and build. Each of these hobbyist-directed magazines had circulations that reached hundreds of thousands of people monthly. But neither *Popular Electronics* nor *Radio-Electronics* spoke specifically to the emerging community of readers who were new to personal computing and wanted to know what they could actually do with their recently purchased machines.

That situation began to noticeably change in November 1974 with the appearance of *Creative Computing*, published by David H. Ahl out of offices in Morristown, New Jersey. Born in 1939, Ahl grew up as a self-described "science nerd" on Long Island where he tinkered with war surplus electronics that could be easily found on Radio Row on Manhattan's Lower West Side. He enrolled at Cornell University, which, at the time, offered undergraduates a five-year program combining engineering with extra classes in the humanities. A summer spent interning for Grumman Aircraft left him unenthused about an engineering career. Instead, he opted for graduate study in management at Carnegie Mellon, where Herbert Simon taught. After finishing school, Ahl started working for Digital Equipment Corporation (DEC) where he helped market machines like the PDP-8 to educational institutions.[79]

Ahl believed educators lacked a proper forum to share best practices when it came to computers. In response, he created a company-published newsletter called *EDU*.[80] Ahl optimistically hoped for a readership of about 2,000 people. Instead, his subscriber list grew to 20,000 as educators who didn't even use DEC's machines went looking for useful information. The recession of the early 1970s pushed Ahl to a new job with AT&T in New Jersey. But, before he quit DEC, Ahl collected scores of programs written in BASIC and assembled them into a compendium, which, after securing publishing rights from DEC, was published as *BASIC Computer Games*. Ahl's book was a huge hit, selling over one million copies, a figure that suggested the size of the largely untapped market for personal computing books.[81]

Fueled by a desire to address the social and educational aspects of computing, Ahl decided to launch *Creative Computing*. He spent months trying to arrange funding from the National Science Foundation and various philanthropic groups but received only encouragement in exchange. In the end, he decided to do it alone using his own money.[82] Ahl assembled a small editorial team and, in the summer of 1974, they contacted advertisers and started a publicity campaign. By October, with only 600 subscribers lined up, Ahl took a gamble and had 8,000 copies of *Creative Computing*'s first issue printed. After meeting obligations to his subscribers, Ahl gave the remaining copies to school principals, librarians, and educational businesses across the United States for free. Soon, *Creative Computing*'s circulation—$2 an issue or $8 for an annual subscription—started growing.

Ahl originally imagined that *Creative Computing*'s articles would be about "the social impact of the computer and technology on our lives" in areas like "the entire educational spectrum. . . . health care, law enforcement, privacy, jobs, elections, credit ratings, etc."[83] However, the introduction of the Altair 8800 and other machines opened up another vista for the magazine. Ahl started recruiting people who could write about how new owners could use their machines. Each bimonthly issue included programs for games and graphics programs written in BASIC that users could experiment with. Ahl and his crew happily parodied other computer magazines as well. *Dr. KiloByte's Creative Personal Recreational MicroComputer Data Interface World Journal* was one memorable spoof; it came with fake advertisements and articles like "Bake Your Own Computer."

Ahl's combination of a user-friendly approach with desirable content worked. By 1978, circulation was over 60,000 and *Creative Computing*, now printed on coated stock paper for a "slicker" look, was attracting more advertisers. Ted Nelson appeared on the magazine's masthead as publication shifted to a monthly schedule. Ahl and his staff eventually spun off a book publishing operation, a mail order book service, and a "Creative Computing Software" branch before Ziff-Davis purchased the whole enterprise in 1981 for an undisclosed but presumably generous price.

If *Creative Computing* showcased an entertainment-based approach to personal computing, *Byte* magazine took members of the nascent home computing community back to its amateur radio roots. *Byte*'s founder, Wayne Green, was born in New Hampshire in 1922 and, like many hobbyists, came to home computing via ham radio. He started a newsletter, *Amateur Radio Frontiers*, in 1951 before becoming the editor of an established magazine for amateur radio operators called *CQ* (this was a hobbyist's radio call when inviting listeners to reply). Then, in 1960, he launched his own magazine, named *73* (it was what radio hams transmitted to signal "best regards"). In 1975, Green decided he wanted to integrate computers into his publishing offices but found little practical advice in the books or magazines he paged through. Unsatisfied with irregularly published hobbyist newsletters, Green launched *Byte* in September 1975.[84] As Ahl had found with *Creative Computing*, Green's timing was spot on, given the commercial ferment among home computer makers.

Although Green was enthusiastic about personal computers, he was no expert on the subject. For that role, he hired Carl Helmers, a computer hobbyist from New Jersey.[85] Helmers learned how to program via FORTRAN classes offered by Sandoz Pharmaceuticals, a Swiss company best known for its discovery of LSD. Entranced, Helmers decided that, like Ted Nelson, he wanted his own computer. After completing a physics degree at the University of Rochester, he intermittently issued a small-circulation newsletter called *Experimenters' Computer Systems*. Its contents, which Helmers put together using a typewriter and clip art graphics, trended toward the technical with articles about building and programming microcomputers.

Green convinced Helmers to join his new magazine venture, which would blend Helmers's technical knowledge with the readership of technically competent amateurs that *73* already enjoyed. Their original plan was to start conservatively with a twenty-four-page magazine and a print run of 1,000 copies. Green and Helmers, however, noticed an unusually high response rate to inquiries they sent to computer equipment companies. As more companies expressed interest in purchasing advertisements, the planned print run increased to 10,000 and then 35,000 before the first issue—50,000 copies at $1.50 apiece—appeared in the fall of 1975. In contrast to the software orientation of *Creative Computing*, the focus of *Byte* was on hardware, with articles such as "Cassette Interface—Your Key to Inexpensive Bulk Memory" and "Build a TTL Pulse Catcher." Eventually, disagreements grew between Green and Helmers over whether the magazine should appeal primarily to the larger community of neophytes that Green favored or to experienced hobbyists. Green was eventually forced out. When McGraw-Hill bought *Byte* in April 1979 for several million dollars, it boasted an international readership of over 150,000 people.[86]

*Creative Computing* and *Byte* were markedly different in terms of intended audience but each helped define the emerging market for personal computers by disseminating important news, opinions, and information, and fostering a community of home computer users. Their business models relied largely on advertising revenue with additional income generated via subscriptions and single-issue purchases. A glut of advertisements helped swell the size of each magazines' issues. By the early 1980s, a typical issue of *Creative Computing* had over 300 pages packed with full-page color ads. Both magazines were quickly snapped up by larger publishing firms as computing magazines of all kinds promised a good investment return. By 1984, over 200 publications were trying to capture some share of the computing community's attention. In addition to magazines aimed at users of specific machines, such as the IBM PC or the Apple II, computer retailers and software purchasers also enjoyed their own specialty publications.[87]

If 1975 marked a point when small microcomputers with limited capabilities could be purchased and assembled for home use by hobbyists, then 1977 stands at an annus mirabilis for the wider availability of personal computers. In that one year, a "holy trinity" of fully assembled, first-generation home computers—the Commodore PET, the Apple II, and the TRS-80 from Radio Shack—all entered the market. By 1978, sales of these three machines approached 200,000 units and, now that consumers could purchase assembled systems rather than kits, MITS and its Altair faded from the market.

The personal computing community's growing excitement was evident at the first West Coast Computer Faire. Held in San Francisco in April 1977, the three-day event was the brainchild of Jim Warren, a former mathematics teacher, social activist, and computer consultant. In 1976, Warren began editing a magazine called *Dr. Dobb's Journal of Computer Calisthenics and Orthodontia* ("Running Light Without Overbyte," the first issue joked), which mixed whimsy with technical material. For the event, Warren initially imagined that maybe 7,000 people would come to the Civic Auditorium to hear talks, swap advice, and sample the hardware and software offerings on display. Instead, more than 12,000 people showed up as lines stretched around the block and computer enthusiasts swamped the exhibition booths.

Some attendees spilled over to the St. Francis Hotel, a posh venue near San Francisco's Tenderloin District, to hear a speech by Ted Nelson titled "Those Unforgettable Next Two Years."[88] Nelson blended predictions about the future of computing with his distinctive views on liberation via computer. With the sudden availability of so many new machines, Nelson pronounced that the computer world was now riven into two camps. There was the "strange coalition of hobbyists" who were already comfortable building their own computers. But a growing community of "straights" simply wanted to buy all-in-one machines that were easy to use.[89] Nelson believed the latter demographic would eventually dominate once the "confusions and oppressions of yesterday's computer systems" vanished and the computer became just another "home appliance, as glamorous as a can opener."[90]

Nelson's new book, *The Home Computer Revolution*, expanded on many of the ideas he previewed in his presentation. By the time it appeared in late 1977, however, it had lots of competition from other books and, as a slight paperback with a conventional cover, it didn't capture as much attention as *Computer Lib/Dream Machines*. Nelson later dismissed the book as a "throwaway" he had dictated to a friend and hastily self-published.[91] Nonetheless, some of its ideas, such as email and "pocket computers," proved prescient if not completely original. As the book's back cover predicted, "The interactive computer will be mankind's new home. The sooner we understand it, the better." Over the next three decades, Nelson maintained his techno-evangelism, especially for his Xanadu system. Reporters who wrote about Nelson often presented him as both visionary technologist and idealistic eccentric, roles Nelson seemed willing, if not always happy, to play.[92]

In 1984, when *Creative Computing* celebrated its ten-year anniversary, the editorial staff considered the people who had influenced and shaped personal computing. They singled out Ted Nelson as the person who had tirelessly preached concepts about computer design that were only slowly becoming accepted. "Always on the fringe of the field," Nelson still managed "to get there before the rest of us. And every time we reach him, he moves on."[93] Despite the recognition, Nelson seemed hardwired to challenge the phenomenal success personal computing enjoyed. He was especially offended by the compromises and concessions computer manufacturers forced on ordinary people with their new machines.

In 1987, Microsoft Press published a much-revised version of *Computer Lib/Dream Machines*, which former RESISTOR Lauren Sarno had patiently edited and reassembled. In reflecting on the current state of computing, Nelson concluded that "NOTHING HAS CHANGED." Computers, he lamented were as despotic as before, just smaller, cheaper, and more widespread. "Now you can be oppressed by computers in your living room," he wrote.[94] Control had indeed shifted but Citizen Nelson still longed for liberation.

When Carl Schlesinger passed away in 2014, an obituary noted his love for two things. One was tap dancing, an activity Schlesinger learned as a boy busking outside theaters on Broadway. The second was printing, something else he started doing as a boy, using rubber stamps to hand make copies of *The Eagle*, a (very) small circulation newspaper for residents in his Bronx apartment building. While dancing remained a lifelong hobby, printing became his life's work.[1]

For thirty-five years, Schlesinger worked for the *New York Times*, first as an apprentice printer and then as a professional typesetter. Each evening, Schlesinger and his coworkers used an aging battery of Linotype machines in the newspaper's composing room to arrange the next day's paper, letter by letter. Invented in Baltimore in the 1880s by an immigrant German crafts-man named Ottmar Mergenthaler, the Linotype employed a "hot metal technique." The process started with the typesetter's individual keystrokes, entered via a ninety-character keyboard. Upper- and lower-case letters had their own keys (there was no "shift" key as with the standard QWERTY keyboard). Commonly used letters in the English language started on the left side and read down and to the right, ETAOIN in one column, SHRDLU, in the next. Typesetters like Schlesinger would sometimes deliberately add those two "words" to fill out a line or indicate errors in reporter's draft.

Those human-generated keystrokes were then converted, via small individual brass "matrices," into larger "slugs" cast directly from molten lead inside the Linotype machine. Once cooled, the lines of type (hence "Linotype") would be used by Schlesinger and his colleagues to fashion a mirror image of a page from which an individual newspaper sheet would be printed. A marvel of mechanical moving parts, Linotypes were later linked to teletypes with letters and sentences inputted not by skilled operators like Schlesinger but via paper tape readers. Experienced typesetters could sense further technological changes to come.[2]

At the *Times*, this occurred on the evening of July 1, 1978, when the last edition of the paper was produced using the old Linotype machines. Future issues would be set in so-called cold type by operators working at computer terminals. The visceral, sensory experience of the *Times'* Linotypes was exchanged for "precisely controlled temperature, pervasive hum, cleanliness and cipher-locked doors" as brass and lead were swapped for digital bits.[3]

The replacement of skilled typesetters like Schlesinger with computers was exactly what Norbert Wiener had warned about (and what automation advocates like John Diebold had encouraged). David Loeb Weiss's 1980 documentary, titled *Farewell, Etaoin Shrdlu*, captured the final hours of the Linotype machines at the *Times*. An homage to a vanishing era of printing technology, the film ended with Schlesinger, now trained in the new methods, quietly tapping away at a computer terminal's keyboard. Beyond the immediate tactile sensations and deep reservoirs of knowledge needed to operate the old Linotype machines, *Farewell, Etaoin Shrdlu* revealed a professional community of men and women who had coalesced around an ensemble of venerable printing technologies.

This chapter starts with the idea of a technological community based on the printing of words on paper and takes it in a digital direction.[4] In 1977, Donald E. Knuth, a computer scientist at Stanford University, started creating programs that would allow him to use a computer to do the typesetting for his own articles and books. He was, in essence, writing a tool for writing. Knuth called his software creation "T$_E$X" with the "E" rendered the same size as the other two letters but placed "out of kilter," that is, slightly lower. (Knuth realized not all systems, including the one I'm using, could render the letters properly so he agreed that "TeX" was an acceptable substitute. I shall follow suit.) TeX, he explained, rhymed not with the first syllable of Texas, but rather the child's cry of "bleccchhh." When a user said it correctly in front of their computer, he observed, "the terminal may become slightly moist."[5]

For Knuth, books represented more than just patterns of words and symbols arranged on a page. Getting them to look right motivated him to invent TeX in the first place. What Knuth originally developed for the "creation of beautiful books," was especially valued by mathematicians and other scientists who needed to be able to write complex formulae.[6] Knuth also wrote a companion program for creating digital typefaces called METAFONT. Developing these programs combined Knuth's expertise in computer science and mathematics with his strong sense of aesthetics and design. He presented his entire digital typography system in a series of books, including *TeX and METAFONT: New Directions in Typesetting* (1979) and *The TeXbook* (1984). With TeX and METAFONT, the practice of printing was central, although not performed in a manner that someone like Carl Schlesinger would have immediately recognized. Eventually, Knuth's books formed the core of Computers and Typesetting, a popular series published by Addison-Wesley.

Preparing a formatted document via a computer with whatever font one desires might seem trivial today but, when TeX was emerging, commercial programs such as Microsoft Word were just becoming available.[7] And, of course, one had to buy and install them. TeX, however, remained free of charge at Knuth's insistence. TeX became the basis of a more user-friendly and widely used software program called L$^A$T$_E$X (hereafter LaTeX) that was also free. Within a decade, LaTeX had become a standard document preparation program used worldwide by physicists, computer scientists, mathematicians, and other researchers for writing articles, uploading them to online repositories, and submitting them to journals for publication. By the end of the 1990s, to be a successful practitioner in these fields required at least some proficiency with LaTeX.

Using TeX in its various forms meant being part of a larger community of writers connected by a common technology. I chose the word "community" quite deliberately. Knuth initially created TeX for his own use. But soon after he introduced his program, a growing group of people, starting with mathematicians, embraced it. They also collaborated with Knuth to improve it, find and fix errors, and encourage other researchers to adopt it. The participants in this technological community originally acquired their knowledge and expertise from reading Knuth's articles, manuals, and books. In time, some members produced their own writings in the form of newsletters, articles, and other books, which expanded the community and solidified its sense of identity.

In the early 1980s, as the TeX community was taking shape, medievalist Brian Stock coined the term "textual community." In his 1983 book *The Implications of Literacy*, Stock used the expression to describe those groups who coalesced around particular medieval religious texts.[8] These textual communities constructed their identities via key written texts that were interpreted for them by charismatic religious leaders.

Obviously, there are worlds of difference between twelfth-century spiritual groups and the community of mathematicians, computer scientists, and typography experts who came together to use and promote TeX. Yet Stock's concept offers a framework for understanding how TeX, METAFONT, and LaTeX became widely established typesetting tools for the broader scientific community. Knuth and his books played an essential role in making and maintaining this community. As TeX's author, Knuth, a highly respected computer scientist, stood as a gentle arbiter as to what was TeX and what wasn't.

Knuth often displayed a playful and punning sensibility in his writings. With that in mind, referencing a "TeXtual community" is quite fitting. In other words, this chapter is about making books (and, for Knuth, making better-looking books) but also how books helped make a community.

### BIBLIOPHILE

In November 1996, Donald E. Knuth gave a lecture, titled "Digital Typography," in connection with receiving the Kyoto Prize, Japan's highest award for work in the arts and sciences. Knuth—then fifty-eight years old, tall, lanky, and bespectacled—declared to his audience, "I have been in love with books ever since I can remember."[9] Knuth noted that his affection for books extended to the actual letters from which words were composed. Knuth displayed pages from an alphabet primer he had read as a young boy. He pointed out how he had marked each serif (the small lines attached to the larger strokes which constitute a letter or number), on pages such as

"K—Kitten" and "P—Pony." Knuth's fascination with letters, words, and books continued throughout his life.

Born in 1938, Knuth grew up in Milwaukee, Wisconsin. The local Lutheran church was a big influence in his life. His father played the organ there on Sundays and classical organ music was something Knuth himself pursued with passion, even having a pipe organ built in his California home. Keyboards of all types fascinated him; as a teen, secretarial classes taught him to type eighty words a minute. The Milwaukee Public Library was another powerful force. A newspaper clipping notes that, at the age of four, Knuth was already a member of the "Ancient Order of Book Worms."[10] In high school, Knuth edited the school paper, assembling issues using a mimeograph machine his father kept in the family's basement to earn extra money. His father also bought a small printing press that he used to typeset material for local groups. Years later, Knuth observed, "I must have ink in my veins."[11]

In high school, physics and math appealed to the scholastically adept Knuth. In his senior year, Knuth submitted two entries to the nationwide Westinghouse Science Talent Search. One was a traditional physics project. But Knuth—a huge fan of *Mad* magazine—also proposed a novel measuring system created around an imaginary unit of length he christened a *potrzebie*. Knuth used this entity to derive an entirely fictional ensemble of units for quantities like volume, temperature, and time. This essay received an honorable mention in the Westinghouse contest and Knuth was thrilled when *Mad* published an illustrated version of it. As of 2025, Knuth's curriculum vitae still lists it as his first scientific paper.[12]

For college, Knuth chose the Case Institute of Technology in Cleveland (today's Case Western Reserve University) where he started as a physics major. This required taking laboratory classes, something that Knuth, a self-described klutz, did not excel at. The precision of mathematics proved more appealing. He started doing extra problem sets found in his textbook, *Calculus and Analytical Geometry*. Written by George B. Thomas, an MIT mathematician, the now-classic book was first published in 1951 by Addison-Wesley. "I loved this calculus book so much," Knuth recalled, "that I chose them later to be the publisher of my own books."[13]

To help cover expenses where his scholarship fell short, Knuth worked part-time for Case's Department of Statistics. The department was located near where a new IBM 650 had just been installed. Knuth soon started spending time with the "wonderful machine with flashing lights." As Knuth read the operating manuals for the machine, he kept discovering more efficient solutions to the examples in them.[14] Knuth soon became interested in a special kind of program called an "compiler." These convert instructions written in one language into another. A compiler might, for example, translate higher-level code written in FORTRAN or BASIC into lower-level "machine code" that the computer directly worked with. When Knuth compared two different compiler programs, he found one "kludgy" (i.e., inelegant) while the other was "beautiful, like hearing a symphony."[15] Knuth rewrote the poorly done program and called it RUNCIBLE, after a nonsense word from Edward Lear's 1870 poem "The Owl and the Pussy-Cat." Professors at Case adopted the manual Knuth had written for RUNCIBLE, putting the young programmer in the curious position of taking a course for which he had written the textbook.[16]

In 1960, Knuth graduated, summa cum laude, with a degree in mathematics. In fact, his academic performance had been so exceptional that Case's faculty took a special vote and awarded him a master's degree in addition to his bachelor of science credentials. Knuth, who already wanted to move to California, thought he might like to teach college math. Knuth had been accepted to Stanford and Berkeley but ultimately picked Caltech. Knuth moved to Pasadena and started graduate studies in mathematics in September 1960. (At this time, computer science departments didn't exist yet; the first one in the United States wasn't created until 1962 at Purdue University.) At Caltech, Knuth set himself a goal of finishing his PhD in three years. He also consulted for the Burroughs Corporation, a rival to IBM, by writing compiler software for the company's mainframe machines.

In January 1962, soon after he and Nancy Jill Carter married, Knuth was approached by a representative from Addison-Wesley. Given his programming experience, would he be interested in writing a book about compilers? The fact that Knuth was only a second-year graduate student says a good deal about both his growing reputation as well as the dearth of quality books on the subject. The idea intrigued Knuth. He respected Addison-Wesley as a publisher and he liked the idea of writing a book he believed would be useful. In the summer of 1962, Knuth wrote up a list of a dozen topics he imagined would provide a framework for a book. Over time, these twelve ideas transformed into what eventually became a monumental series of books for Addison-Wesley.

In 1962, Knuth used the notes for what would become the first three chapters of his book to teach one of the first computer science courses at Caltech. A year later, Knuth received his doctorate degree and Caltech hired him as an assistant professor of mathematics (he was promoted to associate professor three years later). While Knuth would later characterize this period as the most creative time in his life, it was also stressful. A typical day saw Knuth helping Jill care for their two toddlers, prepping lectures, and pursuing a full research agenda. This, combined with the expanding ambitions for his book, put him in the hospital with a bleeding ulcer. After a doctor convinced him to pare back his activities, Knuth had an epiphany of sorts.

In the fall of 1967, at a mathematics conference in Santa Barbara, California, one of the attendees asked Knuth to describe his research. The question made Knuth realize that he wanted to be a computer scientist, not a mathematician. But, given that the discipline of computer science was still developing, what kind of computer scientist? One essential topic of study in the field is algorithms. Broadly defined, an algorithm is a sequence of rules that are rigorously followed when, for example, a human solves a math problem or when a computer performs an operation.[17] Given their importance in computer science, Knuth decided that analyzing algorithms mathematically, so as to determine which ones were better than others, would be both intellectually interesting and valuable to the discipline. He began to refer to this as the "mathematical analysis of algorithms."[18] Because some programs were more elegant than others, Knuth's new research agenda also matched his sense of aesthetics.

But there was still the book and it was expanding far beyond what Knuth had originally imagined. By late 1965, Knuth had produced thousands of handwritten pages. When he shared the first chapter, typed on an IBM Selectric typewriter, with Addison-Wesley, the editors pointed out that, at this rate, the resulting book would be some 2,000 pages long and quite unpublishable.[19]

Eventually, Addison-Wesley agreed to publish what Knuth had written not as a single book, but as a projected set of *seven* volumes. This would be done by taking Knuth's original twelve ideas and have each volume explore one or two of them. As a result, Knuth's first book would not address compilers—that would happen in a later volume—but would instead tackle basic concepts such as how information is represented, stored, and structured inside a computer as well as fundamental programming techniques. He and the publisher agreed that the series would be called The Art of Computer Programming. Knuth chose the word "art" quite deliberately. It reflected his belief that programs could be beautiful and making them an activity that retained an artisanal character even as practitioners worked to make it more "scientific."[20]

The first volume in The Art of Computer Programming series—sometimes abbreviated TAOCP—was a 634-page work titled *Fundamental Algorithms* that Addison-Wesley sold for $19.50.[21] It appeared in 1968, the same year that Knuth accepted a full professor position in Stanford's recently formed Computer Science Department. Addison-Wesley's promotional brochure described Knuth's book as the first in a series that would, in toto, summarize "all present knowledge of computer programming techniques." Given how fast the field was changing, this was, of course, quite a moving target.[22]

Years later, Knuth observed that "you can't have a good piece of writing if it doesn't have a little bit of your soul in it."[23] Dedicated to the IBM 650 machine with which he had "spent many pleasant evenings," the tone of Knuth's book was encouraging and accessible. Knuth opened with a 1963 passage culled from *McCall's Cookbook*: "Here is your book, the one your thousands of letters have asked us to publish. It has taken us years to do, checking and rechecking countless recipes to bring you only the best, only the interesting, only the perfect." Knuth noted that programming could "not only be economically and scientifically rewarding" but, if done well, also be an "aesthetic experience much like composing poetry or music."[24]

Throughout his book, Knuth sprinkled quotes from other authors, including this one taken from a Sherlock Holmes story, that he placed near the end of his tome: "You will, I am sure, agree with me that if page 534 finds us only in the second chapter, the length of the first one must have been really intolerable."[25] The joke was that Knuth's massive 1968 book only had *two* chapters and the first one clocked in at over 220 pages long.

Knuth imagined his average reader as someone familiar with computers and programming—perhaps they had written some programs already—but still a relative novice. (Some reviewers questioned this assumption, noting that Knuth "created the impression that his subjects are simple when they are not."[26]) Algorithms, he wrote, were like instructions in a cookbook. A recipe, however, "notoriously lacks definiteness" as two cooks could interpret "add a dash of salt" very differently. In comparison, algorithms "must be specified to such a degree that even a computer can follow the directions."[27]

As an example of an algorithmic procedure, Knuth described (over several very detailed pages) an imaginary program that simulated the elevator system in Caltech's Mathematics Building as it responded to user input. There was a logistical issue, however. Many readers of his book would be learning how to write their "beautiful programs" on very different kinds of computers. To get around this issue, Knuth deployed a clever pedagogical work-around. He based his instructions on a "mythical computer" he called MIX and an accompanying MIX language. This imaginary machine was, Knuth said, very much like "nearly every computer now in existence, except that it is, perhaps, much nicer."[28]

Reviewers praised Knuth's first book (and the ones that followed) as major steps toward putting computer science on a solid foundation of mathematical principles and theories. As such, it offered not only a "fine blend of humor, finesse, and scholarship" but also a "truly positive step towards eliminating the existing breach between mathematicians and computer scientists."[29] Martin Gardner, author of a popular column in *Scientific American* called "Mathematical Games," complimented Knuth's book for its "wealth of exciting and fresh material . . . written with a grace and humor that is, as you know, exceedingly rare in books on mathematics."[30] When the Association for Computing Machinery awarded Knuth its Turing Award in 1974, the citation highlighted his "major contributions to the analysis of algorithms . . . in particular his contributions to the 'art of computer programming' through his well-known books."[31]

Readers also responded enthusiastically to the first volume in Knuth's series. A second printing was ordered in July 1969 and Knuth produced a completely revised second edition, which appeared in December 1973. After he became one of the world's best-known (and wealthiest) computer nerds, Bill Gates provided a blurb for volume one, which challenged "really good programmers" to send their résumé to Microsoft if they understood all of Knuth's book. Meanwhile, volume two (subtitled *Seminumerical Algorithms*) appeared in 1969 with volume three (*Sorting and Searching*) coming out four years later. Programmers, novices and experienced alike, showed their appreciation by purchasing tens of thousands of copies of the books.[32] Knuth often noted with pride that, carved over the entrance of the house he and Jill shared, was a "grook" written by Piet Hein, the Danish physicist and friend of Norbert Wiener, which read:

The road to wisdom?
Well, it's plain
and simple to express:
Err
and err
and err again
but less
and less
and less.

In keeping with this sentiment, Knuth challenged readers to find legitimate errors in his textbooks, for which he would send bounty hunters an ego-boosting check for $1. As Knuth's fame grew, most of these checks were framed as trophies.[33]

Knuth's perfectionist tendencies, combined with community-based "fact-checking," improved his books' contents. But these couldn't address a problem that increasingly worried Knuth. In 1973, Addison-Wesley sold off the hot-metal machinery used to print Knuth's books and switched instead to a newer photographic-based method. Knuth deeply disliked the new fonts his publisher used as well as how the final products, prepared via phototype-setting, looked. Knuth started thinking that he could solve the problem himself, working in a manner in which he was supremely comfortable: sitting at a computer keyboard and writing programs.

## TEACHING TeXNICIANS

In February 1977, Knuth was chairing a committee that chose the books graduate students studying computer science at Stanford had to read for their exams. He encountered an advance copy of Patrick Winston's *Artificial Intelligence*, a soon-to-be influential textbook which Addison-Wesley was about to publish. The book's typography impressed Knuth, even more so after he learned it had been produced using high-resolution digital typography. "I realized that a central aspect of printing had been reduced to bit manipulation," Knuth later said. "Letters made of little dots—that's computer science! I didn't need to know about metallurgy or optics or chemistry. All I had to do was construct the right pattern of 0s and 1s. Put a 1 where you want ink, put a 0 where you don't want ink, and you can print the pages of a book!"[34]

Compared to other authors, for whom typesetting choices are a matter of aesthetic preference, people writing scientific papers need specific fonts and text placement to accurately convey their message. Publishers faced extra challenges when typesetting books and articles written by mathematicians and computer scientists as the manuscripts are often chock full of subscripts, superscripts, equations, and other special symbols. As a result, printers had long referred to these works as "penalty copy," something they grudgingly worked with at extra cost, hence the "penalty." A University of Chicago Press style manual from 1969 noted, for instance, that "mathematics is known in the trade as difficult" because authors' text was "slower, more difficult, and more expensive to set in type."[35] This problem was exacerbated by the fact that manuscripts in computer science often use *two* fonts simultaneously. One of these would typically be for the author's prose. But another would be reserved to indicate specific programming instructions for the computer.

Knuth had been thinking about typesetting for some time. In 1976, he assigned students the not-as-easy-as-it-sounds problem of breaking paragraphs into sensible-looking lines of text.[36] Then, in March 1977, Knuth received the galley proofs for the second edition of *The Art of Computer*

*Programming*, volume two. That evening he wrote in his diary: "They look typographically awful." Knuth later said, "I had spent 15 years writing those books. I didn't want to write anymore. How could I be proud of such a product?" Knuth had planned to spend an upcoming sabbatical year in South America where he would work on volume four of TAOCP. Instead, he decided to stay at Stanford and start an entirely new project: digital typography. In his diary, Knuth wrote, "I have to solve the problem myself."[37] He imagined the project wouldn't take more than a year. In the end, some thirteen years would pass before Knuth completed his contributions to the digital typesetting system he invented.

Knuth began writing his program in early May 1977. Ever fond of words, Knuth explained that its name, TeX, stood for "technical text." But it was also the upper-case form of τεχ, the "Greek word [which] means art as well as technology."[38] On the first page of his handwritten programming notes, Knuth set out one of his underlying premises. "A character," he said, "by itself is a box," or, in other words, a virtual rectangular space in which one can create a shape using ones and zeros.[39] A week after he started, Knuth, after pulling an all-nighter, finished an initial version of TeX.

However, Knuth soon realized he faced a classic chicken-and-egg dilemma. Knuth could not set type digitally until he had a collection of digital fonts. But Knuth couldn't design the fonts until he could begin to digitally set type with them. In other words, he needed *two* complementary programs—one to set type and another to generate fonts. So, Knuth wrote a second program, called METAFONT, that would define fonts digitally "using purely mathematical formulas under my control." Knuth reasoned that "with my own computer program controlling all aspects of the 0s and 1s on the pages, I would be able to define the appearance of my books once and for all."[40] Where TeX was intended to place characters on a page, METAFONT was about the design of the characters themselves.

To better understand the esoteric world of font design, Knuth and his wife Jill started studying old books. Knuth immersed himself in his university's large collection of rare books where he studied fonts made by accomplished designers such as Frederic W. Goudy and Hermann Zapf. He also called on a nearby lab that Xerox operated where people were interested in using personal computers for desktop publishing. Xerox offered to let Knuth use their equipment but stipulated that anything he designed would belong to the company. Knuth rejected these terms on the premise, as he later said, "Mathematics belongs to God."[41]

At Knuth's request, Addison-Wesley shared the original hot-metal page images, printed in Monotype Modern 8A font, which were used to prepare the first edition of Knuth's first TAOCP volume. He experimented with television equipment borrowed from Stanford's artificial intelligence laboratory to magnify the letters and capture them digitally. Another unsuccessful effort involved Jill photographing the pages and then projecting them on a wall so that Don could trace the enlarged outlines with pencil and paper. Neither attempt provided consistent images with high enough resolution. Knuth then had a breakthrough idea. Instead of trying to directly copy the letters, he would move to a more abstract level. "These letters were designed by people," he reasoned. "If I could understand what these people had in their minds when they were drawing the letters, then I could program a computer to carry out the same ideas." In other words, it should be possible to mathematically describe and then digitally define the curves and lines that would constitute an entire new family of fonts with regular, bold, italics, and so forth.

In taking this turn, Knuth, who had read widely about the history of type design, was returning to much older attempts done, for example, in seventeenth-century France, to make typography both rational and quantifiable.[42] To Knuth, this blend of aesthetics and mathematics proved quite pleasing. Achieving it took considerable work, however. In writing TeX, one of the biggest challenges was to determine how text in a paragraph should be divided into lines of roughly the same length (what is called "justification"). If done well, readers should not notice breaks in words. But when poorly done, the results can distract the reader. The problem is exacerbated by the fact that not all letters in a given font have the same width. When it came to mathematically defining a new font, the letter "S" proved especially challenging. One entry in Knuth's diary reads, "Tried again for eight hours to finish the letter 'S.' Not feeling very good." In the process of solving the problem, Knuth studied work by long-dead European typesetters, such as

Francesco Torniello who, in 1517, had applied geometric analysis to letters in the Latin alphabet.[43] Eventually, Knuth developed a prototype font family called "Computer Modern," which was based on an old Monotype font from 1896.

When Knuth started drafting early versions of TeX, he had no intention of writing a book on digital typography or creating computer programs around which an entire community would coalesce. In his preliminary description of TeX, he professed more modest goals. "I am preparing the system primarily for use in publishing my series The Art of Computer Programming," he noted, explaining that the initial program would "be tuned for my books."[44] Besides himself, he imagined his longtime office assistant, Phyllis Astrid Benson Winkler, would be the only other user. TeX, Knuth said, "would be something I would make so that Phyllis would be able to take my handwritten notes and go from there. I used her as the model for the language I was developing."[45]

Likewise, in the late 1970s, writers did not have easy access to user-friendly "What You See Is What You Get" (or, WYSIWYG) graphical interfaces later found in programs such as WordStar (1981), MacWrite, or Microsoft Word (both 1983). Instead, Knuth built TeX around hundreds of "control sequences." To use TeX, an author would write their text and insert the commands directly into it as markup language. For instance, the command {\bf words words words} would make "words words words" appear in bold face as **words words words** in the final document. Essentially, a TeX file contained two things: the text an author wanted to publish along with the specific commands as to how it should be formatted and displayed.

Knuth's ideas regarding typesetting and font design existed in relative isolation until January 1978 when he delivered the American Mathematical Society's (AMS) prestigious Josiah Willard Gibbs Lecture. Named after a nineteenth-century American physicist and mathematician, the AMS lecture was aimed at both specialists and the general public. Besides Albert Einstein (who offered a new derivation of $E = mc^2$), other presenters included John von Neumann, Norbert Wiener, and Herbert Simon. Rather than talk about his accomplishments in optimizing computer algorithms, Knuth decided his Gibbs lecture would show mathematics "as a printer might see it."[46]

To do this, he displayed representative samples of changing typeface drawn from the society's flagship journal and explained the printing technology used for each. Knuth argued that font quality peaked in the 1920s and, when he showed examples from the early 1970s, he confessed that it was around this time he stopped submitting articles to the AMS. "The finished product," he said, "was just too painful for me to look at."[47] To be fair, Knuth admitted being "somewhat of a stickler and a perfectionist," joking that, as a Wisconsin native, "I refuse to eat margarine instead of butter."

Knuth went on to describe two main areas that his new work would benefit. The first was the intellectual content of mathematics itself. A lot of sophisticated mathematical work was needed to analyze older efforts at typography and develop new systems. Digital typography, in other words, meant "applying mathematics in new ways."[48] The second payoff was institutional. Advances in typography like TeX could help the AMS by providing new tools for publishing works of higher quality at a lower cost. Knuth envisioned a future where the office typewriter would be replaced by a small computer with a keyboard and television screen attached to it. In this scenario, a standard typesetting language like TeX would enable authors to transmit their manuscripts—already formatted—to journal editors and book publishers "from computer to computer via phone lines."[49] Galley proofs would no longer be necessary as referees or copyeditors could put their suggestions directly in the digital text and return the file to the author for revisions. Knuth's vision was essentially what became known as "desktop publishing."

Knuth's lecture marked a point where he went from thinking about TeX as a program for his own use to one that might interest a wider community. Knuth sent copies of his talk to curious editors at major publishers.[50] Colleagues at schools beyond Stanford began experimenting with copies of TeX that Knuth shared via the ARPANET, a precursor to the modern internet. One challenge was getting TeX to run on other computers, in part because Knuth had originally written it in SAIL, a computer language developed in 1970 at Stanford. At MIT, for instance, a colleague modified TeX so that it worked on a PDP-10 machine at the school. Digital Equipment Corporation, which made the PDP line of machines, also expressed interest in exploring how the company might "incorporate TEC [*sic*] into DEC" and help "promote this new technology in the printing business."[51]

The biggest boost to the spread of TeX as a typesetting system for writers came from the organization whose publications Knuth had rejected years earlier for aesthetic reasons. After Knuth's Gibbs lecture, the American Mathematical Society, which had around 20,000 members in 1978, began to consider adopting TeX for its publications. The idea started with the group's "Standing Committee on Composition Technology," which was chaired by Richard Palais, a mathematician at Brandeis University. Palais recalled the 1970s as a "grim one for mathematical publishing." With hot-metal printing on its way out, AMS leaders faced a dilemma of whether to increase the price of the society's journals or abandon high-quality typesetting. AMS had tried both approaches and "no one was happy."[52]

TeX offered an alternative. In October 1978, Palais described the possibility of "typesetting by authors" to the AMS governing board. Palais noted that the journal *Proceedings of the AMS* contained about 2,000 published pages per year. This translated to about $82 per page (or $164,000 annually) with half of that going solely to produce camera-ready copy.[53] By significantly reducing costs, TeX could would allow for less expensive subscriptions, perhaps drawing in new readers, while speeding up the publication process. The obstacles to adopting computer typesetting, Palais reported, were not technical in nature but instead were social, that is, "convincing authors and typists that they should make the initial effort" to learn a new system.[54]

In 1979, the AMS dispatched a small group from its headquarters in Providence, Rhode Island, to Stanford to see how authors could learn how to "typeset their own papers at the preprint stage."[55] Barbara Beeton was one of the people who learned from Knuth and his students. Trained in math and German literature, she took a position with the AMS after college where she worked on various typesetting and computer-based composition projects. Publications and newsletters about TeX are full of Knuth's puns, with words like "TeXpert" and "TeXercise" appearing regularly. After a month of intensive tutorials at Stanford, Beeton, now christened a "TeXnician," returned to Providence with a copy of the program on a data tape. Soon, the society was testing TeX for nonpublication purposes such as preparing membership lists and catalogs. By the end of 1979, Palais reported that the AMS was planning to shift "all of its composition over to the TeX system" within the next eighteen months. The actual adoption of TeX, not surprisingly, took somewhat longer.[56] What would help accelerate this transition, Knuth realized, was a book.

## SPREADING THE WORDS

In the fall of 1978, Stanford's Computer Science Department printed several hundred copies of a manual by Knuth. Titled "Tau Epsilon Chi: A System for Technical Text," it described TeX as a "new typesetting system" especially suited for writing and publishing "books that contain a lot of mathematics." Knuth wrote it for novices as well as those he called "wizards." To do this, he deployed visual cues throughout the book, like a road sign symbol that warned readers that a "dangerous bend in the train of thought" lurked ahead.[57] Staying in character, Knuth sprinkled jokes throughout the manual along with programming puzzles and entertaining references to other mathematicians and popular culture.

In mid-1979, the AMS published 1,000 copies of a revised version of this manual with a cover designed by Jill Knuth.[58] This was the same year that Knuth received the National Medal of Science for his research on the mathematical analysis of algorithms. Later that year, the AMS collected Knuth's Gibbs lecture on mathematical typography along with his manuals describing TeX and METAFONT and bundled them into a single book.[59] More than 15,000 copies of *TeX and METAFONT: New Directions in Typesetting* sold for $12 each, a sign that Knuth's system was attracting growing interest among mathematicians as well as other researchers.

As a student of books and printing, Knuth was naturally interested in incunabula. Relatively rare today, these books were produced in the earliest stages of European printing technology when movable type was just starting to be widely used. While Knuth had used TeX to write his manuals, he was just as proud of what he saw as the "first *real* book printed with TeX." *Lena*

*Bernice: Her Christmas in Wood County, 1895* was a modest-size work, written by Elizabeth Ann James, typeset via TeX, and illustrated with Jill's handmade prints. Knuth was also pleased when Addison-Wesley put out a new edition of Patrick Winston's book *Artificial Intelligence*, which stimulated his initial interest in digital typography, and the publisher typeset it using TeX.[60]

Knuth spent considerable time in 1980 using TeX for its original purpose: preparing a new edition of volume two in his series.[61] His diary notes that he finished this task in late July. About five months later, Addison-Wesley sent him a bound copy of his book. To mark the occasion, the publisher also prepared a limited edition of 256 copies where each was hexadecimally-numbered—something significant to computer scientists as a byte, defined as eight bits, can have 256 different values—and bound in leather. Few of these sold, Knuth recalled, as computer enthusiasts didn't care about the leather binding while book lovers were not drawn to works on programming. Nonetheless, Addison-Wesley sold some 10,000 copies of Knuth's (traditionally bound) book in its first year.[62]

After Knuth had published his initial books and manuals on TeX and METAFONT, he continued refining the programs. In the early 1980s, one of his graduate students converted TeX to Pascal, an efficient and popular programming language, which increased its portability to other machines. Knuth later rewrote his program in WEB, a programming system he created to foster what he called "literate programming." Instead of code providing only the instructions for a computer to execute, Knuth proposed that programmers should "concentrate on explaining to *human beings* what we want a computer to do."[63] By including better descriptions and documentation, Knuth believed that one day computer programs might be considered works of literature, perhaps even worthy of the prizes given to traditional authors. He made revised versions of his programs available for testing via the ARPANET or data tapes his office sent through the mail.[64]

In response, Knuth received hundreds of pages of comments and suggestions about how to further improve TeX.[65] Letters, often written in TeX, and emails piled up as Knuth's colleagues sent their advice and opinions. Many suggestions were highly technical and addressed issues like commands and macros.[66] Other readers noted that Knuth should also include more instructions for novices who wanted to start using his typesetting system. Knuth logged all the suggestions, noting that TeX had "become one of the most thoroughly checked computer programs ever written."[67]

Perfecting TeX and METAFONT absorbed a good deal of Knuth's time. It also required a considerable amount of research funding. Knuth's articles credited a range of patrons, including the National Science Foundation—its computer science division was interested in Knuth's research on the mathematics of paragraph breaking—and the Office of Naval Research. IBM and Addison-Wesley contributed support as well but probably the most generous benefactor was the prosaically named System Development Foundation. Created as an offshoot of the RAND Corporation's software development activities, a series of tax and legal decisions resulted in the foundation having to disburse its considerable assets. Computer science was one of the areas the foundation's trustees decided to support and they provided Knuth with about $1 million for his digital typography research.[68] This allowed Knuth to support about a dozen graduate students who wanted to work with him on TeX-related topics.

In June 1983, Knuth finished the preface for his new guide, called *The TeXbook*, and sent the manuscript to Addison-Wesley. He kept his symbols for "dangerous bends" (i.e., challenging programming sections) and added "road signs" to mark passages that were "so esoteric" most users probably could ignore them. Knuth liberally included quotes throughout *The TeXbook* from authors he respected, such as Plato, Jonathan Swift, and Francis Bacon as well as one from his wife, Jill ("Don't use footnotes in your books, Don"). Attentive readers could still discover the occasional hidden joke. For example, there is an entry in the Index for "Derek, Bo." The actress's breakthrough film, the romantic comedy *10*, appeared in 1979. The index therefore directed readers to page 293 where they found an exercise on the "Powers of Ten."

To illustrate *The TeXbook*, Knuth turned to Duane Bibby, a cartoonist based in the San Francisco area who shared the author's fondness for *Mad* magazine. Bibby read Knuth's manuscript and, although not a computer expert, found it user-friendly.[69] Bibby had recently adopted a Maine Coon cat and suggested to Knuth that an anthropomorphic lion and lioness would be good characters for his typography books. Knuth "instinctively felt this was right" and, years later, when he was visiting the Boston Public Library, he saw that the building's grand staircase had lions. "That's it!" he recalled, "TeX and METAFONT try to be like these lions, fixtures that support a great library. I love books, and lions represent books!"[70]

*The TeXbook*, with Bibby's lions on the cover, appeared in late 1983. Demand for it was such that Addison-Wesley ordered a second printing for October 1984, a sixth one in 1986, and a twentieth print run in 1991. Translations into Japanese, Russian, and Polish followed as well. TeX was, of course, just one part of Knuth's larger digital typography ensemble and, by 1986, Addison-Wesley had published five books that in toto described the entire system. Besides *The TeXbook*, there was a companion manual for how to use METAFONT. Two other volumes contained the source code for TeX and METAFONT while a fifth book detailed Computer Modern. Knuth had designed and refined this typeface in collaboration with type designers such as Charles Bigelow, who he brought to Stanford from the Rhode Island School of Design, and Hermann Zapf, a German designer. Zapf, with input from Knuth, had also created a new font for the American Mathematical Society called "AMS Euler."[71] Not including the publications that contained the source code, Knuth's books ran to nearly 1,500 pages of prose, instructions, descriptions, and exercises.

When Addison-Wesley advertised what it called the "Computers and Typesetting series," its marketing staff described Knuth's five tomes as both a "work of genius" and a "work of art."[72] They had reason to be so laudatory. Knuth's bestselling books provided a profitable foundation for Addison-Wesley's entire computer science catalog, selling hundreds of thousands of copies.[73] Peter Gordon, Knuth's editor, noted how publishers, including his own, were adopting TeX as were "academic departments and research laboratories throughout the world." As a "standard language" for typography, TeX offered publishers the opportunity to produce books "more quickly and more cheaply than ever before possible" while giving authors "increased convenience and facility in developing their words."[74]

Addison-Wesley may have been profiting from Knuth's books on digital typography but the computer scientist wasn't. This was no accident. Early in the development of his typography programs, Knuth, who was doing quite well with his TAOCP series, decided that it "didn't matter to me whether I got anything financial" from TeX. The system was largely a labor of love stemming from his personal attachment to fonts, typesetting, writing, and the appearance of books, especially his own. Interestingly, Stanford University, with a reputation for aggressively commercializing its professors' intellectual property, did not do likewise with Knuth's programs. Knuth was also motivated (or, more accurately, demotivated) by what he saw happening in the professional typesetting industry where he believed proprietary claims held back technological advances. The history of software development provided Knuth with a better example to follow. Programmers at IBM had developed the FORTRAN language in the 1950s and the company encouraged its use on non-IBM machines. As a result, FORTRAN became what Knuth called the "lingua franca" for scientific computing applications.[75] He hoped TeX would achieve similar status for digital typesetting.

Knuth used his publications to place his typographic system in the public domain. Readers could freely use algorithms and code found in his books so long as they did this correctly. To maintain control over his creations, Knuth created a "torture test" that any software presenting itself as a legitimate version of TeX would have to pass. Ever fond of wordplay, Knuth called this test program TRIP.TEX (i.e., "triptychs"), which would "trip up" unworthy programs.[76] By freely offering his programs, publishing the underlying code, and allowing users to tinker with them, Knuth had effectively created one of the first examples of "free and open source software." However, unlike the developers of programs such as Linux, which debuted in 1991, Knuth expressed few of the revolutionary sentiments espoused by open-source advocates.[77] TeX's raison d'être was not programming in and of itself. Rather, as Knuth had stated at the outset, TeX was "intended for the creation of beautiful books."

In 1990, Knuth announced that he would make no further changes to his programs except to correct "extremely serious bugs." He still maintained his offer of a personal check (these started at $2.56 in 1984 and doubled every year) to talented readers who spotted any mistakes still lurking in his code. In 1990, Knuth branded the latest version of TeX as 3.1.

For modifications made after that, the next version would be 3.14, then 3.141, eventually "converging to the ratio of a circle's circumference to its diameter."[78] Knuth stipulated that, after his death, whatever was the current version of TeX would be relabeled as "Version $\pi$" with any remaining "bugs" accepted as "permanent 'features.'"[79] Anyone who wanted to make a better or different typesetting program was free to do so. However, Knuth stressed that no future "authors"—the choice of this word over "programmer" is revealing—could call their systems TeX or METAFONT unless they conformed to his own programs as specified by the various torture tests he had devised. Thirteen years after he first typed "a character by itself is a box," Knuth declared his work on digital typography at an end.

## COMMUNITY ORGANIZER

For medievalists, a "textual community" was something organized around specific authoritative texts. Textual communities in medieval Europe often had an influential and respected figure who offered interpretations of these sacred texts. Members of these communities correspondingly oriented their spiritual and social lives around their foundational texts. These works provided a source of solidarity as well as a means to distinguish the community's members from the outside world.[80] (Of course, heretics—the distinction was in the eye of the beholder—likewise claimed their own texts.)

Parallels between medieval textual communities and the expanding group of users, editors, and authors that emerged around TeX are striking. Knuth's books, publications, and programming code provided a corpus of central texts for acolytes to access, learn, and argue over. Knuth himself was obviously the group's most prominent teacher and key arbiter of what was TeX and what wasn't. And, of course, membership in the growing "TeXtual community" was readily apparent to journal editors and book publishers whenever anyone submitted a manuscript. Someone's writing, right there on the page, identified them as a TeX user.

The TeX community started small at first with just Knuth, his secretary, and his students. The arrival of Beeton and other staff members from the American Mathematical Society at Stanford for their summer tutorial sessions expanded the enterprise. The AMS group helped TeX migrate to the East Coast as did Knuth's own willingness to share it with colleagues at places such as MIT. When *TeX and METAFONT: New Directions in Typesetting* appeared, it contained a postcard readers could mail to the AMS if they wanted to join a "TeX users' group." Soon, an official "TeX Users Group" (TUG) formed.[81] The group's primary goals were to spread the word about TeX, advise accomplished and potential TeXperts, and share news about upcoming events.

To facilitate this exchange of information, TUG started publishing—using TeX, of course—a regular newsletter. Because Knuth had originally written TeX using the SAIL language, the handsome-looking newsletter was, in classic Knuthean humor, christened *TUGboat*. Besides introducing TeX, the first issue introduced leaders of the group, including Beeton ("Wizard of Format Modules"), Hermann Zapf ("Wizard of Fonts"), and, of course, Knuth himself (the "Grand Wizard of TeX-arcana"). It featured a traditional printing press on its cover along with an eminently suitable epigraph from the nineteenth-century British logician Augustus de Morgan: "The subject of mathematical printing has never been methodically treated, and many details are left to the compositor which should be attended to by the mathematician. Until some mathematician shall turn printer, or some printer mathematician, it is hardly hoped that this subject will be properly treated."[82]

At first, just over a hundred TeX enthusiasts paid $10 annually to join TUG. By 1984, when *The TeXbook* was published, the community had grown to over 600 people. More striking was the rapid spread of TeX to a wide variety of institutions. In the mid-1980s, some 220 universities, labs, and companies in the United States and overseas, including major publishers such as Elsevier and John Wiley & Sons, had installed TeX on their machines. By the end of the decade, the TUG community consisted of some 4,000 individual members who paid $45 a year, with institutional memberships costing as much as $1,300.[83] Issues of *TUGboat*, which Beeton assumed editorship of in 1983, grew until each one was 200 pages or more. Meanwhile, a regular lineup of conferences, workshops, and intensive short courses introduced new people to TeX-based writing. Perusing issues of *TUGboat* reveals an active and helpful group, albeit one quite fond of wordplay and puns.

To many hopeful TeXperts, using Knuth's program to place text and white spaces on a page exactly where one wanted could be intimidating. A simplified and streamlined version would broaden its appeal and help the TeX community expand even more. In the early 1980s, a computer scientist named Leslie Lamport started developing just such a program. Born in Brooklyn in 1941, Lamport was working at the time for SRI International, a nonprofit research organization in Palo Alto, California. Lamport was already using another program, called Scribe, to produce documents but it was neither free nor amenable to user modification.[84] Then, in the early 1980s, Lamport decided he wanted to use TeX to write a book. When he inspected Knuth's program, he uncovered some shortcomings with its existing "macros." These are automated input sequences a computer follows to produce specific outputs. (A very simple example of a macro on my own machine is to simultaneously press COMMAND + B, which renders highlighted text from plain to **bold font**.) Developed before the era of commonly available graphical user interfaces, TeX relied on hundreds of such typed commands. Lamport proceeded to assemble a package of standard macros that would help make it easier for authors to use TeX.

The resulting program, which became available in 1983, was called LaTeX (the "La" presumably comes from Lamport). Both TeX and LaTeX were software systems designed to do high-quality typesetting but, compared to the often wild complexity of TeX, LaTeX appeared as a more domesticated species to many users. To help guide users, Lamport wrote a manual for LaTeX that caught the attention of Peter Gordon at Addison-Wesley. Although Lamport later said he doubted anyone would buy a manual for software that was already free, Gordon thought otherwise and saw LaTeX as the potential core of an entire ensemble of software applications and books. In 1986, Addison-Wesley published a revised version of Lamport's manual as *LaTeX: A Document Preparation System*. More than 200,000 copies, illustrated with Duane Bibby's ubiquitous lions, were sold.[85]

Physicists were the group that most fervently and quickly embraced LaTeX. The roots of their enthusiasm could be traced back to conventions that predated the digital computer. Traditionally, physicists had circulated preprints (i.e., papers not yet published and perhaps not yet even reviewed), among their colleagues for decades. Sending and receiving them through the mail was an essential part of functioning effectively as a member of the physics community.[86] In the 1920s and 1930s, American physicists working in quantum mechanics were at a comparative disadvantage vis-à-vis their counterparts in Europe, where much of the cutting-edge research was being done. It took a few extra weeks for journals such as *Annalen der Physik* and *Zeitschrift für Physik* to arrive in physicists' mailboxes at Berkeley or MIT, leaving them playing catch-up in that fast-moving research area. As one American scientist lamented, "By the time we can get at it, probably somebody in Europe has already done the same thing."[87]

After World War II, theoretical high energy physicists, a relatively small and close-knit community in terms of membership but remarkably well-funded by such entities as the Atomic Energy Commission, were especially keen to share research results with colleagues around the world. The fact that their computer-literate community was well-defined and socially interactive yet produced little that would pose questions of intellectual property made them especially eager participants in a well-developed "preprint culture."[88] The practice became so pervasive as to be declared pernicious by one journal editor who suggested that physicists should be issued tape recorders so that "all their statements about physics, even those uttered in their sleep, would be preserved."[89] Sarcasm aside, the preprint system was not necessarily democratic, as scientists who were not part of "the in-group," as one physicist griped, had to wait until colleagues' research appeared months later in paper journals.[90]

In the late 1980s, three technologies combined with physicists' historical propensity for preprints. The result was a fundamental change in scientific practice and daily work routines that also brought a new cohort of people into the expanding "TeXtual community." All three technologies were computer-based. The first was electronic mail, which made it quick and easy to circulate research results. Scientists in general were among the occupations whose members had the easiest access to what became the internet. Second, disk storage grew dramatically in size while decreasing in cost, making it possible for researchers to download and amass ever-larger numbers of papers. Finally, LaTeX and TeX provided a standardized (and free) digital document preparation system that was especially amenable to papers with lots of formulae and equations.

What eventually became a deluge of electronically circulated preprints started as a trickle in a small corner of the physics community. In 1989, Joanne Cohn, a string theorist based at the Institute for Advanced Study at Princeton, was preparing to attend a workshop at Rutgers University. Paul Ginsparg, a colleague from Los Alamos National Laboratory, sent her a selection of preprints, prepared in TeX, in advance of their meeting. After Cohn shared these files via email with colleagues, more and more people asked her for copies. Even after the Rutgers meeting had come and gone, the list of physicists on Cohn's email list kept growing until, by mid-1991, it contained some 180 people from over twenty countries.

Ginsparg suggested automating what was still a bespoke system of sending emails to interested colleagues.[91] With Cohn's approval, Ginsparg wrote some computer code to automate her email list and, on August 14, 1991, his system, named "hep-th" for "high-energy physics theory," started running on a Los Alamos computer. Soon, some 2,000 users routinely accessed the system via the electronic address "hep-th@xxx.lanl.gov." (The site predated widespread public access to the World Wide Web, hence it did not include the familiar "www" in its address.) Ginsparg's system could detect an incoming preprint, most likely prepared in LaTeX or TeX, separate the title and abstract from the main text, and then place the paper in the growing computer database. Users could then log onto the Los Alamos machine, see what new work was posted, carry out searches, and, when interesting papers were found, download them for later reading. In time, Ginsparg added other subfields of physics, such as "condensed matter theory" and "astrophysics."[92]

By the end of 1992, what was initially known as the "LANL preprint archive" received some 7,000 papers annually that more than 8,000 subscribers accessed. In 2001, when Ginsparg moved himself and his preprint server—now known as "arXiv," with the "X" standing for the Greek letter *chi*, as in TeX—to Cornell University, annual submissions surpassed 30,000.[93] Recent statistics show that over 2.5 million papers are available at arXiv, with more than 20,000 new articles added every month. Over two billion copies of papers, almost all written in some variation of TeX or

LaTeX, have been downloaded. The community of users came to include not just physicists, mathematicians, and computer scientists but also researchers in biology, finance, statistics, economics, and engineering. For tens of thousands of scientists, checking their computers in the morning to see what had been posted overnight became a daily routine and the time stamp of a paper posted at arXiv could even serve as way of establishing priority for a discovery.

Frustrated by the vagaries of traditional paper publishing, Donald Knuth had set out to build a digital (and personal) typesetting system to prepare publications that met his own aesthetic standards. His authoritative texts about digital typography provided the essential foundation for the building of a global "TeXtual community." By the early twenty-first century, TeX had become a common language that served an entire publication and distribution system used by authors, publishers, and journals around the world.

Amidst the formulae, exercises, and instructions in *The TeXbook*, Knuth had included a quote from a 1889 black comedy written by Robert Louis Stevenson and his stepson Lloyd Osbourne called *The Wrong Box*. "How very little," the two writers noted, "does the amateur, dwelling at home at ease, comprehend the labours and perils of the author."[94] As the creator of both computer code and computer books, Knuth had managed to make those labors a little less perilous.

# 6 DESIGNING CULTURES

In the late 1960s, an anthropologist at Berkeley began suggesting that she and her colleagues direct their ethnographic attention not only at the disempowered and marginalized communities they traditionally studied but also at the people who wielded influence in modern Western societies. This included exploring the lives of scientists and engineers. Laura Nader called her new methodology "studying up."[1] As a graduate student at Berkeley, Lucy Suchman heeded Nader's call to action. For her ethnographic fieldwork, she decided to examine work practices at Xerox PARC, an internationally renowned laboratory in nearby Palo Alto.

Short for "Palo Alto Research Center," PARC was started by Xerox in 1970 as the company's executives anticipated the transition from a world of paper documents to one where digital files and electronic communications dominated. The goal was to have a place, far away from company headquarters in Connecticut, where engineers and scientists could collaboratively explore the "office of the future." Throughout the 1970s, PARC's researchers developed a host of new information technologies including the Alto, an innovative small computer that featured a mouse and graphical user interface, as well as laser printers and computer networking.[2]

In March 1980, Suchman found herself in a PARC office with a tape recorder and notebook. Lynn Conway, a forty-two-year-old engineer, sat across from her. The result was a remarkable set of conversations as the two women—a technologist who had worked in the electronics industry for more than two decades and a Berkeley graduate student groping toward a dissertation topic—discovered they shared similar interests and ideas. Suchman was becoming increasingly attentive to the nexus of design and technology. For engineers, designing things is a central activity in their professional lives.[3] Conway, meanwhile, had just completed a multiyear collaborative project in which she helped reinvent how future computer chips would be designed.

A book was central to Conway's accomplishment. In 1980, Addison-Wesley published the prosaically named *Introduction to VLSI Systems* (VLSI was technical shorthand for "Very Large Scale Integrated"), a textbook Conway had written with an electrical engineer at Caltech named Carver Mead.[4] Their book articulated a new design methodology for building future generations of computer chips. Conway would sometimes refer to what she and Mead had done as a "design of design."[5] *Introduction to VLSI* was soon being used by students at scores of computer science and electrical engineering programs around the world.

What especially animated Conway's conversations with Suchman in 1980 was her realization that her book had helped bring a whole new "clan" of chip designers into existence.[6] Their talk was replete with anthropologically-inflected terms (e.g., token, ritual, tribe), as the two women discussed the creation of a community oriented around new ways to design computer chips. Suchman's curiosity about engineering cultures paralleled Conway's familiarity with classic works in the field of science and technology studies by Thomas Kuhn, Ludwik Fleck, Harold Garfinkel, and Bruno Latour, among others.

Authors write books for many reasons. Edmund Berkeley wanted *Giant Brains* to explain computers to as wide an audience as possible. Norbert Wiener wrote *The Human Use of Human Beings* as a warning about the dangers computing and automation technologies posed. Meanwhile, Joseph Weizenbaum and Ted Nelson—while displaying very different authorial styles—used their books as argumentative tools to persuade readers about the dangers and opportunities that computers (and computer scientists) posed.

Lynn Conway and Carver Mead's book differed significantly from these other works. First of all, the two engineers were established professionals working at the peak of their careers. In contrast, Weizenbaum (and Wiener to a lesser extent) had largely retreated from scientific publishing when they wrote their books. More significantly, the primary purpose of *Introduction to VLSI* was to help students learn a specific set of technical skills and build a communal infrastructure around this knowledge.[7] In other words, *Introduction to VLSI* functioned as a "paper tool" that Conway and Mead purposefully created out of words and images to do a specific job. The information their book presented was specialized, standardized, and highly portable from place to place. Once captured on the page, this information could then circulate and be shared as students migrated from their classrooms to positions in industry.

Traditionally, textbooks offer readers knowledge that is familiar and accepted by the professional community. Generations of students, for example, used *College Physics* (now in its fifteenth edition and retitled *University Physics with Modern Physics*), a textbook Addison-Wesley introduced in 1947.[8] As tools for teaching and training, textbooks transmit information that is an established part of a disciplinary canon.[9] As a reader of Kuhn, Conway understood the role textbooks played in doing "normal science," maintaining "scientific paradigms," and creating a community with a shared understanding of a subject. Textbooks are not meant to subvert the existing epistemological order.

But this is *exactly* what Mead and Conway set out to do with their book. Rather than just present established knowledge, their book helped establish new ways of doing things and new forms of practical knowledge. Because their book provided instructions about a particular activity—designing chips, in this case—we can situate *Introduction to VLSI* somewhere between a traditional textbook and a laboratory manual.[10] And, as all engineers know, a design typically travels an iterative path. The methods Conway and Mead presented in the first version of their book were systematically changed, revised, and perfected based on students' experiences in the classroom.

Donald Knuth created TeX because he was determined to print characters on paper in a manner that matched his aesthetic sensibility. With their VLSI methodology, Conway and Mead focused on a different form of printing. In the late 1950s, engineers developed photolithography, a set of print-based techniques for making computer chips. In time, they could print hundreds and then thousands of components that together formed a single integrated circuit. By the 1970s, these devices—some of the most complex things made by humans—were being mass-produced by the hundreds of millions.

Conway and Mead saw the fabrication of computer chips as being about more than just printing patterns of extremely tiny switches and wires onto silicon wafers. Instead, they wanted to change the entire process of designing and fabricating state-of-the-art circuits. The methodology they developed for chip design arose from the needs of a fast-growing yet anxious multibillion-dollar industry, one that was heavily capitalized and resistant to change. But executives at electronics companies were also imagining a future when silicon chips would be found in all sorts of consumer devices besides computers. Stimulating the necessary changes across the semiconductor industry would, as Conway described it, require new methods to "design the design methodology."[11] It would require a book.

## TRANSITION STATES

Lynn Conway sometimes claimed fellow engineer Charles Proteus Steinmetz as a role model. Steinmetz had emigrated to the United States in 1889 under duress (he was a committed socialist) before becoming one of the most celebrated engineers of his time. While working for General Electric, Steinmetz promoted the transformation of electrical engineering from a craft-like activity to one based on mathematical equations and the latest scientific theories. His research helped resolve contentious debates over the relative merits of alternating versus direct current (think Tesla vs. Edison). Steinmetz's textbooks were just as influential. In 1920, McGraw-Hill published the Steinmetz Electrical Engineering Library, a nine-volume set that offered students the "foundation of modern electrical engineering knowledge."[12] Steinmetz's career exemplified to Conway how a person's research and teaching could reshape an entire industry as well as engineering education.

But Lynn Conway also saw another, more personal parallel in his life. Steinmetz was born with kyphosis, a class of spinal conditions that result in the excessive curving, rounding, or hunching of the back. Steinmetz's physical differences were of significance to Conway, who had been born in 1938 as what is now known as an AMAB, or "assigned male at birth." Starting at an early age, Conway increasingly began to self-present as a girl. ("Robert Sanders" is the moniker both Conway and journalists have used to protect family privacy. Out of respect for Conway and the transgender community, as well as to minimize confusion for you the reader, I will use "Lynn Conway," "she," and "her" throughout this book, as she did in her own writings.) To Conway, Steinmetz's success meant that someone "perceived as different" might still "become liked, even honored" if they made significant contributions to their research community.[13] As Conway's engineering career developed, this became both a goal and her lodestar.

Since childhood, Conway was drawn to physics, math, and especially, electronics. Profound inner conflicts about gender identity, however, left her feeling socially awkward and isolated. In 1955, Conway enrolled at MIT as a physics major and quickly became intrigued by the rapidly changing world of digital electronics. As she recalled, "Instead of seeing electronics as an infrastructure for doing physics, I glimpsed a vast world for exploration, abstraction and meta-architectural innovation."[14] Reading Norbert Wiener's works on cybernetics heightened Conway's enthusiasm for computers and electronics. Academically successful at MIT, Conway's "cross-gender feelings," as she called them, became more powerful and her personal life "slid into serious and obvious difficulties."[15]

In 1959, Conway left MIT without finishing her degree and took work designing and building digital circuits. A few years later, she enrolled at Columbia University and completed her BS and MS degrees in electrical engineering. While at Columbia, Conway also took courses in cultural anthropology. As she later described it, her "outsider" status left her curious about how ethnographers such as Margaret Mead and Ruth Benedict had written about people and communities who departed from conventional norms.[16]

While Conway's studies at Columbia went well, her personal life remained turbulent. In 1962, Lynn (who still presented publicly as "Robert") met "Sue." The two of them married in 1963 and a daughter arrived a year later. With a family to support, Conway accepted a position at IBM's prestigious new research center in Yorktown Heights, New York. The lab's main building, with a curved, concave facade of dark tinted glass designed by Eero Saarinen, suggested the professional prestige that electrical engineers enjoyed at the time. By the mid-1960s, almost 20 percent of America's engineering community worked on electronics of some sort as Cold War military needs and the growth of computer and microelectronics companies drove demand for their skills.[17] In short, to be an electrical engineer in the 1960s was to join a booming professional community with rising economic prospects and plentiful job opportunities.

At IBM, Conway's main area of expertise was "computer architecture." This refers to the manner in which a computer's components, such as its central processing unit and memory, as well as the operations that guide their function, are organized. A classic example is the "von Neumann architecture," proposed in 1945, in which a computer's memory holds its programming instructions as well as the data on which the machine operates. In 1965, Conway was assigned to work on IBM's new Advanced Computing Systems (ACS) initiative.[18] The ACS effort grew out of the company's concern that it was falling behind in building ultra-fast computers for scientific and military applications. When IBM relocated its secret supercomputing project to California, Lynn and her family moved with it.

The ACS project faced a serious technical bottleneck. The initial architecture of IBM's design only allowed for one command to be executed at a time, such as multiplying two numbers together. Conway devised an innovative solution called "dynamic instruction scheduling." It allowed the prototype machine to hold pending instructions in a queue and issue multiple commands in orders other than sequential. It was, she later said, a "turbocharger" for pushing more commands through the computer in the same amount of time. The IBM team incorporated Conway's work-around into their design and the technique proved valuable for subsequent high-performance computers.[19] Her experience with the ACS project helped Conway realize that the design process itself could be optimized. In other words, *how* one designed something was as important as the actual design itself.[20]

Despite the successes Conway's team had, IBM's upper management disbanded the ACS project in mid-1968. This professional blow could not have come at a worse time for Conway. "Sue" and "Robert" had two children by now but their relationship had steadily deteriorated. Conway was monitoring progress in experimental surgical and hormonal treatments carried out at Johns Hopkins University Medical Center, as well as other health care institutions. One prominent news article described procedures by Harry Benjamin, an endocrinologist and author of *The Transsexual Phenomenon*.[21] Benjamin's medical textbook provided Conway with both essential information and a sense of hope. After much additional research and accompanying mental anguish, Conway started gender reassignment surgery via clinics in San Francisco and Mexico. "Sue," meanwhile, agreed to delay divorce proceedings until her partner's medical transition was complete.

Conway informed IBM, a company known for its conservative corporate culture, about her personal situation. While her immediate colleagues were supportive, IBM's senior executives were not. In 1968, they fired Conway on the grounds that her presence would cause "extreme emotional distress" for her coworkers. Conway's gender reassignment surgery proved successful but came at considerable professional and personal cost. "I suddenly lost my job, my reputation, my extended family, and all my loved ones and friends," she later wrote.[22]

Now legally known as "Lynn Conway" (she had used the first name privately for years and adopted the surname after reading *The Salzburg Connection*, a 1968 spy novel by Helen MacInnes), she reentered the workforce in what she called "stealth" mode. Several companies expressed interest in hiring an obviously talented engineer, only to renege once they learned of Conway's background. The exception was Memorex, a relatively new venture based in the Bay Area. In 1970, the company's main activity was making disk drives and other peripheral devices that were designed (ironically, as it pertains to Conway's story), to be compatible with IBM machines.

As Conway began to settle into her new job at Memorex, she kept up with the rapid changes in digital computing. A big event was Intel's introduction to the market, starting in 1971, of a series of increasingly more powerful and complex microprocessor devices. She began to imagine how these new devices would be used in future electronics. One problem, however, was that electrical engineers were becoming increasingly specialized. At the same time, computer scientists who worked on more abstract topics such as architecture design were typically not well-versed in the finer points of semiconductor physics. There must be, Conway reasoned, a way to make a bridge between what she viewed as different tribes of technical experts.

By 1972, Conway wanted more than the career options Memorex offered. A headhunter informed her of two openings. One was at Fairchild Semiconductor, an established firm in the burgeoning technology region exploding around Palo Alto. Conway, however, hesitated to take a job "blocking out simple architectures" for chips that others would then implement. She also had doubts about working in an industry with a "famously macho disdain of women."[23] The second option was with Xerox's newly opened Palo Alto Research Center, which was rapidly expanding. For Conway, working at PARC would mean not just designing chips but also having some say in how they were used. And in 1973, she joined PARC's Systems Science Laboratory.

Soon after Conway started at PARC, she began to realize there was a "gap between the sort of systems we could visualize and what we could actually get into hardware."[24] However, people outside the chip making community (like Conway) were limited to working with integrated circuits that were mass-produced and available off-the-shelf. Designing, making, and experimenting with specialized, custom-produced chips wasn't a readily available option unless one had inside access to one of the semiconductor factories (referred to in the business as "fabs"). As she later described it, using an authorial analogy, "Only 'writers' working for the 'printing plant' could become 'published.'"[25]

There was a cultural aspect at play as well. The professionals who understood and designed computers had become detached from the people who understood and designed integrated circuits. As microprocessors— programmable chips that function as miniature computers—became smaller and more capable, the distinction between chips and computers was blurring. As one expert remarked, "There is something of a subconscious fight going on to establish which group will be doing the creative work" when it came to designing the computers of the future.[26]

The problem wasn't just how chips themselves were designed. Rather, the challenge was reinventing the entire system, from the design of chips and how they were integrated into electronic devices all the way up to university courses, instructor guides, and an overarching "design methodology." Changing this, Conway explained, using language that an anthropologist such as Lucy Suchman would have appreciated, would require the "cultural integration of new methods" in order to create a common "design culture."[27] Instigating the cultural shift Conway wanted would require collaborating with someone who was both tightly connected with the semiconductor industry and held deep reservoirs of technical expertise in the subject.

## A TALL, THIN MAN

In talks with colleagues, Carver Mead sometimes mentioned a fictional character he called "the tall, thin man." Mead used the curious term to describe an imaginary engineer who was proficient in all aspects of chip design. This professional could work at every level of electrical engineering—the "tall" part—whether it was semiconductor physics or the physical layout of circuits. However, they might not be a leading expert in *all* of these areas (hence, "thin").[28]

In his physical appearance, Carver Mead actually was tall and thin. And he had been living and breathing transistors and integrated circuits for decades. Born in 1934, Mead was a native Californian who had grown up surrounded by electrical technologies. His father worked for a rural power company outside of Fresno and collected surplus equipment for his son to experiment with. A self-described Caltech lifer—BS '56, MS '57, and PhD '60—Mead was hired and quickly tenured in the school's Division of Engineering and Applied Science.[29]

When Mead first arrived at Caltech, transistors were still relatively rare and temperamental devices. One might liken them to the bespoke books produced by the hands of scribes and monks in the pre-Gutenberg era. However, by the 1960s, the manufacture of these devices had been extensively automated and the resulting products were akin to books churned out by printing presses: standardized, mass-produced, and widely disseminated. This alone was a remarkable transformation in an equally extraordinary short period of time. But to understand the significance of the additional technological changes Carver Mead and Lynn Conway undertook to introduce, it helps to have an understanding of how integrated circuits were made in the late 1970s. (Note to the reader: Thus far, I've endeavored to keep technical details at a user-friendly level. The physics and engineering of computer chips is, however, remarkably complex and varied. What follows, while technically accurate, is a much-simplified version of many technological developments that unfolded over the course of two decades.)

In the 1970s, electrical engineers referred to the densely packed micro-circuits their companies produced as "large-scale integrated circuits," or "LSI." In these devices, the smallest features—sometimes called "linewidths" for the width of metal interconnections between circuit components—were on the order of about three to five microns. (A micron is one millionth of a meter.) Engineers expected these linewidths to shrink to dimensions of one micron or less. To put this in perspective, the wires of the first transistors made in the late 1940s were easily seen by the unaided eye.

An integrated circuit circa 1970 typically had thousands of extraor-dinarily miniscule parts—transistors, diodes, resistors, and capacitors—all connected together on a piece of semiconductor material, such as silicon. During manufacturing, hundreds of separate integrated circuits might be built up simultaneously on a single silicon wafer some five inches or so in diameter. The entire process required a carefully choreographed sequence of a dozen steps or more, in which solid state physics, materials science, and electrical engineering combined to make integrated circuits by the millions.

Despite the ability of companies to mass-produce chips, their actual design went back to an older, artisanal tradition. An engineer might spend weeks or months using rulers, straight edges, and colored pencils to design a circuit on graph paper. This schematic would then be handed off to a tech-nician who would transfer the design onto a piece of red masking material called Rubylith. Wielding a fine-pointed razor knife, another technician carefully cut the masking material and peeled away sections where wires, transistors, and other components would be placed. These handmade mechanical masters would then be miniaturized to create so-called masks to fashion the actual circuit.[30]

Once the masks were ready, a first step in the actual manufacturing process was to chemically form a protective layer of silicon dioxide over the entire wafer. The next stage was photolithography, a quasi printing technique first developed in the late 1950s that imprinted circuit patterns on the silicon dioxide substrate. This was done by applying a special type of organic polymer called a "photoresist" onto the wafer. Ultraviolet light would be then projected onto the wafer through the handmade masks that contained the design for the desired circuit pattern. By 1980, typical integrated circuits might require half a dozen or more different masks to generate the desired final pattern. "Etching" happened next, as the photoresist, its chemical structure now altered by exposure to ultraviolet light, was washed away, leaving behind mask patterns "printed" on the semiconductor wafer. In the final stages, the wafers might be "doped" as careful amounts of trace elements that altered the semiconductor's electrical properties were added in a specially designed furnace.[31]

At Caltech, Carver Mead's research agenda in the late 1960s focused on both the basic science and the practical engineering of transistors and semiconductors. When Gordon Moore—a Caltech alum and cofounder of Fairchild Semiconductor—visited campus, he and Mead would regularly meet. Moore had recently written an article that predicted the continued miniaturization of electronic components. Using just four data points, Moore argued that this pattern, if extrapolated into the 1970s, foretold a routine doubling of transistors on a single silicon chip every year or so. "Moore's plot," as it was initially known—only later did it acquire the better-known moniker of "Moore's Law"—predicted that chips would have some 65,000 transistors by the mid-1970s. "The future of integrated electronics," said Moore, who would leave Fairchild in 1968 to cofound Intel with Robert Noyce, "is the future of electronics itself."[32]

Like Moore, Mead was fascinated by engineers' ability to steadily stuff more and more transistors onto ever-smaller silicon chips while the overall cost of the devices steadily decreased. Mead enthusiastically spread the message about the future of electronics—smaller and cheaper devices which nonetheless offered more computing power—to the corporate executives and industry engineers he regularly talked to.[33]

As transistors continued shrinking, Moore wondered if basic physics might someday limit just how small they could be made while still working properly. This prompted Mead to investigate the basic science behind the exponential predictions inherent in Moore's eponymous law. Mead's research suggested that physical features of integrated circuits precisely printed via photolithography could be made as a small as a quarter of a micron. At the time, the industry standard was around ten microns or about forty times larger. In other words, according to Mead, engineers still had lots of opportunities to make components on their chips even smaller before the basic laws of physics started to interfere.

The downside was that the complexity of devices threatened to become unmanageable. To explain this, Mead resorted to a comparison acutely familiar to someone commuting back and forth between Caltech and Palo Alto. When they were first developed, the features and connecting wires on a typical chip could be analogized to the buildings and streets in a single city block. By the mid-1960s, this intricacy had grown to be like "the street network of a small town."[34] Jump ahead another decade, however, and chip complexity was comparable to the street network in the *entire* Los Angeles region. Buildings and freeways, in Mead's analogy, were like the transistors and wires on a chip and now severe congestion on those "roads" posed an engineering dilemma.

Tied to this was a transition in how electrical engineers prioritized the components from which integrated circuits and microprocessors were constructed. With engineers packing more transistors into ever-smaller spaces, Mead predicted that the wires themselves would soon become the critical bottleneck.[35] Despite ever-smaller distances between components, it still took a finite amount of time to move electrons around in an integrated circuit. But doing this more efficiently (i.e., speeding up the traffic), would result in faster and more powerful computers. As Lynn Conway later explained it to a journalist, the next generation of computer chips "forced you to think like an urban planner. You had to think hard about where the roads go."[36]

As integrated circuits became more complicated, it was increasingly hard for one company, let alone one person, to manage the intricacy. A hierarchy of experts and a distinct division of labor emerged. System architects worked out the abstract levels of design, which logic and circuit experts then translated into block diagrams that showed logic paths, interconnections, and component placement. Layout specialists would then transform these schematics into specific geometric patterns. Then the final design could be sent to the company's fabrication facility where technicians would see it printed as a silicon chip. The final prototype would then have to be evaluated and tested by yet another engineering group. This process was neither quick nor cheap. Creating a new chip design might cost as much as $150,000 and require months of effort from an entire team of engineers.[37]

Mead anticipated a time when microprocessors would not just be found in computers but also "deep down inside our telephone, or our washing machine, or our car."[38] All of these new chips—hundreds of millions of them—would have to be designed and manufactured quickly, efficiently, and inexpensively. In the old system, big companies like Motorola and Intel designed and fabricated chips in-house. In contrast, Mead promoted a new "foundry model" that would separate design from manufacture. In this scheme, one company's engineers would do the design and then send their plans to an independent "foundry" which would fabricate the actual silicon devices. Given the cost and complexity of making chips, many firms at the time rejected the model Mead was advocating.

To convince naysayers, Mead drew on another analogy familiar to Californians: the motion picture business. Chip designers, he said, were akin to "the producers, the stars, script writers, and photographers." The microscopic patterns they created would ultimately be turned into reality by a different ensemble of experts. Mead likened this second group to the "film manufacturers and film processing laboratories" scattered all around Hollywood. The challenge was to create a "clean interface" between these two groups—"those creating designs and those printing them"—with agreed-upon standards and processes.[39]

To this end, in 1971, Mead started teaching EE 281, a "Semiconductor Devices" course for Caltech students (later renamed "Integrated Circuit Design"). Mead designed the class so it would attract students from computer science as well as electrical engineering. By spanning disciplinary divides, he wanted to create a "new class" of technologists who understood computers and microelectronics but were not restricted by inherent "shibboleths and taboos."[40] When students left Caltech, they would take this knowledge into the larger corporate world. But Mead's class was very small—only about twenty students—and he only taught it once a year. For the sort of industry-wide transformation he envisioned, this pedagogical process would have to be massively scaled up. This is where Lynn Conway's vision for the "redesign of design" via a book became an essential ingredient.

## BOOTSTRAPPING

The collaboration between Lynn Conway and Carver Mead was brokered by two prominent scientists who happened to be brothers. William R. "Bert" Sutherland had been instrumental in developing technologies central to the ARPANET, a forerunner of today's internet, before accepting a position at PARC. Meanwhile, his younger brother Ivan, an expert in computer graphics, joined Caltech in 1974 to help start a computer science program.

In the fall of 1975, when Mead was visiting PARC, Bert Sutherland introduced him to Conway. A few months later, Ivan Sutherland proposed a formal partnership between PARC and Caltech to work on what they started to call "Very Large Scale Integration." It would address how integrated circuits could and should be built over the next decade by creating a "series of design tools" and a new "design philosophy." This "school of thought" could then be transferred to other schools and companies, which would adopt the new approach. Bert shared Ivan's report with Conway who enthused that the "proposed project is an <u>excellent</u> idea. Hope I can be of assistance in getting it started."[41] Soon Conway, Mead, and a small team that formed around them were studying techniques to design new chips.

One initial challenge the group faced was that they didn't have the tools needed to change the way chips were designed in what was a large and diverse industry. It was akin to starting to build a birdhouse and realizing that you would have to first make your own hammer and nails. This was something that an engineer might call "bootstrapping," a simple and self-starting process that proceeds toward greater complexity without outside input. Fortunately, what the team *did* have was the latest computing technology PARC could offer. This included the lab's innovative Alto mini computers as well as ready access to the ARPANET. These tools allowed members of the team to create and share digital files between themselves and researchers at other sites.

For example, Douglas Fairbairn, one of Conway's PARC colleagues, and James Rowson, a Caltech graduate student, used the Alto to create a new software system called ICARUS (an acronym for "Integrated Circuit Artwork Utility System"). Unlike more complex, computer-aided design systems, ICARUS was "within the reach of almost everybody who has an interest in integrated circuit design." A would-be designer could sit at an Alto keyboard and use "a pointing device called a mouse" (now ubiquitous, these devices were still a novelty in the mid-1970s) in conjunction with screen commands and prompts. This allowed a user to sketch a circuit directly on a computer display. Their files could be saved on the system and sent to a networked printer for inspection or shared over the ARPANET. Fairbairn and Rowson were especially keen to see that new users could "become familiar with ICARUS in just a few hours." In the future, "personal design systems" like ICARUS would lead to a "tight coupling between system designers and integrated circuit layout" and "radically change" the design process itself.[42] By 1978, Rowson and Fairbairn had used ICARUS to design prototype chips with as many as 6,000 transistors.

While ICARUS was being developed at PARC, Carver Mead started the "Our Machine" project (or, as it became known, OM). Mead and a group of Caltech students focused their experimental efforts on the de novo design of a particular kind of microprocessor called a "data path chip," a primary processor in computers at the time. After sketching out chips, Mead used his industry connections to get them fabricated. The project offered Mead and his students a way to test "design possibilities" and obtain "insights into future constraints that will be placed on all large chip designs" as the complexity of circuits continued to increase.[43]

A third path that the Caltech-PARC collaboration explored was how to standardize the data files the group members and external collaborators shared. Ivan Sutherland and Ron Ayres, a Caltech student who would earn a trifecta of degrees from the school, developed a file format called "Caltech Intermediate Form" (CIF). It allowed the interchange of data between, for example, a design program such as ICARUS and a printer or video display. By converting data into a basic text file, CIF enabled the easy sharing of designs with others and it could merge several designs to create a single, larger circuit.[44] What made CIF especially powerful was that it offered users a standard definition of a chip's geometry that wasn't tied to any specific design system or fabrication technology. It simply served, as the name suggested, as an intermediary between the chip designer and a manufacturer. It didn't matter whether the design was a modest-size student project or one of extremely complex circuits that the VLSI team expected to see produced in several years' time.

Together, ICARUS, OM, and CIF provided Conway and Mead's team with essential tools for designing, making, and sharing data about new circuit designs. While this engineering work was happening, Carver Mead and Ivan Sutherland began promoting the new design methodology to a wider audience. In an unclassified report to DARPA, they chided US electronics manufacturers for continuing to think that future computer chips would just simply follow a predictable path of decreasing feature size and increased complexity. This, they said, was "disappointingly conservative." One of their major recommendations was to encourage "efforts aimed at understanding the system design implications of very-large scale integrated circuits."[45] This was, of course, exactly the path that the Caltech-PARC collaboration was already taking. The hope was that program managers at DARPA would get on board and provide funding.

Mead and Sutherland also argued that pedagogical practices in the larger discipline of computer science needed to change. In the pages of *Scientific American*, they described how computer science was too focused on "mathematical reasoning" (recall how Knuth established his reputation by analyzing algorithms while artificial intelligence researchers had long ago drifted toward logical abstraction). Because computer scientists too often ignored their machines' materiality, they didn't appreciate how "physical limitations" constrained the "complexity of the computing tasks" that computers could perform. Part of the problem was historical. The discipline of computer science had emerged when "wires were cheap and switching elements were expensive." By the late 1970s, that situation had reversed and now, as Mead and Sutherland described it, there was a disconnect between theory and practice in the classroom. Mead and Sutherland's article persuasively argued that the design of future chips needed a new approach to circuit layout that was more standardized and rule-driven.[46]

However, the inexorable march of Moore's Law posed challenges to creating a new design culture. While presented by advocates like Mead as an immutable scientific rule—one grounded in the past yet looking perpetually to the future—Moore's Law relied on huge amounts of money and labor.[47] Engineers at scores of companies worked long hours to see that the number of transistors and other components on a typical integrated circuit or microprocessor doubled every few years. "How could," Conway recalled, "such complex rapidly changing geometric layout rules be encoded, applied, and checked given the increase in circuit density anticipated in the coming years?" Another challenge was standardization. Engineers at different companies all had their own proprietary sets of design rules. In his Caltech courses, Mead taught design methods that were useful for making prototype circuits where, unlike in industry, packing components onto a chip as densely as possible wasn't essential. But, as Conway recalled, Mead's rules were "ad-hoc" and "had to be redone as new processes came online."[48]

Conway resolved the dilemma by looking to the longer history of electrification.[49] When Charles Steinmetz started consulting for General Electric in 1892, the established practice of transmitting electricity via direct current was gradually being supplanted by alternating current concepts. Steinmetz realized this process would be accelerated by teachable methods that combined electrical theory and mathematics with system design and practical implementation. Steinmetz developed basic mathematical methods instead relying on the more complicated differential equations or graphical techniques engineers had previously used. Steinmetz popularized his pared-down approach in works such as his influential 1897 textbook *Theory and Calculation of Alternating Current Phenomena*. Steinmetz's simplified methodology transcended specific engineering situations and could instead be applied to an entire range of circuits and machines.[50]

Conway saw parallels between what Steinmetz did and the VLSI team's need for a "coherent but minimalist set of methods." Rather than try to address a future of ever more complex devices, Conway persuaded Mead that they needed a more basic design methodology. "This wasn't about engineering new things," she later explained, "it was about the engineering of new knowledge."[51] With the realization that simplification, not complexity, was the answer, "something magical" happened in the spring of 1977.[52]

Conway's epiphany was to suggest that these minimalist design rules should not be based on specific physical dimensions because these would keep changing à la Moore's Law. Instead, the rules would be built around *ratios* of dimensions using a constant she designated as "lambda" (λ). The beauty of using ratios was that it allowed for the generation of dimensionless design rules, which could be scaled up or down as needed. As a result, circuit patterns could be laid out in a "timeless form" that could later be changed as circuits inexorably became smaller. This also meant that people with only a basic understanding of the physics behind a computer chip could still design one by following a modest set of rules. "Suddenly a clean separation between chip design and fabrication was possible," Conway later wrote, "with extremely simple rules providing the interface."[53] Conway's insights impressed Mead and the Caltech-PARC group quickly adapted her scalable design rules. "We ended up with just two pages of very basic, easy to understand design rules," she recalled, "whereas most chip design processes had about 30 pages with all sorts of arcane stuff in them."[54]

Despite the team's rapid progress, the most pressing question still remained: How could the group efficiently disseminate its new design methodology out into the larger electronics industry? Conway didn't want to "publish bits and pieces" in the pages of academic journals or "scatter fragments into commercial products."[55] Instead, drawing on her familiarity with Thomas Kuhn's work, she recognized that textbooks were an ideal tool for creating a new research paradigm. Her own experience with gender transition offered another data point. She recalled how Harry Benjamin's "paradigm-shifting" book, *The Transsexual Phenomenon*, had "packaged a collection of avant-garde but serious research results" into a "comprehensive, compelling, tutorial textbook that had changed lots of minds."[56] Ultimately, Conway believed the main challenge facing the VLSI team was not scientific, but cultural. In June 1977, she suggested that they present their new design rules in the guise of a textbook. "Mead let out a big 'Yeah!'" Conway remembered. "The decision had been made, and off we went."[57]

The Caltech-PARC team had already acquired lots of technical knowledge over many months as they developed a whole new ensemble of chip-making techniques. But much of this know-how was based on personal experiences and skills that were not yet written down in any sort of formal manner. *Introduction to VLSI Systems* collected this knowledge and presented it in a legible manner for engineering students. Conway hoped their book would reshape the design culture and pedagogical practices around chip making. Like making any tool, Conway and Mead constructed their book in an iterative fashion, adding and dropping elements based on their experience as they went along. As their methods evolved, so did their book's contents and expectations for its success.

Mead and Conway initially planned to self-publish their textbook. Ready access to PARC's Alto computers allowed Conway to draft chapters, draw schematics, and share them as digital files with Mead and other colleagues. Paper copies could be easily created using the latest laser printers connected to Alto machines. And, of course, there were plenty of Xerox-built copiers available. This gave Conway and Mead the ability, perhaps unmatched anywhere else in 1977, to rapidly create and refine the evolving book the group was producing. There was a cultural factor at play as well, one with an interesting parallel to the creation of books. Conway recalled how she wanted chip "architects and designers to be brought out from under the dominance of the fabrication technology and facilities" and "make them 'authors and writers' who got their stuff 'printed in silicon'" while also "gaining visibility for the intellectual work."[58]

The working relationship between Mead and Conway was turbulent at times—they were, one journalist said, "prickly teammates"—and writing a book together exacerbated these tensions.[59] Conway certainly felt some insecurities, especially given her personal history, as an outsider (and a woman) in the world of semiconductor manufacturing whereas Mead was in his element. There was also the question of assigning credit, especially given the potential for securing patents and starting new companies. Conway recalled having an argument with Mead when she discovered his students were attributing the design rules she had invented solely to him. At the same time, Conway willingly accepted a less visible role. "It was actually OK in a way—after all, he was already famous, knew tons of key people and could open any door—while I was less credentialed and had 'a past' that needed to be closely guarded," Conway said. "We were uneasy but incredibly productive collaborators, each playing our respective roles."[60]

By October 1977, Conway and Mead had prepared a draft of what at this point they were calling *Introduction to LSI Systems*. (A few months later, Conway, "in a rush of enthusiasm," changed their title to *Introduction to VLSI Systems*.)[61] It was a concise document—only 118 pages—which Mead would test in a Caltech classroom that autumn. Despite its brevity, Conway and Mead clearly signaled their publication's purpose. Whereas previous textbooks provided too many details about "some very narrow horizontal segment" of the subject, they wanted students to learn a "small set of carefully selected key concepts" with the "least amount of unnecessary mental baggage." After absorbing design methods based on Conway's λ rules, students could then map her scalable concepts onto their "own space of application and technology."[62] In terms of audience, Conway and Mead wanted to reach both electrical engineers and computer scientists. This would reconnect what had, as they saw it, segregated into two different cultures with one group taking courses in "device physics and integrated circuit design" and the other focusing on "digital system architecture and computer science." By thinking of the subject matter as "integrated systems rather than integrated circuits," Mead and Conway thought their book could speak to students from two increasingly specialized and distinct disciplines.[63]

Using the feedback from courses taught at Caltech as well as schools such as Berkeley and Carnegie Mellon, Conway and Mead revised and improved their book draft until, by 1978, it had doubled in size. As the textbook evolved, their plans for publishing it changed as well. In May of 1978, William B. Gruener, an editor of computer science books for Addison-Wesley, offered to publish their book.[64] By this point, the draft included a large number of full-color diagrams essential for teaching Conway and Mead's design methodology. After the publisher agreed to accept the higher-than-usual production costs, the authors signed Gruener's contract.

Up to this point, however, Mead and Conway's "book" had only been used for selected parts of more general electrical engineering courses. An entire course based around the book itself had yet to be taught. Moreover, it was unclear how Mead and Conway's approach would fare in an interdisciplinary classroom setting where engineering students mixed with nascent computer architects and digital system designers.[65] Bert Sutherland, Conway's supervisor at PARC, proposed an experiment. Sutherland, who had connections with MIT's Department of Electrical Engineering and Computer Science, suggested that Conway teach a pilot course there on VLSI design.

The opportunity both thrilled and terrified Conway. A successful offering at an elite school such as MIT would demonstrate to skeptics that the methods she and Mead had created actually worked. At the same time, returning to MIT brought up a host of unpleasant memories for Conway. "Shy among strangers, fearful of public speaking," she recalled, "I lived in dread of being 'outed.'"[66] Despite her fears, Conway accepted Sutherland's challenge. (She later dedicated *Introduction to VLSI Systems* to him.) When Conway arrived in Cambridge, she learned that thirty-two students, including two women, were enrolled in her "Introduction to VLSI Systems" class. (MIT listed the class in its catalog, using that school's system, as 6.978.)[67] Nine MIT faculty members had also signed up to audit the course. While this heightened her anxiety, it also meant that a well-taught course would give the Conway-Mead design methods even greater pedagogical traction.

Conway divided her course into two parts. For the first half of the semester, she lectured on all aspects of VLSI system design, from basic circuit theory and digital logic to computer architecture. But, for the second part—what Conway later called a "techno-socio-political" move—she upped the ante significantly by having her students design actual circuits. Most of the students came to the class thinking they would be learning established design techniques already practiced in Silicon Valley companies. A few of the more perceptive ones began to suspect that they were engaged in some sort of "exotic MIT hack" (in MIT-speak, this referred to a prank, albeit an elevated one) that would target the semiconductor industry. Over several weeks, the MIT students, some working alone and some in teams, began what Conway called their "creative writing" experiment.[68]

Chalkboard and paper cutouts helped Conway and the students keep track of their designs and how they could be configured on a silicon wafer to minimize wasted space. Then, on December 6, 1978, with the students' designs encoded into the CIF format, Conway transferred their designs to PARC via the ARPANET. Colleagues at her lab merged all the student projects—nineteen in all—into a "multi-project chip file" and sent it to a local company that made the requisite masks. The entire package of designs and masks was then transferred to a Hewlett-Packard laboratory that fabricated the students' projects.

Six weeks later, Conway and her students opened up a package that contained their chips. While some had some fixable glitches and wiring bugs, most of the student projects worked straight out of the box. So far as Conway was concerned, the whole project was an unqualified success. Where it had once taken teams of designers and engineers many months and tens of thousands of dollars to design and make a new chip, the MIT group devised (admittedly simpler) chips using a fraction of those resources. With students' projects having validated VLSI design methods, the Conway-Mead textbook, and her pedagogical approach, Conway headed back to PARC in February 1979, sensing that "something profound had happened."[69]

Once back in California, Conway worked to complete the final draft of *Introduction to VLSI* as Addison-Wesley moved the book toward publication. The company's goal was to have it available by autumn so it could be adopted for other university courses. Carver Mead had already primed the pump by informally sharing draft copies with colleagues in the United States and overseas (interest in Japan was especially strong). A sense of anticipation started to build among both university professors and industry engineers who regularly wrote Mead for updates.[70]

Not all readers returned positive evaluations, however. One anonymous reader who reviewed the manuscript for Addison-Wesley praised the textbook's appearance with its "polished graphs, multi-font layout, and other eye-appealing tricks." But the compliments stopped when it came to the book's prose. Besides Conway and Mead, several other colleagues had contributed sections to the book. As a result, the reader found the result "disjointed" and "poorly integrated" (an ironic comment given the subject) with "varying levels of writing skill" on display. Addison-Wesley, they suggested, should hire a skilled copyeditor to remove the "wretched grammar and hackneyed phrases." And, although endorsing publication, the anonymous reader did not envision "much of a textbook market for the book" with Addison-Wesley likely to sell only "a few hundred [copies] per year" to a limited audience of corporate labs and engineers.[71]

With Conway and Mead making last-minute edits and revisions, Addison-Wesley shifted to promoting *Introduction to VLSI Systems*. Conway agreed that Mead (a "well-connected full professor" and known figure in the semiconductor field) should get first-author billing to "enhance the book's credibility."[72] When the publisher asked Mead about other competing books, the Caltech professor was emphatic about the book's novelty. "There are no other books on VLSI systems," he wrote. "It is a totally new field."[73] (Norbert Wiener had said the same thing about cybernetics three decades earlier.)

Addison-Wesley took advantage of this information when it featured Conway and Mead's volume in its "Books About Computers" catalog.[74] An accompanying two-page advertising flyer pitched VLSI as the "new technology too important to ignore" and one that would, within five years, "revolutionize the computer industry." Instead of trying to keep up with new developments by "reading every computer- and electronics-related journal," Addison-Wesley offered (for just $25.95) to give readers "all the design information about VLSI in one book, written by two pioneers in the field." "Do it today," the publisher encouraged, "and invest in the future."[75]

Conway and Mead had made their book well-suited to do the job they had in mind. Leafing through the pages of *Introduction to VLSI*, one immediately notices the large number of color illustrations, some fifteen full-page plates in all. Expensive to produce and publish, these were not included for aesthetics but rather to fulfill a pedagogical purpose. The types of circuits Conway and Mead used as examples were composed of several layers built up on a silicon wafer. The use of color allowed them to highlight these layers and how they would interact with other as they overlapped. "Colors meant something," Conway recalled. "Students using the system could look at a figure and tell right away what it did. It was mathematical abstraction reduced to a diagram."[76] Later, once conferences and workshops on VLSI chip design started to take place, the use of similar color diagrams served to identify who was part of the "VLSI movement," as Conway called it.

The marketing, word-of-mouth buzz, and the book's contents all paid off. Addison-Wesley described the response to *Introduction to VLSI Systems* as "overwhelming" and, by April 1980, only a few months after it appeared, the book got a second print run. Eventually, over 70,000 copies sold with editions translated into Japanese, Italian, and French (there was also an unauthorized edition in Russian produced by the Soviet government).[77] At universities and companies, a growing community of engineers and computer scientists relied on Conway and Mead's book to reenvision the design of computer chips as well as the entire way the industry was organized.

One of those people was Morris Chang. For years, Chang, who had been born in China in 1931, had worked as an electrical engineer and business executive at Texas Instruments. In the late 1970s, the company shifted from making semiconductor devices to churning out inexpensive products (e.g., digital watches and calculators). Chang, who been with the company since 1958, found his career at an impasse. However, reading *Introduction to VLSI Systems* convinced him that the future of electronics would be based on designing chips in one place and then having them made someplace else. Chang soon left the United States and, in 1987, founded Taiwan Semiconductor Manufacturing Company (TSMC).

Over the next three decades, TSMC produced tens of billions of chips as it followed the model Mead and Conway promoted. The company's success also transformed Taiwan into a country with extraordinary geopolitical importance. When interviewed in 2023 at his Taipei office, the ninety-two-year-old Chang "pulled out an old book stamped with technicolor patterns" and "held it up with reverence." *Introduction to VLSI* was, he recalled, the catalyst that altered not just his career but the entire industry.[78] A tool made of words, equations, and diagrams, Conway and Mead's book did not move the world, but certainly provided a lever to nudge it a bit.

## SOUND METHODS

One of the most unsettling scenes in Francis Ford Coppola's now-classic 1979 film *Apocalypse Now* is the confrontation between Colonel Kurtz, a renegade army officer, and Captain Willard, who has been sent to assassinate him. Asked why he is there, Willard replies, "They told me you had gone totally insane and that your methods were unsound." Partly hidden in shadows, Kurtz asks, "Are my methods unsound?" Choosing his words carefully, Willard answers, "I don't see any method at all, sir."

When Lynn Conway returned to PARC, she learned that some of her colleagues regarded her methods likewise. "Rumor was," she later wrote, "that somebody named Conway had gone off the reservation, slipped up the river into Cambodia, and was spreading 'unsound methods.'"[79] In papers she authored circa 1980, she speaks of "taking methods that are new and perhaps considered *unsound methods*, and turning them into *sound methods*."[80]

What she later described as "pushback" came from two directions. One was the culture of PARC itself. Conway was a member of Bert Sutherland's Systems Science Lab. This group was often at loggerheads with another division, the Computer Science Lab (CSL). What might have been productive rivalry between ambitious and technically competent people sometimes degenerated into competition, arrogance, and hostility. The sense that the resources Xerox provided to PARC amounted to a zero-sum game exacerbated the tense situation. "I didn't like what Lynn Conway's group was doing," one CSL scientist remarked, "and I didn't think it was very productive."[81] The kind of circuits Mead and Conway were developing using their design methodology also proved divisive. Conway's $\lambda$ scaling rules were designed to be simple and elegant. Consequently, the integrated circuits they yielded seemed rudimentary compared to the complexity of commercially developed chips. "Critics in the competing Computer Science Lab," Conway noted, "looked askance at what they saw as our 'toy' design tools. Not surprisingly, they questioned what our tiny effort could possibly bring to the huge semiconductor industry."[82] While Conway's MIT course had been successful, it was only one data point. She decided something bolder needed to be done to win over (or discredit) VLSI skeptics.

Conway proposed scaling up what she had done at MIT several-fold by taking her VLSI course to several schools simultaneously. A key to doing this was Conway's easy access to the ARPANET via the Alto computer in her office. This affordance would allow her to communicate with distant collaborators and transfer design files back and forth. Conway emailed a plan to colleagues at ARPANET-connected universities. Her idea was simple—they would collectively offer a VLSI design course using materials honed in Conway's MIT course. As a finale, students could send their design files over the ARPANET to PARC, which would see that they were fabricated into actual chips. Conway called the experiment "MPC79" (for "multi-project chip") and twelve schools joined the initiative.

Unfortunately, Conway's ambitions ended her collaboration with Carver Mead. She had generally deferred to the more voluble Mead when it came to promoting their VLSI design ideas. When Mead learned of Conway's plans for MPC79, he expressed "intense disagreement." Because the effort would rely so much on the ARPANET, he worried that the Pentagon-based agency that had built the network would take credit if the effort succeeded. Conway's plan to have PARC coordinate multiple design projects also conflicted with Mead's advocacy for small, independent silicon foundries. His vision rested on companies and schools making their own arrangements for masking and fab services with different foundries. Conway disagreed, arguing that the foundry model was not yet well developed and that it relied too much on "undocumented personal experience" and hence wasn't yet scalable. "Uneasy collaborators from the start," she recalled, "these sharp differences pretty much ended our interactions."[83]

Conway ran the MPC79 project on a tight schedule. Using *Introduction to VLSI*, students quickly learned the basics of transistor theory, solid state physics, and circuit engineering before turning their attention to a design project. By early December 1979, 125 student groups had electronically transferred eighty-two different circuit designs to Conway's colleagues at PARC. Less than a month later, the custom-made chips were mailed back to their respective designers for inspection and testing. Conway repeated the program the following spring (as "MPC580") and, this time, 250 people from fifteen universities submitted 171 design projects. Taken together, these scaled-up experiments demonstrated to naysayers that Conway's methods were indeed sound.[84]

Some chip designs were especially bold. For example, a group led by Jim Clark, a professor of electrical engineering at Stanford, had built a high-performance chip prototype called the "Geometry Engine." Clark was especially interested in taking geometric models and rendering them in real time as two- and three-dimensional images on a computer screen. Until the advent of Conway and Mead's design methodology, however, it was very expensive to design and fabricate custom microprocessors for computer graphics.[85] After participating in the MPC79 effort, Clark and a small team of Stanford colleagues founded Silicon Graphics Inc. and developed a series of graphics-oriented computer workstations, making Clark extraordinarily wealthy in the process.

The Conway-Mead design methodology opened up other commercial possibilities as well. In 1980, Douglas Fairbairn and James Rowson—they had designed the ICARUS system together—unveiled a new publication called *LAMBA*. Some skeptics questioned the wisdom of adding yet another electronics publication to an already saturated market. But since integrated circuit design "is actively practiced by a large number of people in diverse companies and universities," they insisted a magazine "devoted strictly to their needs is required." The first issue was slight in size but, because Fairbairn had access to the latest in Xerox printing technology, it featured slick-looking graphics produced on thick, expensive paper.[86] *LAMBA* initially published articles from colleagues active in the VLSI field but the magazine's topics soon expanded along with its page count and circulation. In 1982, Fairbairn sold the magazine—now renamed *VLSI Design*—to CMP Publications for $650,000. By this point, Fairbairn had left PARC and become involved with Rowson in another successful commercial effort, VLSI Technology Inc. Designing and making microchips translated into megabucks. By the mid-1980s, their company was posting revenues over $100 million before it was acquired in 1999 by Phillips Electronics for some $1 billion.[87]

Whatever doubts detractors might still have harbored about the soundness of Conway and Mead's methods were quashed in 1981. *Electronics*, a venerable industry magazine, gave the two engineers its annual Award for Achievement. The magazine praised Conway and Mead for "structuring the methodology of the design of very large scale, integrated circuits" and encapsulating that knowledge in a "truly monumental" textbook. By this point, over 100 universities in the United States, Canada, and overseas were offering courses based on the Conway-Mead design methodology. The article ignored Conway's personal history and presented the two authors as equal contributors. In the end, *Electronics* declared, any discussion about the success of their methods "comes back to the book."[88]

More accolades followed for Conway. Xerox made her the manager of the "VLSI Systems Area" before promoting her again to Research Fellow. In 1983, Conway left PARC for a management position in DARPA's lavishly funded Strategic Computing Initiative. By this point, spurred by the success of the Conway-Mead methods, DARPA was heavily invested in spreading VLSI research using methods modeled on Conway's teaching experiments.[89] After staying at DARPA for a few years, Conway took a faculty position at the University of Michigan and was elected to the National Academy of Engineering.

After retiring, Conway interpreted her VLSI work as "meta-level exploration in 'applied anthropology'" where she was trying to understand and change the culture around chip design.[90] Some of her insights also came from conversations she had with Lucy Suchman at PARC. As the anthropology student and the engineer bantered ideas about, Conway repeatedly referred to the creation of a new "shared culture" among chip designers. When students designed their prototype chips using Conway and Mead's approach, they were, she explained, participating in "rituals" or "initiation experiences." In the process, they became members of a new "insider clan."

This membership came with a certain materiality as well. When teams of students received their fabricated chips, they would sometimes hang images of them on laboratory walls. "If you walked around EECS [electrical engineering and computer science] departments," Conway noted, "you could almost spot it from the end of a corridor. That would be where these VLSI gangs had got some territory." But the focal point for this clan remained *Introduction to VLSI Systems*. Conway, who once described herself as "a weed growing in a very, very tailored garden," told Suchman, that "people talk about 'The Book.' And if you don't know what book they're talking about you're not in the clan." The anthropologist agreed. "Membership," Suchman said, "is no small thing."[91]

As the year 1984 approached, journalists and news anchors displayed an increasing interest in George Orwell and his book *Nineteen Eighty-Four*. When Walter Cronkite hosted a CBS special that revisited the dystopian novel, the famed newscaster discussed, among other topics, the use and misuse of computers. In one scene, the editor of the *Encyclopedia Britannica* showed Cronkite how to delete a digital file on Orwell and render the deceased author a "nonperson."[1]

Steve Jobs, the twenty-eight-year-old chairman of Apple Computer Inc., also noticed the swell of interest in "Orwellian" futures. In October 1983, Jobs gave a pep talk to Apple salespeople gathered on Oahu, Hawaii. In Jobs's telling, IBM had serially and cynically neglected consumers until, in late 1981, it finally introduced its first personal computer model. Jobs warned his employees that IBM was a corporate bully with neither vision nor courage. And now the company sought to "dominate the entire computer industry" and thereby monopolize the "information age."

However, sales figures certainly didn't support Jobs's claim that the Apple II, first introduced in 1977, was the "world's most popular computer." Both Atari and Commodore easily bested Apple in terms of consumer purchases. And there was also Jobs's continued insistence that Apple had brought the personal computer to the masses. Nonetheless, when Apple's chairman asked if George Orwell was "right about 1984," hundreds of Apple employees shouted "No!"[2] Apple's charismatic leader then played a short video for them that introduced the company's newest product: the Macintosh personal computer.

Three months later, some seventy million television viewers watching Super Bowl XVIII saw the video that Jobs had recently previewed. It featured a legion of passive, gray-clad drones and a menacing figure on a telescreen who spewed corporate nonsense at them (e.g., "the first glorious anniversary of the Information Purification Directives"). An athletic blonde woman (played by Anya Major), clad in bright red shorts and a white top, bursts into the dismal forum. Eluding the truncheons wielded by visored security forces, she heaves a sledgehammer at the screen. As it explodes in a shower of debris, a voiceover intones: "On January 24th, Apple Computer will introduce Macintosh. And you'll see why 1984 won't be like *Nineteen Eighty-Four*." Advertising executives and marketing experts praised the one-minute advertisement, which aired only once on national television, as the "greatest commercial ever made."[3] MIT's Joseph Weizenbaum, not surprisingly, saw the advertisement quite differently. "The notion that a personal computer will set you free is appalling," he said.[4]

Apple's now-classic advertisement was a prominent signal in a swarm of social, cultural, and economic messages about the state of computing circa 1984. Taken together, this evidence showed just how rapidly both computing technologies and Silicon Valley had moved to the forefront of public awareness. Even a glance at simple metrics, such as the number of times "Silicon Valley" and "personal computer" appeared in published books, points to a significant inflection point occurring in the early 1980s. Part of this interest can be attributed to the scores of publishers who produced a rapidly expanding assortment of books and magazines that introduced readers to Silicon Valley and the "high-tech" (another phrase that joined the vernacular in this era) companies based there.

In 1984, there were more than 250 computer book publishing imprints in the English language alone. The sixteen biggest firms—including companies such as Addison-Wesley, Simon & Schuster, and McGraw-Hill—put out over 1,100 new titles in that year alone. Many were textbooks or specialty books written for computer professionals, but some were trade books for general readers. All told, computer books amounted to a $270 million market, making up about 3 percent of all books sales in the United States. And if computer-oriented magazines and newspapers were considered, publishing revenues easily topped $600 million.[5]

Instead of focusing on a single book or specific author, this chapter is about a particular time—1984—when there was a marked change in both the writing and selling of books about computers and computing. Part of this story is also about the emergence of a new kind of author: the "technology reporter" or "tech journalist." By the mid-1980s, a growing cohort of technology-oriented writers had become essential conduits for generating awareness about Silicon Valley, its companies, and their products for an audience of technology enthusiasts, computer consumers, businesspeople, and policy makers. Taken together, their output stoked public recognition not just of Silicon Valley but a host of social and economic issues the region is still grappling with more than four decades later.

However, readers who wanted to access this abundance of information faced some challenges. Big bookstore chains such as Waldenbooks and B. Dalton tended to have relatively small selections of computing books compared to the actual number of available titles. To address growing consumer interest, a new store, called the Computer Literacy Bookshop, opened for business in 1983 in Silicon Valley. Although it only sold computer books, it soon boasted an extensive inventory of titles with some 60,000 books stocked and available for purchase.[6] The formula worked fantastically well and that first store evolved into four venues as customers made purchases in-person, by mail or phone order, and, eventually, online.

Although the reality of 1984 didn't resemble what George Orwell (or Steve Jobs) predicted, the early 1980s proved a volatile time for both computer users and the computer industry. Rapid technological changes coupled with increased competition from Japan and the worst economic recession in a half century catalyzed boom-and-bust cycles for the electronics industry. And, as the decade progressed, more and more computer users were exploring "cyberspace," a term popularized by science fiction writer William Gibson via his 1984 novel *Neuromancer*. These early adopters read texts and interacted with one another online via computer screens and dial-up modems. When it came to learning about computers and computing, words printed on a paper page still very much mattered. Meanwhile, the computer itself started to gradually fade into the background.

## TIME AFTER TIME

When *Time* magazine first featured a computer on its cover in 1950, Russian American illustrator Boris Artzybasheff drew a military-looking machine monitoring its own input and output. No people were in his illustration. Fifteen years later, *Time*'s editors placed a computer on the cover and, again, Artzybasheff did the artwork. This computer—a six-armed machine, greedily consuming punch cards—was now surrounded by people. They were neither scientists nor military officers but ordinary looking office workers. A woman served up a platter of data cards, for instance, while her coworker gestured at a flowchart. No vacuum tubes or messy tangles of wires were visible; the computer was now just a streamlined set of boxes. In both cases, Artzybasheff's anthropomorphic cover illustrations complemented lengthy articles that explained "thinking machines" and "electric brains" for *Time*'s readers.

In the 1980s, over four million people in the United States alone consumed *Time*'s tightly worded articles to get a middlebrow perspective on national and international news along with specialty pieces covering developments in business, culture, and the arts. But before 1980, computers had appeared on the cover of *Time* only a few times. Each of these issues typically focused on the machines themselves: what they did, how they worked, and their predicted "impact" on society (the metaphor of violent collisions between people and machines had long been a standard journalistic trope).

Around 1982, this pattern started to change. In a single year, *Time*'s editors placed computing on three different covers with accompanying feature sections. While the surge of attention from the nation's largest news magazine was itself notable, these issues saw a marked shift in reportorial focus. Instead of computers and their workings, people—specifically, computer users and corporate entrepreneurs—had become the main subjects. Consider the issue from February 15, 1982. The issue's cover featured "Steven Jobs of Apple Computer" with text that blared "Striking It Rich." A zig-zagging arrow zapped forth from a stylized Apple II computer to pierce the red fruit the illustrator had placed on Jobs's head. (This was the first of eight times Jobs appeared on a *Time* cover). With a 23 percent share of the personal computer market—equal to that of rival Radio Shack—*Time* revealed that Apple was about to release a new machine that "already has the computer world abuzz."[7]

An even lengthier piece in the same issue focused on the new cohort of "bold and brassy risk takers" who were "betting on the high-technology future" and "leading the US into the industries of the 21st century." Even as these entrepreneurs' gambles produced handsome dividends for themselves and the venture capitalists who backed them, the main beneficiaries were particular geographic regions. Chief among them was California's Santa Clara County. While some reporters had been referring to the area as "Silicon Valley" for several years, enough people remained unfamiliar with the term that *Time*'s editors still placed it in quotes.[8] This relatively small place was now home to more than 750 electronics firms that, in toto, employed tens of thousands of workers and produced nearly $9 billion worth of goods.

Much of *Time*'s reporting on computing from this period came from the keyboard of Michael J. Moritz. Born in 1954 in Wales, Moritz studied history at Oxford University before moving to Philadelphia for graduate work in business at the University of Pennsylvania's Wharton School.[9] A voracious consumer of newspapers and BBC broadcasts as a teen, Moritz realized he was more interested in journalism than corporate management. A letter of inquiry to *Time*'s bureau chief in London landed him a posting in Detroit as a correspondent.

Covering the automotive industry's ups and downs (mostly the latter) prompted Moritz to see Detroit as a "terrifying example of what happens when innovation dwindles."[10] In 1981, Doubleday published his first book (coauthored with *Time* reporter Barrett Seaman) on Chrysler's near-bankruptcy, smartly titled *Going for Broke*.[11] By the time it appeared, however, Moritz had left Detroit for *Time*'s San Francisco office where he started writing about the region's technology-oriented companies. Having spent time with Lee Iacocca, Chrysler's charismatic CEO, Moritz saw something similar in Steve Jobs's abilities to sell a product to consumers and a vision to investors. Moritz decided that writing about Apple's rise would make a compelling new book project. "Jobs was obviously a large personality," Moritz noted, "and that made it easier to write by telling the story of a company's evolution via the people associated with it."[12]

Moritz started interviewing Apple employees and, with Jobs's approval, attending company meetings. This offer of behind-the-scene access was not altruistic but rather motivated by the appearance of another book. In July 1981, Tracy Kidder's *The Soul of a New Machine* had appeared on best seller lists.[13] Now regarded as a masterpiece of narrative nonfiction, Kidder's book provided an evocative and bittersweet account of life inside Data General Corporation, a scrappy Massachusetts-based firm. Kidder recounted how the company's engineers poured their time and energy (giving the book its title) into creating a new minicomputer. *Soul* also laid bare the vicious internal competition among Data General's managers over which new design the company would bet its fortunes on.

Before *Soul* won both a Pulitzer Prize and a National Book Award, its author was a struggling freelance writer for the *Atlantic* with only one other book (and this not terribly well received) to his credit. Richard Todd, his editor, suggested that Kidder write something about computers and told him to contact J. Thomas West, a computer engineer for Data General that Todd knew from college. Kidder was initially hesitant ("I was afraid of science and math. . . . The prospect of looking into computers seemed daunting and drab") but, lacking better ideas, accepted the suggestion.[14] Three years later, Kidder had finished a manuscript for what became *Soul of a New Machine* and West had become the book's main protagonist.

Ironically, given the author's initial doubts, book critics praised Kidder's ability to convey the details of coding, digital logic, and Boolean algebra. Kidder combined this vividness with compelling vignettes of West's managerial machinations and the working lives of Data General's young engineers. By the book's end, when Data General finally unveiled its new computer, the machine "no longer belonged to its makers." Despite the bittersweet ending, engineers and programmers appreciated Kidder's depiction of them as "eccentric knights errant" who solved difficult problems in the face of managerial pressure.[15] Apple's cofounder Steve Wozniak was one of the people captivated by Kidder's depiction of engineers seeking self-fulfillment through their labors. "I was very sad at the end," Wozniak recalled (perhaps reflecting on his often fraught relationship with the publicity savvy Jobs), because the "engineers were not better-respected and marketing folks were credited with the computer."[16]

Steve Jobs liked how Kidder's book captured the risk, notoriety, and fame that came with building an innovative new computer. He also saw obvious parallels to what was happening inside Apple. Part of his company was focused on bringing a new personal computer, called the Lisa, to market. But after management booted Jobs from that project, he commandeered another Apple group that was working on a competing machine, the Macintosh. "That was what he wanted," Moritz recalled, "a book about the Macintosh that was like *Soul of a New Machine*."[17] While Jobs wanted the reporter to lionize his flair for innovation, Moritz had a different goal. Where his first book had explored the workings of Chrysler, a decades-old car company, Moritz now wanted to describe "how you go from a little garage to selling your first machine and then hundreds of thousands of them."[18]

As he did his research, Moritz still continued his reporting for *Time*. "It was there," he recalled, "that my troubles began."[19] As 1982 drew to a close, Steve Jobs, despite his recent appearance on a *Time* cover, anticipated being named the magazine's "Man of the Year."[20] He continued to encourage Apple employees to cooperate with Moritz. But this support came to a crashing halt in late December 1982 when *Time* snubbed Jobs and instead named the personal computer itself as "Machine of the Year."

If Jobs was annoyed to be passed over, *Time*'s profile of him in the "Machine of the Year" issue left Apple's chairman incensed. Besides chronicling his now-infamous management style—one employee opined that his boss would have made "an excellent King of France"—the article, titled "The Updated Book of Jobs," revealed that the wealthy entrepreneur refused to pay child support to his ex-partner. The published article was officially credited to Jay Cocks, a writer for *Time* who typically covered popular culture. But, as the article's byline made clear, it also included some information from Moritz. However, *Time*'s New York offices had "siphoned, filtered, and poisoned [it] with a gossipy benzene" to produce what Moritz called a "grotesque caricature" of Jobs. Years later, Jobs confessed that the "hatchet job," which he still blamed on Moritz, was "so awful that I actually cried."[21] Not surprisingly, after the "Machine of the Year" issue appeared, Jobs banished Moritz from Apple.

However, Moritz still had a book to write. Having lost his access to human sources, Moritz had to rethink his project. One task almost all nonfiction authors invariably face is deciding what their book is fundamentally about and what is doable with the materials they have. In the introduction to what became his 1984 book *The Little Kingdom*, Moritz noted that he had initially enjoyed a "carefully circumscribed freedom" at Apple but "the company I saw in 1982 was very different from the little business that filled a garage in 1977." Given the challenges of writing about a rapidly changing company in the midst of equally fast paced technological change, Moritz decided that he would instead focus on the past and narrate "Apple's road to its first one billion dollars."[22]

Moritz organized his book around chapters that explored Apple's founding while also offering insights into recent business and design decisions. For his book's tone, Moritz found inspiration in Joan Didion's and John Gregory Dunne's take on the film industry.[23] In his 1969 book, *The Studio*, Dunne used observations collected in studio lots and conference rooms to create an account of life at 20th Century Fox where "dreams are converted to stock dividends."[24] Didion conveyed her own surreal experiences writing for the film business via characteristically detached reflections about deals, big payoffs, and the industry's financial logic (or lack thereof). Hollywood, she wrote, "makes everyone a gambler" but it remained a place where acquiring money is never "the true point of the exercise."[25] One can imagine how Dunne's and Didion's observations resonated with Moritz as he watched Jobs bet both his reputation and his company on the Macintosh computer.

Moritz secluded himself in a rented cabin near Lake Tahoe and hammered out his book. "It wasn't a happy experience," he recalled. "You are writing about a company where you have been considered persona non grata. If you are writing nonfiction, you are at the mercy of people you don't control."[26] Moritz secured a contract with William Morrow and Company ("they paid the highest price, which was small") but had to persuade his editors that Apple would still be in business when the book appeared. The title of Moritz's book—*The Little Kingdom*—was inspired by Antoine de Saint-Exupéry's 1943 classic *The Little Prince*. Meanwhile, the subtitle, *A Private History of Apple Computer*, allowed the author to declare some independence from his subject.

Unlike *Time*'s "unfortunate magazine article," Moritz believed that his book offered a "balanced portrait" of both Jobs and Apple.[27] Nonetheless, despite an endorsement from Chrysler's Lee Iacocca, *The Little Kingdom* did not secure the acclaim that *Soul of a New Machine* had received. A brief review toward the end of the *New York Times*' book section questioned the author's decision—following Kidder's approach, to a degree—to interrupt a "compelling narrative with vignettes of present-day Apple executives at work." A few other mentions appeared in other venues but, as Moritz recalled, "quiet broke out all around."[28]

Largely a tale of how Jobs and Wozniak founded and grew Apple, by the closing pages of *The Little Kingdom*, a sense of decadence had crept into Moritz's narrative. Apple's initial public stock offering had made Jobs and Wozniak fabulously wealthy but perhaps not very happy. Moritz recounted how Wozniak lavished and lost millions of dollars on an ill-advised "Disneyland version of Woodstock" called the US Festival.[29] Jobs, meanwhile, continued his relentless campaign of pitting the Macintosh against two other Apple products as well as IBM's personal computer. One of the final vignettes Moritz included places Jobs at a Stanford dormitory where he assumed a lotus position before speaking to a coterie of undergraduates. Questions from students ranged from the financial (when would Apple's stock rise?) to the imperial (what was it like to run a corporate empire?). As the event drew to a close, Moritz overheard one young woman say, "Well, at least he's not a jerk."[30] Even so, *The Little Kingdom*'s narrative ended with Jobs, having regained power, laying off scores of Apple employees.

In the end, banishment from Apple was not the career reversal for Moritz that it might have seemed. While doing research for *The Little Kingdom*, Moritz met Don Valentine, an entrepreneur who was one of the original investors in Apple. In 1986, Moritz joined his firm, Sequoia Capital.[31] Jump ahead three decades and Moritz had been knighted by Queen Elizabeth II and was a billionaire. Although no longer a writer, in 2019 a charitable foundation Moritz operated with his wife Harriet Heyman became the prime patron for the Booker Prize, a prestigious award for literary fiction. When the BBC asked Moritz about their philanthropy, he responded, "Neither of us can imagine a day where we don't spend time reading a book."[32]

### "HELLO, WORLD!"

When students learn computer programming, a standard exercise is sometimes to write a short snippet of code that, when done properly, displays the message "Hello, World!" Students doing this using the BASIC language in the 1970s, for example, would enter `10 PRINT "Hello, World!"` while their counterparts learning Python a few decades later would type `print ("Hello, World!")`. Starting in the 1970s, journalists performed an analogous function by transmitting an increasingly powerful "Hello, World!" message that introduced people to Silicon Valley.

At the time, "technology reporter" was not a readily apparent career for aspiring journalists. But, by the time the Soviet Union dissolved in 1991, there were hundreds of writers, working for scores of magazines and newspapers, filing stories every day about what was happening in "technology." Throughout the 1990s, this once broad term was truncated until it came to mean "information technologies" or, narrower still, "information technologies developed in Silicon Valley."

The surge in journalistic coverage devoted to Silicon Valley produced two effects. What started as a trickle of reportage became a veritable flood of articles and books that fostered wider public awareness about the region. Closely coupled to this phenomenon was a second result. As newspapers and magazines devoted more attention to Silicon Valley, a rapidly expanding cohort of writers and reporters provided specialized coverage of computing, electronics, and other information technologies. By the end of the 1980s, "tech journalism" had become an established subgenre, its growth amplified by the appearance of dozens of magazines, including *InfoWorld* (1978), *PC World* (1983), and *Macworld* (1984), whose publishers needed plenty of content to complement the hundreds of profitable advertisements found in each issue.

Reporters had actually been filing stories about what used to be called the Valley of Heart's Delight—the moniker given to Santa Clara County when agriculture, not electronics, was the region's dominant "tech sector"— for some time. More often or not, however, these journalists came to their subject through serendipitous paths. For example, Donald C. Hoefler studied electrical engineering at New York University before working in the radio broadcasting business and writing a series of technical manuals and books.[33] In 1960, Hoefler relocated to Palo Alto and started producing publicity copy for Fairchild Semiconductor before writing articles for *Electronic News*. The frontpage of the January 11, 1971, issue announced Hoefler's interests with the now-famous term he coined: "Silicon Valley, USA." Interestingly, Hoefler didn't use his own phrase anywhere in the actual article. Nonetheless, it was Hoefler, a reporter for a trade newspaper, not a regional booster or businessperson, who introduced the term into the vernacular.[34] In 1975, Hoefler started a newsletter he called *Microelectronics News*. Each gossipy issue—one writer gleefully called the publication "cruel, petty, occasionally libelous"—tracked trends, finances, and personnel shifts in the industry, interspersed with Hoefler's own colorful opinions. When he died in 1986, Hoefler's obituary noted he was "liked by some, disliked by many, and read by all."[35]

Hoefler's approach to business writing skewed orthogonal to much of the genre. Before the 1970s, newspaper editors tended to favor stories that informed readers about important events in public life. Reporters disinclined to question statements made by the people they wrote about or scrutinize their motivations or beliefs.[36] The result was a narrow definition of what was "newsworthy" compared to what would become the norm in the era marked by Watergate and the Vietnam War. By the 1970s, journalistic values were markedly different compared to just a few decades earlier. Instead of a recounting of facts and quotes, readers had come to expect interpretation and analysis. Meanwhile, journalists exchanged deference to political and business leaders for a more adversarial and skeptical approach when dealing with both their subjects and sources.[37]

These changes were especially notable in business journalism. Prior to the 1970s, a newspaper's business section—if it even had one—was limited

to earnings statements, changes in the financial markets, and laudatory articles on business executives and their companies' new products.[38] During the
1970s, this dynamic started to change as articles became critical of business
to the point of antagonism. Arthur O. Sulzberger, the publisher of the *New
York Times*, noted in 1977 that business coverage was now "more analytical,
more skeptical" with reporters aggressively asking "how?" and "why?" rather
than repeating information provided by a company's public relations staff.[39]

Sulzberger made his remarks as business journalism in the United States
was starting to dramatically expand in scale and scope. For example, in May
1978, the *New York Times* introduced a new daily section that gave in-depth
profiles of corporations, analyses of business strategies, and detailed coverage of markets and other economic news. The best examples of business
journalism exhibited what one expert called "interrelatedness" as reporters
looked beyond the financial aspects of a story to bring "economic, eco-
logical, technological, and personal values" to their writing.[40] Other major
daily newspapers followed suit while new magazine ventures, such as *Money*
(launched by Time Inc. in 1972), joined established venues like *Fortune*,
*Forbes*, and *Business Week*.

This expansion produced a surge of writers covering the business beat.
Some estimates suggest that the number of business journalists tripled after
the 1970s to some 12,000 people.[41] Many members of this new group
arrived at newsroom desks with higher levels of education and journalistic
ambitions. By 1980, reporters' attitude toward business was no longer one
of "automatic hostility but of sophisticated curiosity." As more "dispassion-
ate pragmatists" replaced the "young soldiers of the turbulent sixties," their
stories relied more on carefully cultivated sources rather than company press
releases.[42] Starting around 1980, some of these journalists started to focus
their attention on the companies and workers concentrated in what Hoefler
had successfully branded as Silicon Valley.

Evelyn Richards was part of this cohort of reporters, formally trained
in the craft of journalism, who made Silicon Valley their distinct beat. Born
in 1952, Richards grew up in Waukegan, a small city north of Chicago. A
self-described "journalism junkie," Richards watched the *Huntley-Brinkley
Report* every night with her parents and started a newspaper called the *Blue*

*Streak* that she distributed around her neighborhood.[43] As a teen, Richards and her family moved to the Palo Alto area before she returned to Chicago and enrolled at Northwestern University. She took both a bachelor's (1974) and a master's degree (1975) there from the Medill School of Journalism. Unlike some of her contemporaries, Richards wasn't drawn to the so-called new journalism pioneered by writers such as Hunter S. Thompson and Tom Wolfe, and instead preferred reading business magazines and newspapers. While at Northwestern, Richards combined advanced mathematics classes with specialty training in business and economic reporting.

After just a few months of working for a small-town Ohio newspaper, where she covered the city hall beat, Richards navigated a poor job market to land what seemed a better position at the *Palo Alto Times*, a Bay Area daily. "It was a horrible experience, the opposite of Ohio, which was actually a real down and dirty newsroom," Richards recalled, "I was put on the night beat with school board meetings and so on. It was pretty depressing." In 1979, her luck changed when the Chicago-based Tribune Company bought her paper and merged it with the *Redwood City Tribune* (another local paper) to create the *Peninsula Times Tribune*. Aware of Richards's expertise in math and economics, the new management encouraged her to take up business reporting. Silicon Valley became one of her specialties. Besides writing a thrice-weekly column, she was made the business editor and supervised a small staff charged with covering an increasingly expansive topic.

Two years later, the *San Jose Mercury News* hired Richards away from the *Tribune*.[44] Her move to what locals referred to as the *Merc* was, for a business- and technology-oriented reporter, like being traded to a major league baseball team. Once seen as a "profoundly mediocre" local paper, by the 1980s, the *Merc* had become the de facto venue for journalism about Silicon Valley. Throughout the 1980s and 1990s, the *Merc*'s circulation numbers paralleled the massive growth of the computer and electronic industries. The *Merc*'s profit margins—30 percent or higher in some years—was driven by classified advertising, especially employment listings for positions in regional high-tech firms. The reporting staff grew likewise, approaching 400 people in the San Jose newsroom alone. Other journalists worked at bureaus in Seattle, New York, Washington, and Tokyo, as some electronics makers shifted their production to Southeast Asia.[45] As Richards recalled, the *Merc* aimed to be the main international source for news and information about Silicon Valley. "All the Wall Street analysts who worked with technology had subscriptions," she recalled. "The paper just went crazy, so successful and rolling in money. The staff grew and grew until we had something like forty people by the late 1990s working on business topics alone."[46]

For her first few years at the *Merc*, Richards wrote about the fluctuations of Silicon Valley companies while also doing occasional freelancing for the *New York Times*. "The early 1980s were definitely an exciting time," Richards said, "but it could also be brutal, especially given the horrible economic contractions with companies like Atari laying off 1,700 workers in 1983." The *Merc* was the first major daily to have a dedicated computer section, which Richards designed and edited while also writing a Sunday column aimed at personal computer users. Equally important were the regular surveys of venture capital firms, which she started assembling and publishing in 1982. The feature proved quite popular and was eventually taken over by Price Waterhouse, a business consulting firm.[47]

The year 1984 was a turning point for Richards. For years, one of the companies she had closely watched was Apple. Even before she started at the *Merc*, she reported on Apple's exceptionally successful initial public offering at the end of 1980. Many journalists who attended Apple's sneak previews were asked to sign nondisclosure agreements and embargo their stories until a certain date. Richards avoided these constraints by relying on unembargoed information and taking advantage of access to anonymous sources inside and outside the company. As Apple prepared to release the Macintosh, for example, Richards paid special attention to the machine's technical specifications. A week before Apple's now-famous 1984 advertisement aired, two articles by Richards appeared on the front page of the *Mercury News* Sunday edition.[48] The first one, titled "A Look at Secret New Apple Computer" presented a slew of details about the Macintosh, including its price and technical capabilities. Richards's second article, presented as "How the Macintosh Computer Grew," detailed infighting within the company as well as Jobs's mercurial management style. Years later, reflecting on the Macintosh's release, Richards noted that journalists, even as they knew they were being manipulated, performed an essential function in the Macintosh's rollout. "In some sense that campaign was a watershed," she noted, "and after that everybody else had to match the hype."[49]

Evelyn Richards's reporting also addressed the human cost of business, such as with her pointed but sympathetic feature article titled "On the Brink of Bankruptcy." It described the "uncomfortable and eerie experience" of workers at Dynabyte, a now-forgotten Silicon Valley company in its death throes.[50] Richards very much saw herself primarily as a business reporter and she maintained a focus on the economic ups and downs of Silicon Valley companies, national industrial policy, and the rapidly intensifying competition between American and Asian electronics firms.

In 1988, Richards became a national technology writer for the *Washington Post*, covering a wide range of business topics at a time when the Washington political establishment was fostering emerging technologies such as high-definition television and the "information superhighway." The *Post* nominated her multipart series "The Software Snarl," which explored US dependence on flawed programming, for a Pulitzer Prize. Starting in 1993, Richards spent three years as an editor in Tokyo at the English language edition of the *Nikkei*, the world's largest financial paper, before returning to California and the *Merc* in 1996. While stocks listed on the tech-heavy NASDAQ exchange soared, Richards supervised a team that produced a series of articles that chronicled not just the staggering wealth in Silicon Valley but also its housing crisis and the corrosive effect of growing income disparities, writing some of the first analyses of problems the region would later become infamous for.

Today, when one flips through the pages of a major national newspaper— or, more likely, points, clicks, and scrolls from article to article—it's quite probable that the publication they are reading employs many "tech reporters." At such papers as the *Wall Street Journal* or the *Washington Post*, there are now reporters who concentrate their attention on just one company (e.g., Google or Apple), or cover a discrete subject (e.g., social media, autonomous vehicles, artificial intelligence). Once considered questionable as a journalistic beat, tech reporting has become a well-established subject accepted and expected by readers, editors, and publishers.

It would be a gross misrepresentation to say that tech reporters made Silicon Valley (although they did *name* it). But journalists were essential, if often underrecognized, elements in what business management types have come to call the "innovation ecosystem."[51] While they were not designing chips or writing lines of code, journalists engineered the metaphor and helped build the get-rich narrative of Silicon Valley. In the 1980s, that same sense of optimism found among Silicon Valley's business community was taken up by the region's most visible booster. President Ronald Reagan repeatedly invoked Silicon Valley's mystique at political rallies and in off-the-cuff comments to voters.[52] "High-tech is spreading across the country like wildfire," he told a crowd in Philadelphia. "Silicon Valley is being joined by Silicon Bayou in Louisiana, Silicon Mountain in Colorado and . . . [here in] the Silicon Valley of the East."[53] Like an accelerant tossed into the flames of that fire, journalists like Richards helped catalyze international awareness about the technological transformations that were happening in a small sliver of California real estate.

## BOOK PEOPLE

Engineers call a situation where electrical input into a system does not match the output "impedance mismatch." Daniel A. Doernberg was no engineer, but in the early 1980s he experienced his own sort of impedance mismatch when it came to books. He wanted to learn more about computers and how they worked but "if I went into a computer store looking for a book," he told a magazine writer in 1987, "they'd try to sell me a machine. And if I went into a bookstore, the clerks didn't know what I was talking about."[54]

Doernberg's problem was compounded by the fact that input into the system, so to speak, had never been stronger. Unlike the early 1950s, when a person could have conceivably owned every book published in English about computing, by 1983 tens of thousands of titles filled publishers' catalogs and warehouses.[55] What was causing the mismatch, Doernberg realized, was not a lack of books but rather their availability and access. But it was a solvable problem.

Doernberg grew up in Cincinnati, Ohio, where he nurtured a love of books. He spent hours exploring the shelves of bookstores and building up a collection of science fiction classics and biographies. "I'd buy a book thinking that if I wanted it someday, I could just pull it off my shelf," he remembered, "and that would be so cool."[56] When it came time for college, Doernberg went to Duke University where he met Rachel Unkefer, his future wife and business partner. She was also from Ohio and, like Dan, a baby boomer who grew up surrounded by books (Spanish literature was a favorite genre).

After finishing their degrees in 1980—Latin American studies for Rachel and psychology for Dan—they relocated to Santa Barbara, California. While Unkefer held a series of temporary jobs, Doernberg took a community college class in BASIC. He also helped build a searchable computer database project for a real estate company. While working nights, Doernberg met another, more experienced programmer. By this point, Doernberg was becoming increasingly interested in personal computers ("obsessive," claimed Unkefer) and their potential to affect people's work and home lives.

This fascination was set more firmly when his colleague loaned him Ted Nelson's *Computer Lib/Dream Machines*. "It was life-changing," Doernberg recalled, "I got the idea that what I want to do is open up a little bookstore that just sells computer books."[57] Not everyone was enthusiastic. "I thought it was a stupid idea," Unkefer recalled. "Why would anyone want to go to a bookstore that only has books about computers? I thought it would be like having a bookstore with books about toasters."[58] Unkefer came from a family that had operated a series of small businesses. But it was becoming hard for her to ignore the growing piles of computing books and magazines accumulating in their apartment. And it was hard to deny that the 1970s and 1980s had seen a marked growth in the number of bookstores specializing in particular subjects and audiences.[59]

Doernberg left the database project for a job at a Radio Shack store in the Santa Barbara area. At the time, the chain store sold personal computers, including its bestselling TRS-80 model, as well as peripheral devices and software. Somewhat inexplicably, the store hired the inexperienced (but enthusiastic) Doernberg to provide technical support for new computer owners. "I was pretty worthless at that job," he admitted, even though he and Rachel soon bought their own personal computer, a Commodore 64 machine. But another key turning point for Doernberg happened when a customer at Radio Shack gave him a stack of back issues of *InfoWorld*. According to John Markoff, an editor at the magazine who later became an acclaimed technology reporter for the *New York Times*, *InfoWorld* "wanted to be like the *Rolling Stone* or *Sports Illustrated*" of the computing world, covering not just news but also the rapidly changing culture around computing.[60] "I loved that magazine," Doernberg said, "It was an introduction to a whole new world. *InfoWorld* was just heaven."[61]

In the spring of 1981, when the couple visited San Francisco, Dan persuaded Rachel to stop in the Palo Alto area so he could check out local computer bookstores. He found surprisingly little activity. Stanford's campus store, of course, had textbooks but other shops around Palo Alto offered little in terms of products for the serious enthusiast or computer professional. "There was no bookstore in the heart of Silicon Valley that had a *lot* of computer books," Doernberg recalled, "this seemed a no-brainer. Except that everyone around me said it was a really dumb idea."[62]

The next year Doernberg and Unkefer moved to Sunnyvale, one of several small cities that collectively compose the sprawling entity of Silicon Valley. Doernberg's first few weeks proved a bit of a reality check. "I had the romantic idea," he told a computer magazine in 1987, "that I'd meet David Packard at the meat counter at Safeway."[63] Rachel got a job designing printed circuit boards (something she had no experience with, which says a lot about the boom-and-bust nature of employment in the region). Dan, meanwhile, worked part-time in local general-interest bookstores while creating a plan for his own business. Rachel remained unconvinced about his bookstore idea but, seeing that her partner was increasingly committed to the plan, told him, "Just do it and get it out of your system."[64] Rachel, who had some business experience, also realized that Dan would eventually need help with essential tasks like accounting, calculating sales margins, and managing inventory.

Doernberg started scouting possible store locations. He ultimately settled on a modest-size space that rented for around $500 a month. It was near their apartment in Sunnyvale and close to a freeway (US Route 101), so it could attract customers commuting between San Francisco and San Jose. The offices of several major companies, including Hewlett-Packard (HP), Atari, and Intel, were nearby too. Despite having "no experience and an amateur business plan," Doernberg convinced the landlord to rent him the space.[65] As he moved in bookshelves and other furniture, Doernberg reserved an area for presentations and author talks. Boxes of books slowly began to accumulate in advance of the store's opening and he set about choosing a suitable name for his venture.

In the early 1980s, computer scientists and education experts frequently used the phrase "computer literacy." In some ways, it was an extension of Ted Nelson's mantra, placed on the cover of *Computer Lib*, that all citizens "can and must understand computers NOW." A major promoter of the computer literacy movement was Arthur Luehrmann, a computer scientist at Dartmouth, where the user-friendly BASIC language had been invented, and the term subsequently appeared in a bevy of books and articles aimed at educators and policy makers.[66] As one of Luehrmann's colleagues noted, "the evidence is overwhelming that computers may appeal to students who have never shown any talent for mathematics and, indeed, may have hated the subject."[67]

The customers that Doernberg and Unkefer imagined coming to their store—computer scientists, engineers, and programmers—certainly didn't hate the subject. But the term "computer literacy" was, he recalled, a "huge buzzword at the time." Doernberg also thought a bookstore with literacy in its name was an inside joke leavened with a personal sense of mission. "Teaching people to use computers was a very Ted Nelson thing to do," he said, "I was super idealistic." He later reflected, "We had the perfect name for the first two weeks of our existence. And then it became almost an embarrassment as our actual business model and customer base—computer professionals—emerged."[68] The Computer Literacy Bookshop opened for business in March 1983 and quickly found success.

At first, Doernberg and Unkefer carried about 1,200 unique titles. A year later, when *PC Magazine* profiled the store, that roster had grown to some 5,000 titles, a number that tripled again in the next few years.[69] Many of these were, for the average reader, rather esoteric books. But works such as *dBase II for Every Business* and *Principles of Computer Speech* were of considerable interest to the region's professional programmers and computer engineers. To add to the store's offerings, Doernberg attended book industry trade shows and combed through publishers' catalogs. Meanwhile, the bookstore's staff—about a dozen people worked at Computer Literacy by the end of 1984—prided itself on being able to find almost any title a customer requested. For people working on time-sensitive projects, Doernberg would get books directly from the publisher and even hand-deliver them. Eventually, companies such as IBM, Apple, and Intel opened corporate accounts to make sure their employees had whatever resources they needed. "We're not a part of the book industry," Unkefer explained to one reporter, "as much as a part of the computer industry."[70]

Just a few decades earlier, computer professionals and enthusiasts had a relatively small selection of books to choose from, especially when it came to specialized technical topics. By the mid-1980s, as Computer Literacy's impressive inventory suggested, this was no longer the case. If anything, readers were spoilt for choice and faced the challenge, as the *New York Times* put it, of "finding a good book in a sea of mediocrity."[71] Because Computer Literacy carried only computer books, Doernberg, Unkefer, and their staff could offer personal guidance as to the best book on a particular topic. Regular interactions with hundreds of customers each week provided additional and valuable feedback that figured into what books they carried and could recommend.

Customers bought books for all kinds of reasons. A typical client might be an engineer who needed to burnish her technical knowledge before a job interview. Some customers, especially if they were overseas, might order a dozen or more books at a time. One patron in Poland, lacking dollars, offered hand-carved wooden dolls in exchange for a particular title. And, of course, besides being a "paper tool" an engineer might need to do their job, books have long been objects to display. Doernberg noted that Donald Knuth's The Art of Computer Programming series was a perennial best-seller. "Whether you ever looked at his books or not, it was in your interest," he noted, "to have them on your bookshelf to show you were a serious card-carrying credentialed programmer."[72]

Within a year of opening, the Computer Literacy Bookshop had received favorable publicity via stories in several computer magazines. To Doernberg's delight, *InfoWorld* showcased the Computer Literacy Bookshop, with a color photograph of the bookshop on the cover, and in a feature article about the rapidly expanding universe of computer books. The *Whole Earth Software Review*, a new publishing venture Stewart Brand launched in 1984, called it a "civilized haven in a savage landscape" and provided instructions for readers to place orders by phone or mail.[73] Customers flocked to the store during lunch breaks and then again after the workday had ended. Soon, Computer Literacy was operating seven days a week.

Part of what made shopping so appealing was the store's obsessive organization. A regular bookstore typically had (if it had one at all) a modest section called "Computers" with titles arranged alphabetically by author. Maybe it would be subdivided into "Computers: PCs" and "Computers: Other." At the Computer Literacy Bookshop, however, books were organized with a much higher degree of granularity. A document outlining the store's "taxonomy," for example, was several pages long with 100 categories and 560 subcategories. "It was like drafting the Constitution," Unkefer said.[74] There were over a dozen subcategories for "Apple" alone, including "MAC-APPLIC," "MAC-TECH," and "MAC-LANG." Books about "IBM" were broken down likewise while "Computer Languages" had at least thirty-five subcategories, with some subdivided even further. Inside the store, book shelving following this schema; a customer who wanted to see everything on the design of "graphical user interfaces" could go straight to that part of the store and then peruse sections like "GUI-GEN" or "GUI-Windows."[75]

The Computer Literacy Bookshop quickly became more than just a place to buy books. Customers chatted about technical matters and exchanged gossip about recent hirings and firings. Doernberg added to this sense of community by hosting a regular speakers series. Legendary computer programmers, such as Donald Knuth and Dennis Ritchie, could easily attract several hundred fans for events that stretched to several hours.

(*The C Programming Language*, a 1978 book by Ritchie and Brian Kernighan, was Computer Literacy's top seller over the years.) Bimonthly bulletins went out to tens of thousands of people on the store's mailing list. Festooned with the store's logo—an anthropomorphic computer reading a book—these newsletters gave short assessments of new books and advance notice of forthcoming titles. To Doernberg, these newsletters were not just commercially valuable but, when combined with short author interviews, represented a parajournalistic contribution to the larger tech community in the region and beyond.[76] As information technologies changed, so did the store's methods of getting people the books they wanted. Computer Literacy set up an electronic bulletin board system that allowed customers to leave book requests outside of store hours. In time, it expanded into an electronic forum where users argued over which book was best for a given topic, exchanged opinions and suggestions, and even left questions and comments for authors.[77]

Doernberg and Unkefer opened their store at an opportune time as the computer book business often reflected what was happening in the larger computer industry itself. For years, the publishing business around computer-oriented books and magazines had been growing at a swift clip of about 15 percent each year with total revenues well over $700 million.[78] By the end of its first year, Computer Literacy's cash registers had rung up more than $1 million in sales.[79] However, in the mid-1980s, the computer industry experienced a brief but severe downturn.[80] This also hit publishers of computer books and magazines as fewer companies bought advertisements. More than 100 computer periodicals vanished in 1984 and 1985 (those aimed at the home computer market were especially hard-hit) while the number of new books declined by about 20 percent.

The Computer Literacy Bookshop weathered the financial headwinds and, by 1986, the computer publishing industry had righted itself. In response, Doernberg and Unkefer expanded their operations, opening two new stores in 1987. Taken together, these locations added another 10,000 square feet of retail space. One of these was initially in Santa Clara but, in the early 1990s, the store moved to Apple's new campus in Cupertino. A third and much larger shop, located in San Jose, became Computer Literacy's new flagship venue. A fourth store, serving the tech-heavy region around Washington, DC, opened in northern Virginia.

By the early 1990s, annual sales at Computer Literacy regularly broached the $10 million mark and it had become a modest-size retail empire with an international reputation among the technology cognoscenti. Doernberg recalled how tour buses would occasionally pull up to their main store in San Jose and discharge scores of tourists, eager to buy something that spoke of "Silicon Valley." All told, more than 100 staff members were needed to ring up sales, chat with customers, operate pop-up bookstores at trade fairs and, do all of the other behind-the-scenes work that kept the business running. As journalists and authors wrote with increasing frequency about virtual communities connected by the internet, the Computer Literacy Bookshops offered real, brick-and-mortar spaces for experts and enthusiasts alike to gather, converse, and buy books about computing.

### THE UNCANNY VALLEY

The author's letter was, by turns, pleading, boastful, and resolute. Michael S. Malone had written it in response to one of the most common yet painful requests an author receives: cut some pages. In this case, Adrian Zackheim, a New York-based senior editor at Doubleday, and Luther Nichols, a Doubleday editor in Berkeley, had asked Malone to trim about a third of the manuscript for his first book. Malone contested their advice. "The strength of *The Big Score*," he insisted, "lies in its detail, the type of information not found in any other book about Silicon Valley." If left alone, *The Big Score* would "make all other books on the subject superfluous." Malone supported his claim by referencing his writing style ("superior to any other") and critical perspective ("as opposed to the usual utopian point of view"). "I'm sorry," he concluded, "but I believe in this book."[81] In the end, the parties compromised. Malone made some of the edits Doubleday wanted but, when it appeared in the summer of 1985, *The Big Score* was still big. It checked in at nearly 450 pages but nowhere near the 1,000 typed pages of prose Malone had originally prepared.[82]

*The Big Score* was by no means the first book about Silicon Valley. For decades, authors had produced romantic memoirs and boosterish reportage about the region. Malone's opus was distinguished by its panoramic scope. The usual names that readers had come to expect (Gordon Moore and Robert Noyce of Intel, Nolan Bushnell of Atari, and, of course, Steve Jobs) were all there. But Malone also included workers' voices he heard on shop floors and sweatshops where immigrant laborers churned out circuit boards. His reportage encompassed the personal tragedies, environmental ruination, and bankruptcies—financial as well as moral—common in any boomtown. If one understood Silicon Valley as an "ecosystem," then Malone was its "digital-age Darwin."[83]

As strange as the social and economic activities of Silicon Valley may have been for some readers of Malone's book, the place was familiar territory to him. Indeed, it was home. Born in West Germany in 1954—his father was a military intelligence officer—the family moved to Mountain View, California, when he was in the fifth grade. What was still sometimes called the Valley of Heart's Delight became a foundation for Malone's personal and professional identity. The Malones arrived when the region was in the midst of a vast transition, much of it driven by the Cold War's command economy. Rapid demographic changes produced schoolyard antagonism as the children of blue-collar workers, many of them still toiling in the area's quickly disappearing orchards, confronted the offspring of well-paid engineers employed by local aerospace and electronics firms.

When he was in junior high, Malone's family moved to the more prosperous town of Sunnyvale and bought a house close to where Steve Jobs and his family had recently moved.[84] Another neighbor of the Malones, Regis McKenna, would start an influential marketing firm and provide professional counsel to Jobs. Malone saw a homebuilt computer-of-sorts that a young Steve Wozniak had built for a local science fair. However, electronics did not interest Malone much at the time. Instead, he focused his teenaged attention on becoming an Eagle Scout and exploring places not yet gobbled up by suburban sprawl and industrial parks. "Computers and electronics were everywhere," Malone recalled. "But then I fell in love with writing in high school."[85]

While attending nearby Santa Clara University in the mid-1970s, Malone produced a long-running, and sometimes controversial column—the author had a penchant for colorful language—for the school newspaper. One of Malone's professors encouraged him to apply for an internship at Hewlett-Packard, a leader in the region's electronics industry. The part-time position metamorphosed into a steady job doing public relations work for the company. While Malone went on to get an MBA, "my real education was at Hewlett-Packard," he said. "Almost everything I know today about reporting and writing comes from them."[86]

Malone produced "an endless stream of press releases" for HP while also observing that the "human side of business wasn't in any of the grad school textbooks."[87] He became increasingly interested in all of the people, technicians and workers as well as executives, behind the products that companies like HP made. At the same time, writing publicity copy taught Malone how to "tell stories about technology for the average person . . . I quickly learned the nature of the sellable hook and the importance of following that up with a really good quote."[88] Now in his midtwenties, Malone disliked seeing his work repackaged by other writers at trade magazines like *Electronic News*. "It tore my guts," he confessed, "I just wanted my own by-line. I was a newsy at heart."[89]

In 1979, Malone left the security and steady paychecks HP provided and started as a business writer for the *San Jose Mercury News*. It was an unusual transition as Malone entered journalism from the public relations side rather than working up the reportorial ranks, as Evelyn Richards had done. When Malone joined the *Merc*, the paper still largely focused on local news and few reporters wrote much about Silicon Valley that went beyond quarterly profit reports. Malone, however, was more interested in what was behind companies' products and financial statements—what he later called the "real story" of Silicon Valley. "I knew I wasn't going to get there writing sales-and-earnings stories for the business section," he recalled.[90]

One of the first big topics Malone dug into was something the region's trade press, which relied on maintaining harmonious relations with local companies, was inclined to ignore. On the surface, the electronics industry seemed like a postindustrial alternative to soot-belching factories and oil refineries, the sort of technological future that Alvin Toffler had imagined. The tacit agreement made by boosters and businesspeople with residents of Santa Clara County was that prosperity would come via "clean" industries like electronics manufacturing. But producing the chips and integrated circuits that made computers possible depended on a vast array of toxic and dangerous chemicals. As one administrator at the Occupational Safety and Health Administration pointed out, "People think of it as wires, soldering, and transistors. But when you get to the semiconductor business, you're really talking about chemical reactions. It's a *chemical* industry."[91]

In 1980, Malone partnered with investigative journalist Susan Yoachum to produce a damning, three-part series they called "The Chemical Handlers." Their reporting showed how the careless use of hazardous materials translated into workers' injuries (some fatal) and the illegal dumping of chemicals. The negligence produced work-related illness, widespread groundwater contamination, and increased birth defects.[92] Malone's and Yoachum's reporting went beyond statistics to give a voice to workers who had been harmed. The series was nominated for a Pulitzer Prize and, in response to concerns it raised, local companies pledged tens of millions of dollars to address environmental and safety issues. "That experience," Malone said, "showed me the power of investigative reporting." By the end of the 1980s, dozens of journalists followed Yoachum's and Malone's lead and were reporting on the toxic legacy created by Silicon Valley's electronics firms.[93]

In 1981, Malone left the *Merc* (Evelyn Richards replaced him) and took up the more precarious position of a freelance writer. By this point, Malone knew there were important stories about Silicon Valley that rivaled the oft repeated tales of successful companies and instant millionaires. For several months, he collaborated with Pete Carey, a *Merc* reporter who later shared a Pulitzer for investigating the Marcos regime in the Philippines. Working with Carey taught Malone how to efficiently search public records, file Freedom of Information Act requests, and talk to people engaged in questionable, if not illegal, activities. Over several months, Malone's and Carey's articles revealed the widespread use of sweatshop-style labor in the electronics industry, drug use in the workforce, corporate espionage, and, given the prominent use of gold in manufacturing integrated circuits, the lucrative market in stolen metals.

By 1982, Malone was starting to consider what to do with all the stories he had collected about his hometown. He got some unexpected guidance from Niven Busch, a San Francisco-based novelist (his 1959 novel, *California Street*, was about the newspaper business). Busch was doing research for a new novel and he asked Malone for a tutorial on Silicon Valley. The two writers got along well and Busch connected Malone with Luther Nichols, Doubleday's West Coast editor and a former book critic for the *San Francisco Examiner*. Besides serving as Malone's main contact at Doubleday, Nichols introduced the neophyte writer to Don Congdon, a literary agent who would represent Malone for two decades. Congdon already had an impressive roster of writers, including William Styron, J. D. Salinger, and Ray Bradbury.[94] For Malone, who had never written a book, the serendipitous contact left him feeling like the "luckiest guy in the world."[95]

Malone convinced Doubleday's editors that the time was right for a book about Silicon Valley's "turbulent history" that would not follow the well-trodden path of "popular technology books of recent years." Instead of focusing just on technology, Malone would tell a "story of people, from overnight tycoons to drug addicted assembly line workers."[96] Doubleday responded with an advance contract and Malone settled down to write. As he wrestled with hundreds of pages of notes, interview transcripts, and anecdotes captured over drinks at Walker's Wagon Wheel, a popular cocktail lounge tech workers frequented, Nichols provided encouragement. "I winced," Nichols wrote in one letter, "on seeing yet another Silicon Valley volume up for review." (The book in question was likely Dirk Hanson's *The New Alchemists* [1982], which focused on the broader history of electronics.) "This book's author didn't get half as deeply inside as you're planning to," Nichols observed, noting, "the added competition makes it all the more desirable that your book be angled a little differently than the rest."[97] While offering reassurances, Nichols was sending a familiar message from editors to their authors: hurry up before you get scooped.

Malone *was* hurrying. But he was also living hand-to-mouth as a freelance writer, at one point taking up residence with his future wife in a former chicken coop that had been converted into a (very small) apartment. Buried in Malone's correspondence is an apology for having to borrow money from Nichols when he was down to his last $20.[98] As he assembled his chapters, Malone kept investigative reporters Bob Woodward and Carl Bernstein, famous for their coverage of Watergate, in mind. He also appreciated the cultural skewerings penned by John L. Wasserman, a prominent critic for the *San Francisco Chronicle* in the 1970s.

Perhaps more than anything, Tom Wolfe's "new journalism" provided Malone with a stylistic model. In 1983, *Esquire* had published one of Wolfe's last major articles, titled "The Tinkerings of Bob Noyce." This monumental, 15,000-word piece profiled Intel's cofounder while also giving readers a visceral sense of how the Bay Area had changed since 1968 when Wolfe had written *The Electric Kool-Aid Acid Test*, his classic chronicle of the counterculture. Whereas the hippies had once claimed they were making the future, in the end it was entrepreneurs like Noyce who won those bragging rights. The revolution was over and the "squares" had won.[99] Wolfe ended his essay on a moralistic tone by contrasting Noyce's understated approach to doing business to that of a newly minted Silicon Valley millionaire who died when he crashed his brand-new Ferrari on the day his company had gone public.[100] "Wolfe nailed it," Malone said. "He got into the soul of Silicon Valley. Bob Noyce and Gordon Moore came from working-class worlds and carried those values with them."[101]

Malone completed his book by the end of 1984 and Doubleday moved *The Big Score*—subtitled *The Billion Dollar Story of Silicon Valley*—into production. He was especially pleased with the book's cover art, which featured a computer chip shaped like a dollar sign. In advance of the book's appearance, Malone penned the brief but essential copy that would go on the book's dustjacket. *The Big Score*, he wrote, "burrows beneath the well-tended image of the high-tech revolution for a hard-hitting look at the reality of Silicon Valley life . . . the view is both tantalizing and startling." Malone identified himself as the "first reporter on high-tech business" and "perhaps the only writer on Silicon Valley to have grown up there." In terms of competing books, Malone noted that another recent book, *Silicon Valley Fever*, actually relied a good deal on *his* freelance writings. More people, Malone insisted, will want to read *The Big Score* to learn about "the bright and dark sides" of a high-tech region. "It's a great fucking read, full of sex, drugs, and millionaires and spies and murderers," he told his editor. "There, that's enough self-aggrandizement."[102]

While promotion (and self-promotion) is important for an author, publishers also recognize the ineffable importance of timing. Unfortunately, when Malone gave his completed manuscript to Doubleday, the company was experiencing difficult financial times. There was even speculation that the company, which was almost ninety years old, might not survive much longer.[103] Malone's book fell afoul of financial and staff changes roiling the company. "At the end, I had a pick-up crew handling the book," he remembered. "It was something which they hadn't signed and knew nothing about. *The Big Score* was born an orphan."[104]

If Silicon Valley was where "heaven and hell sit at adjoining workbenches," then Malone promised to guide his readers just "as Virgil led Dante through a strange and often terrifying land."[105] Reactions from critics and readers were positive if somewhat muted as Malone's book fell somewhere between Michael Moritz's *The Little Kingdom* and Tracy Kidder's *Soul of a New Machine*. Newspapers and magazines in San Francisco and Silicon Valley, not surprisingly, were especially interested in *The Big Score* and their reviews were uniformly positive. The *San Francisco Chronicle* called it a "first-rate history" and, although his "very strong book" still included "admiring portraits of Valley tycoons," the newspaper praised Malone's willingness to look past laudatory hype and call out the region's environmental, economic, and social woes.[106] The highest profile venue to review *The Big Score* was *Business Week*. Placing it among the "lengthening shelf of books" about Silicon Valley, the venerable weekly called *The Big Score* "the best of these" and "the most ambitious" as its author tried "to get at the motivations and emotions of the Valley's people." The magazine's reviewer likened Malone's depiction of Silicon Valley to a "speedboat driven by greed and egomania" and *The Big Score* appeared on its list of the year's top-ten books.[107]

With Doubleday's operations in disarray, Malone arranged his own publicity for his book. He gave numerous radio interviews about *The Big Score* and sent copies to leaders of the technology community, including Robert

Noyce and his old boss, David Packard—both replied with enthusiastic letters of encouragement.[108] One of Malone's friends, meanwhile, traveled around Silicon Valley taking photos of *The Big Score* in bookstore windows, including at the Computer Literacy Bookshop. Malone's self-promotion produced dividends. Doubleday announced a second print run and also sold the rights for a Japanese translation.[109]

While executives such as Noyce appreciated Malone's tales of Silicon Valley's past, politicians and policy makers were more keenly concerned with its present and future. Malone sent *The Big Score* to Ed Zschau who represented California's 12th Congressional District. Zschau had earned degrees in the philosophy of science and business management before starting a company that made printers and disk drives for computers. Known in the early 1980s as "Silicon Valley's Congressman," Zschau advocated for tax cuts that would bolster high-tech entrepreneurship.[110] A colleague at the Capitol noted that Zschau "now has members of Congress pronouncing it 'Silicon' instead of 'Silicone' Valley" (while including a tasteless reference to Carol Doda, a famed Bay Area dancer and performance artist).[111] Zschau complimented Malone—"a master storyteller"—for his "definitive history of Silicon Valley" and even penned a blurb for Malone to use.[112]

By 1985, Silicon Valley was established in the lexicon as a distinct and powerful technological region. More than just denoting a place, it suggested a state of mind with a bullish attitude toward entrepreneurship, risk, and reward. As the 1980s acquired a reputation as the "decade of greed," the lifestyles of the rich and famous were extolled.[113] Silicon Valley, with its growing population of recently minted millionaires, fit this mindset perfectly. The region became, in the minds of the public and policy makers (less among the workers toiling in the sweatshops Malone helped document), a synonym for financial success and technological innovation.

By no means did writers and journalists invent the economic phenomenon of Silicon Valley. But they gave it a name and crafted compelling narratives about it. Just as critically, their articles and books drew readers' attention toward Silicon Valley and the technologies it was indelibly associated with. This public awareness coincided with the emergence of the tech reporter as a new kind of writer. Despite their diverse professional backgrounds, these writers still had one thing in common. Their journalism was presented via the modern world's oldest medium: words printed on paper. As the Cold War ended and Clinton-era rhetoric about an internet-enabled "new economy" reached rhapsodic heights, many journalists started to see the technologies about which they had written hundreds of thousands of words with fascination and foreboding. A digital tsunami was coming that would transform the very nature of writing, reading, and selling books, along with the computer itself.

# 8  MATERIALISM, MAPS, AND MANUALS

Their mark on this land is still seen and still laid
The way for a commerce where vast fortunes were made. . . .
—THE POGUES, "NAVIGATOR," FROM *RUM, SODOMY, AND THE LASH* (1985)

When the 1990s began, only a few companies were registered for an internet domain. In fact, in 1992, as Bill Clinton and Al Gore campaigned for the White House, one could, if compulsively inclined, write out a list of the internet's entire commercial sector—all the dot-coms—on a few pages of a single legal tablet.[1] Today, of course, the situation is quite different with more than 160 million such registered entities.

One of the businesses that *did* have a registered dot-com domain name in 1992 was Computer Literacy Bookshops, Inc. In August 1991, Dan Doernberg registered clbooks.com and encouraged readers to submit book orders via email.[2] But Doernberg remained leery of doing business over the internet due to legal restrictions and cultural norms. In the early 1990s, using the internet for commercial purposes was still a gray area and missteps might offend his customers. But soon after he filed the paperwork, the bookstore's staff started processing electronic orders sent to clbooks.com, making Computer Literacy the first bookstore to stake out some cyberspace on the internet.[3]

One customer recorded in Computer Literacy's business accounts was "Jeffrey P. Bezos." In the mid-1980s, Bezos had studied electrical engineering and computer science at Princeton before taking his quantitative skills to Wall Street. In 1996, not long after Amazon.com started selling books online, Bezos placed an order with Computer Literacy for a book offering "answers for computer contractors."[4] Was Bezos interested in this topic? Maybe. Or perhaps the entrepreneur was testing if Computer Literacy could secure a rather esoteric publication from a very small press.

Obviously, Computer Literacy Bookshops sold books because it was a bookstore. But why did Bezos choose *books* as the vehicle to explore the new world of online commerce? To be sure, Bezos did not choose them because of a deep commitment to print culture. Books were only one of about twenty different products he considered offering for sale. As he told a gathering of undergraduates in 1998, books offered a "huge diversity of products" allowing one to "build a store online that simply could not exist in any other way."[5] By the end of 1996, Amazon's sales figures easily surpassed the money recorded by the cash registers and credit card imprinters and electronic payment terminals of all the Computer Literacy Bookshops. That same year, the *Wall Street Journal* profiled Bezos as the "whiz-kid programmer" with "thinning brown hair and frayed blue jeans" who was disrupting "the tweedy book-publishing business."[6] The rest, as they say, is history.

But how does *that* history relate to *this* bookish history of computing? Rivers of ink have chronicled and critiqued Amazon's trajectory to becoming a mammoth multinational company with hundreds of billions of dollars in annual sales. Oceans more have flowed forth in an effort to account for the massive changes the internet and the World Wide Web brought to the publishing business, as well as the nature of being an author—and reader—in the age of the internet.[7] While relevant to this chapter, I want to focus instead on two topics in computing (and books about computing) that emerged during the early 1990s.

The first topic is, oddly enough, the "disappearance" of the computer. The digital computer gobbled up a host of other technologies throughout the twentieth century, including typewriters, punch cards, card catalogs, slide rules, and calculators. In the 1990s, this trend accelerated as various types of media—music, maps, film, photographs, and books—vanished from physical form and reappeared in the ephemeral-seeming territory of cyberspace. As George Gilder, a Republican speech writer and supply-side economist turned technology evangelist, phrased it in his 1989 book *Microcosm*, this represented the "triumph over materialism."[8] While computers still mattered, their ubiquity ironically led to a form of invisibility. Physical commodities such as mainframes, machines, and personal computers mattered less and less as stuff was supplanted by services and software. For Gilder, this digital transformation contained an importance that transcended economics and technology. At the level of quantum physics, which governed how computer chips worked, he claimed to have found an underlying transcendent, even religious, significance.

Now, of course, the computer qua machine never disappeared. Companies such as Dell and Gateway—both were founded in the mid-1980s and expanded markedly in the 1990s—made and sold millions of desktop and laptop machines. The number of households in the United States with personal computers, in fact, more than doubled between 1990 and 1997.[9] Even so, computer hardware itself became increasingly camouflaged as more and more media and public attention was directed toward software applications.

We can see the computer's "disappearance" by looking again to the covers of *Time* magazine. Between 1993 and 2000, *Time*'s editors featured some aspect of information technology an astonishing twenty-eight times. But unlike earlier decades, the focus of this cover art and accompanying articles was no longer the machines themselves. What excited politicians and entrepreneurs more were the digital pathways that would transport millions of citizen-consumers to the new virtual spaces and places where information, entertainment, and, perhaps, internet-fueled fortunes awaited. As one well-placed observer of the "digerati" explained, "It's not about computers. It's about human communication."[10]

As a result—and this is the second main topic of this chapter—there was massive public interest in all things internet. *Time*'s colorful covers highlighted cyberpunk, cyberwar, and cyberporn while lengthy stories appeared in print that addressed "online shopping," "online news," "growing up online," and "Jesus online." *Time* even published an entire special issue in 1995 titled "Welcome to Cyberspace"; its cover featured a stylized digital doorway receding into infinity but with no machinery in view that one might use to enter this virtual world.

Even as everything seemed poised to disappear into the digital, the traditional medium of printed paper books remained essential tools for understanding the internet. People were keen to learn about cyber-everything but found themselves overwhelmed when they ventured into local book-shops and the large chain stores that came to dominate bookselling in the 1990s.[11] Who and what would serve as a trusted guide to the internet?

O'Reilly & Associates started the 1990s as a publishing company noted for its technical manuals written primarily for computer professionals. In 1992, however, the company departed from its usual fare with *The Whole Internet User's Guide and Catalog*. Profitable as well as popular, the book also provided the foundation for another O'Reilly product. The Global Network Navigator (GNN) was both an online magazine and a gateway to the World Wide Web. GNN shined briefly in the early 1990s until it was eclipsed by more highly capitalized entities such as Yahoo and Netscape Navigator.

A lot of money could be made by publishing guides written for novices. In 1991, International Data Group launched its hugely popular series of "For Dummies" books. The first book to appear was *DOS for Dummies*. Written by Dan Gookin, it sold millions of copies, as did *Windows for Dummies* (1992) and *The Internet for Dummies* (1993). These books, with their bright yellow covers and easily accessible prose, were instantly recognizable at bookstores around the world. While some readers groaned at the books' cornball humor, International Data Group's executives laughed all the way to the bank.

When Edmund Berkeley wrote *Giant Brains* in 1949, he wanted to explain how computers worked. By the 1990s, few readers needed such knowledge. As Ted Nelson had predicted, the ubiquitous computer was now as "glamorous as a can opener."[12] The books discussed in this chapter all shared a common feature. They treated the computer less as an "information machine" than as a portal *to* information, another realm where adventure and riches, perhaps even redemption, seemed to await just beyond the modem's screech and howl.

## DISAPPEARING ACT

In May 1988, Ronald Reagan traveled to Moscow to meet with Mikhail Gorbachev. By all accounts, it was a remarkable encounter, especially given

Reagan's denunciation of the Soviet Union as an "evil empire" just five years earlier. While he was in Russia, Reagan addressed students at Moscow State University, Gorbachev's alma mater. The speech was classic Reagan. He spoke about cowboys, Hollywood, and his tenure as president of the Screen Actors Guild. While most journalists focused on Reagan's admiration of religious and civic freedoms, he also referenced the transformative power of computers.[13] Standing in front of a painting filled with flags from various political revolutions, Reagan talked about another kind of upheaval. An "information revolution," he said, was "quietly sweeping the globe" but "without bloodshed or conflict." These new technologies, many of which were made in Reagan's home state of California, were exemplified by the enormous power of the "tiny silicon chip." As a result, American society was, Reagan said, a "chrysalis" about to transform after decades of being "confined to and limited by the Earth's physical resources."[14]

Reagan's rhetoric drew on ideas lifted from an obscure book written by a conservative newspaper columnist named Warren T. Brookes. In *The Economy in Mind*, Brookes argued that America's economic growth was less and less dependent on traditional industries. Instead, American business was poised to "mine the rich resources" of a new "mind-based" economy. An equation scribbled on a blackboard could quickly become a new company. Ephemeral computer code could become a bestselling new product. Even though some economists, like a cadre of obsolete Marxists, remained stubbornly "materialistic in their perspective," the "real economy" was in fact inexorably becoming "more metaphysical."[15]

Reagan, of course, didn't write the speech he delivered in Moscow. That task was done by Joshua Gilder, a journalist and critic. But how did Joshua Gilder learn about Brookes's book—it was no bestseller—so as to include it in what is regarded as one of Reagan's best speeches? The likely answer is hidden in plain sight. The cover of Brookes's book announced it included a foreword by George Gilder, Joshua's older and better-known cousin.

George Gilder experienced what a *New Yorker* writer would later characterize as a genteel, patrician, yet penurious childhood. Born in 1939, he grew up on a "marginally profitable" Massachusetts farm—albeit one with a collection of objets d'art made by his great-grandfather, Louis

Comfort Tiffany.[16] After George's father was killed in World War II, David Rockefeller—grandson of the oil tycoon—promised to look after the young man. Schooling at Exeter and Harvard was followed by stints writing for progressive Republican magazines and penning speeches for Nelson Rockefeller and George Romney. In 1973, Gilder achieved his first dash of celebrity with his book *Sexual Suicide*. Decried by feminists and homosexuals alike, Gilder's book championed the traditional heterosexual family. *Time* magazine, itself no bastion of radical leftist thought, branded Gilder the "nation's leading male-chauvinist-pig author," a label that probably boosted book sales among some demographics.[17]

Gilder used his next book to praise the new spirit of anti-Keynesianism, tax slashing, and entrepreneurial fervor that underpinned the ascendant conservative political philosophy. Basic Books published *Wealth and Poverty* in 1981 just as Reagan took office. Besides offering readers opinions about business cycles and economic policy, Gilder's book infused Reagan-era neo-liberalism with a certain moral rectitude. Capitalism, he wrote, was not just about the selfish accrual of wealth. It also had a spiritual (i.e., Christian) aspect, presenting a path toward altruism (an idea that infuriated Ayn Rand) and a renewal of religious faith (ditto). Before he pleaded guilty in 1986 to insider trading, Ivan Boesky told the commencement audience at the School of Business Administration of the University of California at Berkeley: "You can be greedy and still feel good about yourself."[18] Gilder took this idea a step further and insisted that greed could actually make you a good person. Selling over a million copies, *Wealth and Poverty* made Gilder, if not good, then at least rich and famous. William F. Buckley blurbed Gilder's book, the *New York Times* called it a "guide to capitalism" for the 1980s, and the *Wall Street Journal* said Gilder's message offered a "key to a better world."[19]

One can draw a straight line from phrases in *Wealth and Poverty* to what Reagan would espouse in Moscow seven years later. In a chapter titled "The Entrepreneurial Future," Gilder praised the risk takers who had made computers, software, and semiconductor chips possible. If historical facts inconvenienced the larger ideological argument, Gilder was quite comfortable to ignore them. For instance, if "the industrial analysts of the Left" had actually been right about "the bureaucratization of invention," then the "computer revolution" should have been led by industry behemoths like IBM. Ignoring decades of Big Blue's innovations, Gilder insisted the real heroes were smaller and ostensibly more nimble companies such as Intel (although, with 16,000 employees in 1980, it was not exactly a garage-based start-up). Economic and entrepreneurial miracles had happened, not because of federal patrons and policies, but because "callow geniuses" had transmuted sand, "the world's most common matter," into an "incomparable resource of mind."[20] Materialism, like Warren Brookes had argued, was passé.

Despite his growing fascination with immateriality, Gilder faced a weighty problem. Although he was extolling the importance of semiconductor chips, Gilder didn't know very much about them. At Harvard, he had studied economics (poorly, by some waggish accounts) and since then had cast himself largely as a mildly misogynistic moralist, a gadfly to the Left, and a spirited advocate for Reaganomics. Gilder resolved his dilemma by writing himself out of it. He had already devoted sections of *Wealth and Poverty* to the ideal of entrepreneurship, depicting it a solution for America's moral and economic decline. In 1984, he focused his new book, *The Spirit of Enterprise*, entirely on business entrepreneurs or, as he put it, the "forgotten heroes of capitalism."[21] Although *Spirit* did not grab reviewers and headlines like his previous book, it provided Gilder with an opportunity to start exploring the semiconductor industry in earnest.

A writing assignment for *Forbes* magazine took him to Boise, Idaho, where he mingled with executives at Micron Technology, a small chip company that was just about to make a public stock offering.[22] A colleague suggested that Gilder should immediately write yet another book, a sequel to *Spirit* that would give a "history and analysis of the semiconductor industry."[23] So, just as *Wealth and Poverty* provided Gilder with a jumping off point for writing about entrepreneurship, *The Spirit of Enterprise* likewise provided opportunities for him to learn and write about information technologies. Notes and draft chapters started to pour forth from his Osborne personal computer.

Gilder originally planned to focus his new book on Intel and its cofounder, Robert Noyce. However, Tom Wolfe, one of Gilder's writerly heroes, had already brilliantly profiled Noyce for *Esquire* so that option vanished. He would need another perspective from which to view the subject. As he set about trying to understand his topic, Gilder took advantage of the rapidly expanding universe of computer magazines, such as *InfoWorld* and *Byte*, as well as books like *The Soul of a New Machine* and *The Big Score*. But to be taken seriously, Gilder needed more than stacks of magazines. He would need immersion and instruction.

Gilder received some of this via a small but influential monthly publication called the *Rosen Electronics Letter*, which Benjamin M. Rosen, a Wall Street analyst, founded in the mid-1970s. Each monthly issue was typically less than twenty pages but the information it provided about the computer and semiconductor industries made it essential reading. When Rosen sold his profitable newsletter, its new owner, Esther Dyson, hired Gilder as a contributing writer.[24] (We'll meet Dyson again in the next chapter.) Writing for Dyson gave Gilder an opportunity to get acquainted with the industry's arcana.[25]

A more important contributor to the education of George Gilder was Carver Mead. In the mid-1980s, the Caltech engineer was still advocating for the design revolution he and Lynn Conway had launched. Gilder learned about Mead while finishing *The Spirit of Enterprise* and he soon started making regular trips to Pasadena where he attended Mead's classes and chatted with his students.[26] Gilder's tutelage started with basic quantum mechanics and extended to the design of integrated circuits using the Conway-Mead methodology. Gilder got, as he said, "neck deep in the science" and Mead "became his sage."[27] Mead's coaching helped Gilder write admirably accurate descriptions of the technological and scientific foundations of semiconductors.

In *Microcosm*, the book that came out of these experiences, Gilder deployed a capacious definition of what he meant by the titular term. He started with key breakthroughs in quantum mechanics made in the first three decades of the twentieth century. Gilder claimed these groundbreaking discoveries had "revealed the inner structure of matter" and thereby "made modern computers possible." Gilder's opening sections referenced the theoretical accomplishments of Albert Einstein, Niels Bohr, and Max Planck. Generally speaking, Gilder was correct. The quantum physics of semiconductors certainly differs from the Newtonian laws that govern the familiar, macroscopic world around us. That said, building computers relies on much more expertise than just quantum physics.

To bring Mead's research activities into the realm of Nobel-winning luminaries, he stressed the engineer's studies on the quantum mechanical phenomenon of "electron tunneling." Mead had researched this topic since the late 1960s at the urging of Gordon Moore, who wanted to know just how small transistors could theoretically be made.[28] As Gilder saw it, Mead's exploration of these physical limits had set the stage for subsequent breakthroughs in chip design. By the time Gilder had finished writing *Microcosm*, Carver Mead had become the book's central character. One of Mead's favorite adages—"Listen to the technology"—provided an epigraph for *Microcosm*'s preface and the engineer's name appeared hundreds of times throughout the book.[29] Gilder transformed Mead into the "prophet of the microcosm," a visionary with technical acumen, entrepreneurial zest, and ample foresight about the technological future.[30]

While Mead's accomplishments were notable, he had, of course, collaborated with many people including, most notably, Lynn Conway. While Gilder admired how their 1980 textbook, *Introduction to VLSI Systems*, had helped train a new generation of chip designers, his book gave Conway relatively little attention. For her part, Conway recalled that "he wasn't interested in really talking with a woman" about the engineering she was doing. Years later, she expressed some disappointment that Mead had befriended a "misogynist right-wing technology futurist" and imagined Gilder's horrified reaction "when he learns [about] my past."[31]

As he considered the past, present, and future of semiconductor chips, Gilder believed he glimpsed something more profound. "The central event of the twentieth century is the overthrow of matter," he wrote. The products of Silicon Valley resembled "books more than steel ingots" because their value was in their "content, not substance." Gilder imagined a future where traditional wealth based on physical resources would fade in importance because the "powers of mind are everywhere ascendant over the brute force of things."[32] It was exactly this thinking that Reagan channeled in Moscow in 1988 when he said "in the new economy, human invention increasingly makes physical resources obsolete."

Seen another way, Gilder argued that technologies like microchips represented the rising value of intellectual property, or what he termed "mind." Once a computer program was written or a chip was designed, its main value, he argued, was embodied in the intellectual labor that went into its creation. "The most valuable capital is now the capital of human mind and spirit," he said, a notion that also found its way into Reagan's speeches. "The laws of the microcosm subvert any attempt to capture, intimidate, confine or overwhelm the exertions of mind by the tyranny of matter," he wrote in one of his book's more extravagant passages. "Mind," he wrote, "flees the corporate traps of the pre-quantum economy."[33]

Gilder's favoring of mind over matter, suggested two broader implications. The first was political. The old economy, he said, relied on ossified hierarchical structures such as unions and regulatory agencies. The "laws of the microcosm" favored a more fluid and decentralized approach. However, leftist professors and entrenched "Silicon Valley patriarchs" opposed the ennobling forces of entrepreneurship and instead sought the safety and security of state-led industrial policy.[34] A chief villain Gilder identified was SEMATECH. Short for "Semiconductor Manufacturing Technology," this entity was created in 1987 as a partnership between the federal government and leading US chip making firms in response to the threat that foreign companies posed. SEMATECH was nothing more, Gilder said, than a "tax-paid collaboration of the established firms . . . [acting] in revolt against the microcosm."[35] Despite decades of funding from the military, Gilder claimed (falsely) that the success of Silicon Valley owed much less to federal patronage than the champions of federally managed industrial policy claimed. Rather, the technological triumphs Gilder trumpeted in *Microcosm* were the result of laissez-faire economics.

Gilder's eloquent, sometimes florid, prose camouflaged the deceptiveness of such claims. In assessing the communities that created the microchip and other computing technologies, Gilder declared they had received little financial input from the government nor contributions from Ivy League elitists with their "Brooks Brothers suits, gentleman Cs, and warbling society wives." Such claims might have surprised Sherman Fairchild who—like Gilder—attended Harvard and whose initial investment arguably launched Silicon Valley. Instead, Gilder attributed America's technological prowess to a certain form of populism. In one of *Microcosm*'s most memorable passages, he extolled the "immigrants and outcasts, street toughs and science wonks, nerds and boffins, the bearded and the beer-bellied, the tacky and uptight, and sometimes weird, the born again and born yesterday, with Adam's apples bobbing, psyches throbbing, and acne galore, the fraternity of the pizza breakfast, the Ferrari dream, the silicon truth, the midnight modem, and the seventy-hour week."[36]

Gilder's second main intervention was spiritual. For Gilder, the overthrow of matter also meant overturning "the great superstitions of materialism," whether these appeared as "a Marxist dialectic or a Midas's hoard." Materialism, Gilder stated, had unjustly reduced "men to mechanism" and subordinated "soul to solid state physics."[37] The rules of the microcosm would reverse these historical trends and elevate a world "in which thought is paramount."[38]

In *Microcosm*, Gilder expressed a particular form of asceticism in which humans could escape "the traps and compulsions of pleasure" that accompanied a preoccupation with the material. In this vision, the computer as a physical "thinking machine" faded against the brilliance of divinely provided human thought and creativity. For Gilder, technology, economics, politics, and religion converged and combined at the microcosm where "free men and women" scaled the "hierarchies of faith and truth seeking the sources of light." Ultimately, at the end of their quest, people would discover a "paradoxical and redemptive cross at the heart of light, radiant in the microcosm and the world."[39]

Reactions to Gilder's *Microcosm*, especially its message of Christian mysticism, varied. Aside from botching the book's title, a review in the *Los Angeles Times* found Gilder's book rewarding even as it encouraged readers to skip over the "tough, boring stuff" and seek out the "entertaining anecdotes."[40] The *Wall Street Journal* gave a similar positive assessment, noting that while Gilder's prescriptions would be "unpopular with advocates of national industrial policies," *Microcosm* would still "enlighten those who reject its economic and political analysis."[41] Langdon Winner offered a strongly dissenting view via the *New York Times*. A professor of science and technology studies, Winner, a former colleague of Joseph Weizenbaum, was perhaps best known for arguing that technological objects embodied particular political beliefs.[42]

Winner charged Gilder with misunderstanding the nature of technological change. By ignoring the "complex relationships" among business, government, and universities, Gilder had reduced new technologies to the "isolated accomplishments by men of rare genius." Perhaps, Winner said, "it is embarrassing for a free-market philosopher to admit that the earliest venture capitalists" were actually the "millions of ordinary taxpayers" whose monies flowed to technology companies through federal support and investment.[43]

In the years that followed, Gilder successfully reinvented himself as technology pundit. Less a fighter in the culture wars of the 1990s, he promoted his views on technology and entrepreneurship via books and articles. *Forbes* even created a special publication, called *Forbes ASAP*, to showcase Gilder's writings. Proclaiming opinions produced material riches for Gilder who was paid as much as $20,000 for a single speaking engagement.[44]

By the early 1990s, the focus of Gilder (and practically everyone else who opined about technology) shifted from microchips to the even more ephemeral empire of the internet. The "telecosm" became Gilder's new neologism for network connectivity, fiber optics, and wireless communication.[45] In an interview with Kevin Kelly, the executive editor of *Wired*, a publication scrutinized and lionized by the "digerati," Gilder predicted that internet bandwidth would soon become "virtually free."[46] While his prediction proved unreasonably optimistic, the number of people accessing the internet did start expanding dramatically. "Surfing the web" entered the lexicon while "cyberspace" became the new global frontier where society, economies, and politics would all be digitally reinvented.[47] But before people could explore this territory, they needed to understand it. Video clips abound of befuddled news anchors and talk show hosts asking, "So . . . what *is* the internet?" or trying to decipher the now-ubiquitous "@" symbol. Help was needed. A guide was wanted. Indeed, a book was required.

### DON'T PANIC

It began with a "Request for Comments" (RFC), one of the mundane documents computer professionals wrote to discuss technical protocols for the ARPANET, a forerunner of today's internet.[48] Programmers and engineers produced hundreds of these documents, stored on the network itself, in which they shared essential technical information. However, RFCs neither circulated much beyond the experts who did computer networking for a living nor were they stimulating to read for their own sake. That changed in September 1989 when a new RFC appeared.

RFC 1118 was written by Edward Krol, a computer scientist at the University of Illinois Urbana-Champaign.[49] Born in 1951, Krol earned engineering degrees at Illinois before taking a job there as a programmer. In time, his duties shifted to overseeing digital infrastructure projects. As part of his job, he attended a workshop in the Washington, DC area in 1989 on the arcane but essential topic of internet protocols. These are the technical specifications that allow for the connection of different, disparate networks. Most of the other students were from military or federal agencies and had little idea what the internet even was. "I realized there was no basic documentation," Krol recalled. "The first instruction a person might read would be 'FTP [File Transfer Protocol] this document and read it.' And the people would be asking 'But what's an FTP?'"[50]

Krol titled his RFC "The Hitchhikers Guide to the Internet," a nod to Douglas Adams's comedy/science fiction classic *The Hitchhiker's Guide to the Galaxy*. In it, he interspersed humor amidst dry sections that explained "Address Allocation" and "ICMP Redirects." Krol hoped his RFC would provide an "indispensable companion to all those who are keen to make sense of life in an infinitely complex and confusing internet." He breezily noted that "where it is inaccurate, it is at least definitively inaccurate" before concluding with reassuring words familiar to Adams's readers: "DON'T PANIC."

One of the people who read Krol's RFC was Michael Loukides. Loukides had double majored in electrical engineering and literature at Cornell University. While attending graduate school at Stanford, Loukides took freelance jobs writing technical manuals for computer-related products. This afforded him access to a Xerox Alto computer, which he used to write a dissertation about William Wordsworth's poetry. After graduating in 1985, Loukides moved back East where, while freelancing as a technical writer, he met Timothy O'Reilly. The encounter set both of them on a new professional course.

O'Reilly was born in Ireland in 1954 but grew up in the San Francisco area. Years later he recalled how his father had read Homer's *Odyssey* to him at night. "I just love books," he reflected, "and I can still smell that book."[51] O'Reilly studied classical literature and philosophy at Harvard. But he was also influenced by a nonacademic mentor named George Simon who advocated "general semantics." This was a quasimystical approach to thinking about language and consciousness that philosopher Alfred Korzybski—known for his adage "the map is not the territory"—had popularized in his 1933 book *Science and Sanity*.[52]

After finishing college, O'Reilly transcribed some of Simon's writings and then, in 1981, wrote a biography of Frank Herbert, author of the science fiction novel, *Dune*.[53] O'Reilly also started authoring technical manuals for companies such as Hewlett-Packard. Since he wasn't a trained programmer, O'Reilly's drew on his experiences with classical languages. "You don't learn ancient Greek or Latin the way you learn a modern language. You're always engaged with a text but you're not speaking it," he explained. "A lot of technical writing is actually rewriting what engineers have given you as you try to turn it into a more human language. I tried to put things in the right order, and tell the story in a way that didn't surprise the reader."[54]

O'Reilly's focus shifted to publishing technical books and computer manuals. In the mid-1980s, he launched O'Reilly & Associates and based it near Boston. O'Reilly packaged many of his books in a plain brown cover and sold them directly at computer expos.[55] They were, in other words, exactly the sort of products that customers flocked to the Computer Literacy Bookshop for. Some of O'Reilly & Associates' most requested publications were its "Nutshell Handbooks," nondescript looking creations with pages held together via staples and sometimes lacking a title on the spine ("Awful," Dan Doernberg recalled, "but only in terms of retail display").[56]

By 1990, O'Reilly relocated his expanding publishing business to Sebastopol, a small town north of San Francisco. As part of this growth, Mike Loukides joined O'Reilly & Associates as a technical writer and editor. Loukides soon convinced O'Reilly that a book based on Ed Krol's *Hitchhiker's Guide* RFC would make for a good book. Speed, however, was essential as other authors were working on similar projects. Moreover, both the subject matter and the number of people accessing the internet were changing very rapidly. Many of these users were starting to explore what Tim Berners-Lee, in 1990, had christened the "WorldWideWeb" (it was initially written with no spaces). In response to these developments, O'Reilly proposed that Krol and Loukides produce a guide to the menagerie of "locations" one could visit via now-obsolete predecessors to the World Wide Web, such as Archie and Gopher.[57]

To explain the internet to the uninitiated, Krol used his experiences working at his university's computer center, where he helped students unravel complex problems using language they could understand. O'Reilly & Associates started to imagine Krol's book might appeal to a wider reading public. The project got a big boost from Brian Erwin. Erwin had done public relations for major publishers in New York, such as William Morrow (one of his clients was Alvin Toffler). He then spent several years helping the Sierra Club expand its membership. At O'Reilly & Associates, Erwin encouraged his colleagues to think less about marketing the book per se and instead get readers enthused about the internet itself. "It resonated with my background," Erwin told one interviewer. "It is very hard getting communities formed. But if a community is already formed and it doesn't know it's a community, that's easy. That's where the internet was at that time."[58]

Erwin sent copies of Krol's manuscript to the usual contacts at magazines and newspapers, as well as congressional staffers who might be interested in technology. But he also sent copies to the people who moderated internet newsgroups and bulletin boards. Many of these online groups were actually mentioned in the book and therefore were pleased to proselytize *Whole Internet*'s pending publication. In other words, Erwin used the internet to promote a forthcoming book about the internet. Erwin's "activism-driven marketing" impressed Tim O'Reilly who embraced the idea that promoting a book's subject matter could be more important than promoting the actual book.[59]

In September 1992, the first edition of *The Whole Internet User's Guide and Catalog* appeared in the (now visually improved) Nutshell book series.[60] The title, of course, riffed on the acclaimed *Whole Earth Catalog*, which Stewart Brand had first published in 1968. *The Whole Internet*'s cover featured an illustration of an alchemist that was adapted from a nineteenth-century engraving. Instead of the traditional glassware and other tools associated with alchemy, however, the sage was shown using a geometer's compass and globe to mark out distant geographic points. The message conveyed to readers likened the internet to an undiscovered land.

A reader, after flipping past blurbs that praised the book's "breezy conversation style," arrived at its intended purpose. Krol assumed his readers were "computer-literate, but not network-literate." The internet, he said, could be a "friendly place to meet people" instead of a "machine-dominated wasteland where antisocial misfits sporting pocket protectors flail away at keyboards." Readers didn't have to be experts in "telephone lines, data communications, and network protocols." "Getting a handle on the internet," he admitted, "is a lot like grabbing a handful of Jell-O." All a person needed was "a spoon" and then they could "dig in and start eating." Krol's book was that utensil.[61]

Over the course of fifteen chapters, Krol explained how the internet worked, what was legally (and socially) allowed on it, and, of course, how to access it. The book—the first edition ran 400 pages—ended with some 100 pages of online resources, assembled by Loukides, which covered everything from "Aeronautics and Astronautics" to "Zymurgy." Attentive readers, meanwhile, discovered nerdish Easter eggs scattered throughout the book. Krol was a fan of *The Adventures of Buckaroo Banzai across the Eighth Dimension*, a wacky science fiction movie from 1984, and his book's internet tutorials included such terms as the "Yoyodyne Corporation" (first coined by Thomas Pynchon, most notably his 1966 novella *The Crying of Lot 49*), and "overthrusters" borrowed from the film.

The formula proved enormously successful. At the Computer Literacy Bookshop, for example, Dan Doernberg watched customers eagerly purchase copies of *The Whole Internet*. It became one of O'Reilly & Associates' big successes as more than 300,000 copies sold within a few years. Krol, meanwhile, gave scores of talks to librarians, congressional staffers, and other groups eager to understand the internet's possibilities. In April 1994, a second edition, with more than 100 additional pages, appeared. Positive press helped spur sales, which eventually surpassed one million copies. At the

*New York Times*, technology journalist John Markoff called it the best way for novices to navigate the "thicket of information known as the internet."[62]

Perhaps the biggest recognition came from librarians. In 1995, the New York Public Library (NYPL) was preparing to commemorate its one hundredth anniversary. As part of the celebration, the NYPL asked its staff to suggest books that "had a significant influence, consequence, or resonance" in the past century. In all, the NYPL created a list of more than 1,100 works—Upton Sinclair's *The Jungle* and Freud's *The Interpretation of Dreams* drew the most votes—before whittling it down to just 159 titles. Naturally, once the list was made public, "gasps, laughter, and righteous indignation" followed.[63] Nonetheless, the ensemble of books formed the basis for an exhibition called *Books of the Century*. Krol's *The Whole Internet* appeared on the final list. Praised for its historical significance, the NYPL placed it in the "Economics and Technology" part of the exhibit, alongside works by Milton Friedman, Max Weber, and Jane Jacobs.

Although the NYPL's celebration spoke to the enduring value of printed works, the exhibit catalog noted that "there is much talk today" of paper books as "threatened species."[64] Krol's book itself was hardly endangered as revised print editions and multiple translations continued to appear in stores around the world. Nonetheless, within a year of its arrival, *The Whole Internet* did a disappearing act of sorts. Like other media content, its paper pages were converted into digital form, absorbed by the computer, and made accessible via the very medium for which it served as a guide. But there would be no driving on the information superhighway if one couldn't get their vehicle into gear. In other words, millions of frustrated computer novices and internet neophytes still wanted friendly advice about how to use their computers.

### MANUALS FOR THE MASSES

In 1958, customers could instantly spot the new books, with their cheery black and yellow covers, when they walked through their local bookstore. The slim volumes offered information and advice, presented in an eminently readable manner by experts in the subject, for navigating, even mastering, seemingly esoteric information. Students, eager for a shortcut to literary understanding, bought *Cliffs Notes* (no apostrophe) by the millions.[65] Jump ahead several decades . . . in 1991, customers could immediately spot a new series of books, with cheery black and yellow covers. They offered information and advice, presented in an eminently readable manner by experts in the subject, for navigating, even mastering, seemingly esoteric information. Computer owners, eager for a shortcut to technical knowledge, bought *For Dummies* books by the millions.

Any confusion that students experienced in trying to decode the literary messages in something like *Pride and Prejudice* paled in comparison to the befuddlement many new computer owners expressed. The computer industry itself didn't make the problem better as its manuals employed a "lexicon bursting with ungainly acronyms, twisted English, and baffling technical terms" that left the "technologically illiterate" suffused with "fear and loathing." As a result, non-experts were left with the sense that they couldn't understand computers. "They're not stupid," one well-placed industry observer noted, "they're just uninformed."[66] Stupid, of course, suggests an innate incapacity to learn. Dumb, on the other hand, is a different but treatable condition.

To understand the *Dummies* phenomenon, one has to appreciate the trajectory of Patrick J. McGovern, owner of International Data Group (IDG), which published the series. Born in 1937, McGovern first encountered computers via Edmund Berkeley's *Giant Brains*, which he found in a Philadelphia library. Berkeley's book inspired the teen to build a rudimentary logic machine that could play tic-tac-toe. The device won a prize at a science fair and helped McGovern secure admission to MIT.[67] While an undergraduate, McGovern responded to an advertisement Berkeley posted seeking an assistant for his magazine, *Computers and Automation*. McGovern rose to associate publisher before deciding to chart his own fortune; he started IDG in 1964 and headquartered it in a Boston suburb.[68]

McGovern originally planned to collect market information on the computer industry and sell it to executives eager to keep up with the rapid pace of change in technology. In 1967, McGovern branched out with the newsletter *Computerworld* and publishing soon became IDG's main activity. McGovern envisioned *Computerworld* as the "first newspaper for the full computer community," offering a weekly assortment of news, information, and gossip.[69] What began as a slim newspaper of less than a dozen pages soon grew into a magazine-like publication of a hundred or more advertisement-filled pages.

By the 1980s, IDG had launched scores of computer-related publications, with accompanying translations, into the global market. These included dependable monthlies such as *InfoWorld*, which IDG bought from Jim Warren in 1979, and *PC World*, which started in 1983.[70] A 1986 market forecast noted that IDG was the "world's largest computer publishing and research company." By 1990, the company's global revenues surpassed $500 million (with a 10 percent profit margin) and its publications reached over four million readers annually.[71] But IDG had fallen behind its main competitor, Ziff-Davis Publishing, which published *PC Magazine* and *PC Week*. Notably, IDG trailed Ziff-Davis when it came to reaching readers who lacked basic knowledge about computing.[72] As a result, IDG was primed for an opportunity to expand its readership.

There is a long history of consumers being baffled by their gadgets. In the early twentieth century, users of bicycles and automobiles experienced a decided sense of "unfriendliness" as people tried to understand, use, and repair the machines. The arrival of personal computers and the myriad array of software packages available for them opened new realms of outrage and frustration.[73] As one journalist remarked, "the 'personal' in 'personal computing' means that once you have bought a machine, the problems are all yours personally."[74]

This problem appeared to be especially acute for new owners of IBM PCs and the various "clones" sold by companies such as Dell and Gateway. In 1990, these machines represented about 85 percent of the home computer market. Just turning the machine on challenged some neophyte users. IBM had located the power switch on some models in nonobvious locations so they would not be toggled off accidentally. Once a PC machine booted up, users were greeted with a "command line" (something like C:\>) and a flashing cursor that, to some, appeared despotic. Entering instructions in MS-DOS—short for Microsoft Disk Operating System—required familiarity with an array of inscrutable typed commands. Copying a file, for example, from one disk to another, required a user to input C>COPY READMEBOOK. DOC A: (no typos allowed). Intuitive, it was not.

In the late 1980s, Dan Gookin, a technical writer living in Southern California, had an idea for a book written for people with no prior experience using DOS. While he was at college, Gookin bought his first computer, a TRS-80 made by Radio Shack, on which he learned programming and wrote fiction. After graduating in 1983, he produced articles and manuals for a series of publishers before starting a company he cheerily called Not Another Writer.[75] With titles like *Hard Disk Management with MS-DOS and PC-DOS*, Gookin's primary audience was so-called power users—people with considerable technical knowledge who wanted to make their work even more productive.[76] "People would describe me as a writer who does technical things as opposed to a technical guy who writes," Gookin noted. "One of my goals was to have empathy for my audience and be able to connect with them."[77] By 1991, his name appeared on the covers of more than a dozen books and he had established himself as a well-regarded author with a flair for humor.

Gookin had a short proposal for his new book idea, tentatively titled *The Idiot's Guide to DOS*. A major inspiration came from another classic how-to book. In 1969, John Muir, a former aerospace engineer, wrote a manual for Volkswagen owners. Muir called his self-published book *How to Keep Your Volkswagen Alive: A Manual of Step-By-Step Procedures for the Compleat Idiot*. It blended practical advice with Zen-like philosophical nostrums. "Taking it easy," Muir explained, was both a lifestyle choice and a reminder not to drive your Beetle too fast. Muir's enormously popular book was exactly the sort of do-it-yourself manual promoted in the pages of *The Whole Earth Catalog*.[78] Gookin's father owned a copy of Muir's book and used it when fixing the family's car. Dan Gookin planned to take the same ideal of self-sufficiency and apply it to personal computers.

At a writer's conference in the spring of 1991, Michael McCarthy, an editor for IDG, heard Gookin lament that too many computer books were boring and humorless. McCarthy had been considering commissioning a book for a wider readership and the two started talking. In April 1991, Gookin sent McCarthy a formal proposal for what became *DOS for Dummies*, noting that it "feels like an 'anti-computer book,' which tickles me."[79]

Buoyed with a modest advance of $6,000, Gookin produced a clean draft in a matter of weeks.[80]

By the fall of 1991, IDG was ready to release *DOS for Dummies* to the wider world. Before that could happen, editors at the company had to circumvent the complaints of Patrick McGovern. IDG's owner thought Gookin's title insulted readers and ordered the project killed. However, the book had already entered its first print run of 5,000 copies. In October 1991, the first copies of *DOS for Dummies* appeared in bookstores, including the Computer Literacy Bookshops, and it quickly sold out. McGovern retracted his doubts and, within a month, Gookin's user-friendly DOS manual was in its third printing as IDG scrambled to meet customer demand.

Gookin's book, which IDG priced at $16.95, had a bright yellow and black cover that was somewhat primitive looking. It was dominated by a cartoon hand holding a (protest?) sign proclaiming "DOS for Dummies." Underneath, in a smaller font, the target audience was clearly stated: "A Reference for the Rest of Us!"[81] After flipping past pages of testimonials ("Thanks! I feel human again," read one), readers were started at the very beginning with instructions on how to turn their computers on and off. From that point onward, Gookin provided clear and concise directions on everything from swapping disks and changing directories to understanding DOS error messages and connecting peripheral devices. Although jokes were interspersed amidst the instructions, he understood levity had a proper time and place. "No one trying to recover lost files wants to hear a joke," he noted, "you need to respect readers' intelligence and you don't want to annoy them."[82]

The formula worked. By 1993, IDG had sold more than 1.3 million copies of the "irreverent primer for the perplexed."[83] The *New York Times* described *DOS for Dummies* as a "light-hearted survey of the operating system everyone loves to hate." Readers too embarrassed to admit they were in fact dummies—but not stupid—could order it directly from the publisher.[84] Recognizing it had a major hit, IDG quickly sought more authors for its stable. For example, Andy Rathbone produced *Windows for Dummies* in 1992. IDG eventually sold some fifteen million copies of it and the guide appeared in multiple versions and translations.

The *Dummies* books were meant to appeal to a broad swath of the population. But other computer guides were written for more specific communities. The LGBTQ+ community, for example, could explore internet resources via the book *Gay and Lesbian Online*, which appeared in several editions in the mid-1990s. Other publications aimed to reach members of the evangelical Christian community who were seeking computer guidance.[85] Eager to demonstrate that African Americans would not be "hoboes and lame hitchhikers" on the information superhighway, Stafford L. Battle and Rey O. Harris assembled *The African American Resource Guide to the Internet and Online Services* that McGraw-Hill published. Like *The Whole Internet*, it contained pages of resources and web links but embraced the maxim that "the New Black Power is knowledge."[86]

Although *Dummies* books were easy for customers to find on bookstore shelves, the books themselves had a short shelf life due to rapid changes in home computer technology. Many of the instructions in *Dummies* books were out-of-date as soon as a new version of a program or operating system appeared. Meanwhile, public interest in the internet and the World Wide Web stimulated consumer demand for even more manuals. In 1993, IDG published *The Internet for Dummies*, which former RESISTOR John Levine coauthored. In his author's biographical note, Levine credited Ted Nelson with the idea that "everyone can and should understand and use computers," a maxim the *Dummies* series embraced.[87] Other companies muscled into the market for manuals. Penguin Putnam, for instance, copied IDG with books like *The Complete Idiot's Guide to DOS* (1994).

IDG embraced Nelson's idea about "everyone" perhaps a little *too* enthusiastically. Eventually some 1,000 titles appeared in the *Dummies* series, explaining everything from iguanas and investing to dating, sex, marriage, and divorce.[88] The success of the *Dummies* series coincided with the popularity of chain bookstores such as Borders and Barnes & Noble. *Dummies* books were relatively standard in tone and organization, highly visible to customers and, given the wide range of subjects in the series, scattered throughout entire bookstores. Despite publishing *Branding for Dummies*, IDG eventually diluted the quality of their own product. "For a while, in the late 1990s, they would put out any piece of shit that had 'For Dummies' on it, and the thing would sell," Gookin complained.[89] In 1998, IDG even bought the venerable *Cliffs Notes* franchise, paying over $14 million for what one company executive extolled was "part of Americana."[90]

Despite their rapid obsolescence and the rampant proliferation, the computer-focused *Dummies* books served an important purpose. They truly were manuals for the masses. Their simplicity and user-friendliness stood out in comparison to the books O'Reilly & Associates published which assumed readers had a fair degree of technical experience. A sense of need, perhaps even desperation, motivated people to buy the computer books in the *Dummies* series. Inside these mass-produced manuals, readers found both advice and assurance.

### NAVIGATORS

In the early nineteenth century, Zadok Cramer operated a printing and publishing business in the then-small settlement of Pittsburgh, Pennsylvania.[91] In the decades that followed, the city would become renowned for manufacturing the sort of industrial commodities George Gilder would later claim had been eclipsed by the more ephemeral products of "mind." But the growing town's location at the intersection of three rivers meant that thousands of settlers passed through it as they moved westward. Seeing an obvious business opportunity, in 1801 Cramer wrote and published a compact travel guide he called *The Navigator*.[92]

Almost two centuries after Cramer's *Navigator* helped white settlers travel to unknown lands, which they hoped to exploit, internet novices wanted a similar tool. One of the best known of these appeared in December 1994. For a time, Netscape Navigator was the most popular way people accessed the World Wide Web. But, before Netscape's much-mythologized browser became such a dominant platform, it had a strong competitor.

With Ed Krol's *Whole Internet* enjoying success, a more audacious plan was taking shape at O'Reilly & Associates. This effort was spearhead by Dale Dougherty, who had joined the company in the mid-1980s after majoring in English at the University of Louisville. Whereas some publishing companies viewed instructional books about computing as paperbound extensions of an established academic subject, Dougherty saw the situation differently. "We saw computing as a practice, not a discipline," he explained, "so we wrote O'Reilly books for practitioners."[93] At the same time, since the company published books *about* computers, it was natural to imagine a future where books might be delivered *by* computers and read *on* them.

However, it wasn't clear what an "electronic book" might look like. Would it somehow replicate the physical form of the traditional printed book?[94] Perhaps it would present a book's content on a computer screen in some other format. This latter approach was the path taken, for example, by Michael S. Hart. As a student at the University of Illinois in 1971, he posted a digitized copy of the United States Declaration of Independence via his school's mainframe computer. Because the computer was connected to an external network, the digital text could circulate and be accessed electronically by other people. Over the next several years, the electronic books Hart "created" in this manner formed the basis of Project Gutenberg.[95] Of course, one could not access Project Gutenberg (or any other digitized books placed online) without an internet connection, something which remained a relative rarity for most computer users prior to the mid-1990s.

Another medium that intrigued Dougherty was the CD-ROM, which was short for "Compact Disk Read-Only Memory." Microsoft Press published a (paper) book about this technology in 1986 that extolled its extensive storage capacity for text, images, and sounds. In an accompanying foreword, Bill Gates suggested that this so-called "new papyrus" was the "heart of a new 'viewer,'" which would be "more similar to the stereo or television you have in your home than to a computer." Consisting of a "CD-ROM player, a screen, and a pointing device" and perhaps connected to a flat display screen, it would "go beyond our traditional ways of being entertained, of learning, and of gathering information." Microsoft, Gates noted, had even created a research group "to focus on this new opportunity."[96] However, it was by no means the only company interested in CD-ROMs as a publishing medium. As Dougherty recalled, a proliferation of proprietary interfaces and a lack of standards inhibited cross-platform use.[97] But because users navigated the multimedia content on CD-ROMs via hypertext—a variation of what Ted

Nelson imagined decades earlier—Dougherty started attending conferences on the subject.

At one of these gatherings, Dougherty met Tim Berners-Lee and learned about computer networking being done at the *Conseil européen pour la recherche nucléaire* (CERN), the physics laboratory in Geneva where Berners-Lee worked. Amazingly, from today's perspective, the conference organizers rejected Berners-Lee's paper submission on the grounds that the work was incomplete and not sufficiently engaged with hypertext scholarship. Instead, Berners-Lee was relegated to presenting his ideas for the World Wide Web via the decidedly low-tech medium of a poster.[98] Undaunted, he used the opportunity to describe features such as "hypertext markup language" (HTML) and "hypertext transfer protocol" (HTTP). Now lauded as tools that have defined the World Wide Web for more than a quarter-century, in 1991 they were untested novelties, not universal standards. Curiously, one influence on Berners-Lee was a "musty old book" called *Enquire Within Upon Everything*, which first appeared in 1856. Years later, Berners-Lee recalled encountering this book as a child and finding it "suggestive of magic" because it offered readers a "portal to a world of information." This was an experience he later sought to re-create digitally.[99]

History, the adage goes, is written by the victors. In standard histories of the World Wide Web, Netscape appears as the big winner, its success secured in 1994 when the company released its Navigator browser. Based on software called Mosaic, which some of Krol's colleagues at the University of Illinois's National Center for Supercomputing Applications had written, it helped fuel the exponential growth of the web. Before this happened, however, it was unclear which program would emerge as the champion. By December 1990, for instance, Berners-Lee had created his own "W3" browser (short for "WorldWideWeb"). Other contenders appeared at Stanford ("Midas-WWW") and Cornell ("Cello"). Then, in May 1991, Dale Dougherty got wind of another very promising contender called "Viola" from the University of California.

Viola was an acronym for "Visually Interactive Object-oriented Language and Application."[100] Its creator was Pei-Yuan Wei who was born in Taiwan and moved to California as a teenager. As an undergraduate studying geography at Berkeley, Wei joined the "eXperimental Computing Facility," a campus group devoted to computer science. In 1990, Wei started writing what eventually became Viola.[101] In doing this, he was influenced by a particular application for Apple computers called HyperCard. This was, in essence, a digital version of notecards that one might collect in an old-fashioned Rolodex. HyperCard users could, for example, navigate from one digital card to another while also searching across a whole collection of them for particular passages of text.

In May 1991, the internet-based newsgroup alt.hypertext announced that a version of Wei's Viola—it was described as a "hypercard-like prototype system"—was available for free public use.[102] In the documentation accompanying his program, Wei noted that, with his graduation approaching, "Viola's future development is uncertain." Dougherty quickly emailed Wei and asked if he wanted to write a book on Viola, perhaps as a "book/software combo." In short order, the two men were discussing whether Viola could become a tool for electronically publishing books that O'Reilly & Associates distributed in their traditional paper form.[103]

A web browser is essentially a program that allows users to read hypertext and navigate from one hyperlink to the next. Although Wei did not set out to create such a program, by the end of 1991, he had produced a working version of one. Using on-screen links, users could point and click their way to websites instead of having to type in instructions at a cursor prompt as with Gopher or Telnet. Viola eventually included features that are standard in all of today's web browsers, including "forward" and "back" buttons as well as a history of sites visited. Wei's work impressed Berners-Lee who called Viola a "very neat browser, usable by anyone, very intuitive and straightforward." CERN encouraged its use and even discussed the possibility of licensing it.[104] "The future is very exciting," Berners-Lee wrote Dougherty.

When Krol's *The Whole Internet* first appeared, it included a short chapter on the World Wide Web. It explained what hypertext was and told readers how to navigate among the burgeoning number of web pages. Although "the Web is very much under development," the book claimed, "ViolaWWW is probably the most feature-rich" way to navigate it. In fact, Krol's chapter used images of Viola to illustrate what readers might encounter when they were online. With the near-simultaneous appearance of *The Whole Internet* and Viola, two threads, one pulled from the world of traditional book publishing and another from the uncharted territory of electronic publishing, started to come together.

The addition of a third thread suggested an even greater synergy. In the early 1990s, the National Science Foundation, which managed a substantial part of the internet, was beginning to loosen restrictions on its commercial use. For example, Congress passed legislation in late 1992 that modified the science agency's "Acceptable Use Policy" to read: "computer networks . . . may be used substantially for purposes *in addition to* research and education."[105] Those three new words would have an incalculable impact on the internet's future. More private companies were beginning to offer internet connections and these firms had their own policies on acceptable use. So, when *The Whole Internet* first appeared, it noted the considerable ambiguity in terms of both legality and user behavior (at the time, called "netiquette"). Krol, in fact, devoted an entire chapter to "What's Allowed on the Internet" to introduce neophytes to computer networking while not destroying existing community norms.

Since O'Reilly & Associates had just published a very successful book about the internet, it was logical to imagine an online version that could be navigated via Viola or some other web browser. After all, *The Whole Internet* book included pages of websites. Why not make these into hyperlinks that a user could visit via point-and-click? Even more compelling was the realization that having *The Whole Internet* available as an online publication would promote the internet itself. So, when Dale Dougherty and Tim O'Reilly saw Viola perform for the first time, they told Wei, "That's not a demo. That's a product."[106]

In the fall of 1992, just after Krol's book was published, something new appeared at the Computer Literacy Bookshop. It was a kiosk where visitors, using a personal computer O'Reilly's company provided, could access an online version of *The Whole Internet*. Using Viola as an interface, customers could explore websites and other resources found in the print version of Krol's book. O'Reilly recalled it as very rudimentary: a "stand with a computer on it, some company branding, and a sign with something like 'Try

The Internet.'"[107] It enjoyed only modest popularity, not surprising given that most customers were already familiar with the internet.

Then, in August 1993, the *Computer Underground Digest*, an electronic newsletter that reported on internet news, announced a "new experiment in online publishing" from O'Reilly & Associates called "Global Network Navigator." Dougherty chose the name as an homage to Zadok Cramer's nineteenth-century guide.[108] Just as *The Navigator* had helped people explore rivers as avenues of adventure, travel, and commerce, O'Reilly & Associates hoped GNN would do likewise for 1990s-era digerati.

Visitors to GNN first landed on a home page. Then, by following hyperlinks, they could travel to an interactive version of *The Whole Internet*, browse the news, or read articles in a new, online-only publication called *GNN Magazine*. In keeping with the deliberately archaic look of *The Whole Internet* book, Jennifer Niederst Robbins, a graphics designer at O'Reilly & Associates, chose an unthreatening-looking hot air balloon, with a map of the world on it, as the main emblem for GNN's home page.[109] *Wired* magazine, a cheerleader for all things internet, called GNN "an editorially and visually sophisticated front end to the Net" and said its design captured "the shift from punch cards and printouts to keyboards and screens."[110]

However, what caught the attention of many observers was the inclusion of something called the "GNN Marketplace." Functioning more like a traditional telephonic White Pages, this was a virtual space where businesses provided links to their company's web pages. GNN's first customer was a law firm that paid $5,000 in 1993 to secure a digital presence on the web. Soon after GNN debuted, other companies, including Digital Equipment Corporation and NordicTrack, signed up as advertisers using what would come to be known as "banner ads." Tim O'Reilly was aware, of course, about community resistance to commercializing the internet. At a trade conference in 1993, he discussed the inclusion of advertising via GNN with Stephen Wolff who managed the National Science Foundation's internet network. According to O'Reilly, Wolff accepted the plan on the theory that GNN had been created by a company that produced instructional books and manuals and thus comported with the science agency's larger mission of supporting research and education.[111]

Despite this tacit blessing, there was still the risk of backlash for violating long-held (but rapidly dissolving) internet norms. Soon after GNN was unveiled, the *Wall Street Journal* reported that O'Reilly & Associates was "hustling to line up advertisers" and predicted that the internet would soon "get hit with ad clutter."[112] Tensions over whether and how one used the internet for business were made apparent in *Time*'s first major article about the internet. According to the news magazine, a pitched battle "for the soul of the internet" was happening. The fight was ignited when two lawyers placed an advertisement that forced its way to thousands of electronic bulletin boards where it was seen repeatedly and unwantedly by millions of internet users. Seen as "a declaration of war," in an instant, the lawyers became "the most hated couple in cyberspace."[113] Nonetheless, bringing commerce to cyberspace appeared inevitable even as long-time internet users expressed hopes it would be done sensitively and selectively.

O'Reilly & Associates took pains to explain how their approach to business differed. For example, GNN's users would only see advertisements if they purposefully navigated to GNN's "Marketplace" section.[114] In this sense, the company was following an established model from traditional periodical publishing. For years, computer magazines included "bingo cards" on which readers could circle numbers that corresponded to advertisers in that issue. Readers would mail these back to the publisher and, weeks later, receive promotional material. GNN, Tim O'Reilly explained, was delivering the same sort of commercial content but instantaneously.[115]

At its peak, about twenty people at O'Reilly & Associates, some with backgrounds as librarians, were working on various features and services for GNN. Given that the web was still relatively small, they could flag interesting websites and offer short reviews of them for GNN's users. By the end of 1994, the hybrid platform, which functioned as a combined publication, web portal, and curated collection of noteworthy websites, attracted some 80,000 registered users who accessed 150,000 pages or more per week.[116] To help neophyte explorers get online, O'Reilly & Associates marketed a new product called "Internet in a Box." For $149, purchasers received Krol's (paper) book, a subscription to GNN, and software that would connect users to the internet.[117]

Both the internet and the World Wide Web were transforming at a dizzying rate. *The Whole Internet* went through multiple printings and editions to keep apace but, by April 1994, Krol's book no longer recommended Viola as a web browser. Wei's creation was eclipsed for good in December 1994 when Netscape released version 1.0 of its Navigator browser for free, prompting millions of people to download it over the internet. A year later, Microsoft released its own product, called Internet Explorer that was based on similar Mosaic source code that underpinned Netscape. Microsoft's entry set the stage for years of "browser wars" and eventual antitrust litigation.

Global Network Navigator itself was bested when Yahoo debuted in January 1995. Its success was boosted by former writer Michael Moritz, now a wealthy venture capitalist with Sequoia Capital. Meanwhile, America Online purchased GNN from O'Reilly & Associates in June 1995 for $11 million. The reprieve was short-lived, however. GNN's new managers were unable to compete with powerhouses such as Yahoo and Excite, which combined search engines with web portals. In the breakneck pace of what the digerati called "internet time," America Online shuttered GNN at the end of 1996. (This, ironically, was the same year that IDG published the quickly obsolete *Global Network Navigator for Dummies*.)

The nineteenth-century travelers who had used Zadok Cramer's *The Navigator* to guide them through uncertain territory understood that all rivers eventually lead to the sea. Ed Krol's book helped neophyte explorers travel to the vast but ephemeral realm of cyberspace. But, by the mid-1990s, once-littoral regions of World Wide Web had become an enormous ocean no single book could even survey, let alone navigate. Millions of voyagers made their way to cyberspace but were still left looking for someone to explain what it all meant. And, just like the explorers who read Cramer's guidebook and then lit out to find their fortune, many cybernauts wondered how they might get rich in this new virtual world.

# 9 ALL TOMORROW'S PARTIES

For one writer, the long party ended with a brief and angry epithet. Through-out the 1990s, Michael Krantz had contributed articles for *Time* magazine, reporting on information technologies, Silicon Valley, and, most notably, the giddy atmosphere of the dot-com boom. Soon after Krantz left *Time* in early 2000 for a well-capitalized internet start-up, the stock market experienced a volatile series of "corrections." These downturns were felt most in the tech-heavy NASDAQ. In a matter of weeks, the index dropped nearly 30 percent as traders, investors, and markets grew alarmed and then panicked. By the time the decline ended in late 2002, some $5 trillion of wealth had disappeared. When Krantz surveyed the financial debris, his reaction was, "New Economy, my ass."[1]

Largely forgotten today, the term Krantz ruefully referred to was practically unavoidable in the 1990s. In 1996, as the Clinton administration won a second term, the phrase "New Economy" appeared in magazines and newspapers about 400 times. Two years later, as the dot-com bubble expanded, references to it more than doubled. By 2000, when the market was poised to collapse, the phrase appeared in mainstream media publications a staggering 22,848 times.[2] Whatever the New Economy actually was—on this point, considerable debate existed—it was undeniably pervasive.

In the early 1990s, "newness" of all stripes was au courant. The demise of the Soviet Union had set the stage, some experts argued, for a less ideologically polarized geopolitical order with capitalism reigning supreme. After experiencing years of stagnant wages, low productivity, and curtailed economic growth, followed by a painful recession in 1990–1991, many Americans wanted something new that promised better possibilities. Bill Clinton's 1992 election signaled a rejection of older Republican candidates who appeared out of touch with economic realities. And, of course, the end of the millennium was fast approaching, bringing with it a certain fin de siècle mindset that combined apprehension with acquisitiveness.

As a concept, the New Economy proved remarkably plastic. When *Business Week* used the term in 1981, it meant corporate restructuring and an adjustment to recent oil shocks. Four years later, the magazine claimed it now referred to the service sector and high technology, particularly computing technologies.[3] This was the sort of postindustrial transformation that business consultants and futurists like Alvin Toffler had long predicted. A decade later and *Business Week* now attributed the "triumph of the New Economy" to the twin forces of globalization and new information technologies. Economic winners would be those people that Robert Reich, Clinton's Secretary of Labor, called "symbolic analysts," another term for people who spent their days working in front of computers.[4] And, finally, by the late 1990s, the New Economy had simply become synonymous with the "Internet Economy."[5] Whatever the New Economy was, networked computers were a keystone. In half a century, computers had gone from being electronic brains that did calculations to something intimately associated with making money and marketing an ideology.

The conflation of the New Economy with information technologies was accompanied by a dizzying sense of possibility as wizened venture capitalists and amateur financiers alike plowed money into dot-com companies in the hopes of making quick profits. *Wired* magazine, an enthusiastic New Economy promoter, predicted a "Long Boom" resulting in a quarter-century of "prosperity, freedom, and a better environment for the whole world."[6] *Dow 36,000*, a bestselling book published just before the internet bubble burst, captured this celebratory zeitgeist, its dustjacket proclaiming that "amazing profits can be made, but the time to act is now!"[7] When the book appeared, the Dow Jones Industrial Average index, which tracks thirty prominent companies, hovered around 10,000 before starting to drop. (The Dow *did* eventually reach 36,000 . . . but not until 2021.)

Of course, many experts harbored skepticism about the New Economy's premises and promises. In December 1996, Alan Greenspan, chair of the Federal Reserve Board, cautioned that "irrational exuberance," due in part to low interest rates he had championed, had "unduly escalated asset values." Greenspan's phrase was later adopted by economist Robert Shiller as the title for his timely book *Irrational Exuberance*, which appeared in March 2000, just as the more narrowly tech-focused NASDAQ began to implode.[8]

A diverse array of books appeared that informed, excited, and, at times, inflamed public imagination about the internet's economic and social potential. George Gilder, for example, boosted the idea of the New Economy and a future that was digital and networked. So did Kevin Kelly, *Wired*'s executive editor, in his book *New Rules for the New Economy*. His book (which Gilder blurbed) served readers a "rapid-fire blur of neologisms and breathless declarations" like "Feed the Web First" and "No Harmony All Flux" in ways that suggested the business and New Age sections of the bookshop had merged.[9]

Esther Dyson stood out among the era's many pundits and promoters. The scion of a prominent scientific family, Dyson built a reputation as an astute industry analyst in the 1980s. By the late 1990s, she was regarded as the "most influential woman in all the computer world." As the "First Lady of cyberspace," Dyson cultivated an image as a peripatetic global traveler who was well paid for her views about information technologies, free markets, and the implications of an increasingly wired world.[10]

After Marshall McLuhan became an academic star in the 1960s with his phrase "the medium is the message," he indulged in some wordplay in a new book called *The Medium Is the Massage*. By this, the media guru meant that new communication technologies manipulated readers. "Media work us over completely," McLuhan said. "They leave no part of us untouched, unaffected, unaltered."[11] By the 1990s, millions of people were having similar thoughts and questions about the internet's implications. Sensing an opportunity, Dyson decided to try to explain the Net (as it was often still called at the time) in her book, *Release 2.0*. Even as more and more content appeared online, Dyson's choice implied that traditional books still had power to move markets and influence minds.

In terms of appearances, Norbert Wiener and Esther Dyson could not have been more different. Reporters had routinely caricatured the author of *Cybernetics* as a roly-poly wunderkind who became an absent-minded professor, frenetic with energy and puffing on his omnipresent cigar. Over time, journalists likewise created a stock image of Dyson: "pixieish, puckish" with a "sartorial style that runs to baggy jeans and promotional polo shirts" albeit accented with pearl earrings.[12] Rare was the article that didn't mention Dyson's eccentricities, such as her legendarily disheveled office and compulsively maintained swim routine. And, as had been the case for Wiener, words like "prodigy" and "guru" made regular appearances in journalists' accounts of Dyson, as did mentions of a personality ("curiously undersocialized . . . brittle and sometimes abrasive," read one description) that didn't always mesh well with others.[13]

Whatever the Net was, it was a big, complex, and abstract thing freighted with portentous implications. But Esther Dyson grew up surrounded by adults who had devoted their professional lives to explaining the implications of big, complex, and abstract things. J. Robert Oppenheimer, hired her physicist-author father, Freeman Dyson, for a lifetime appointment at the Institute for Advanced Study in Princeton. As a child, she met Albert Einstein, and John von Neumann frequented the Dyson's dinner table. While Freeman did pathbreaking work in theoretical physics and consulted for the government, Esther's biological mother, Verena Huber, earned a doctorate for her work at the nexus of algebra and logic.

Esther was born in 1951, the same year that the IAS Machine, a groundbreaking computer designed by von Neumann, entered service. She later played with parts leftover from its construction. Verena and Freeman divorced seven years later.[14] In journalists' portrayals, Dyson appeared as a self-sufficient child who, according to her younger brother George, was "totally in control" and "fascinated with money."[15] She hand-produced the *Dyson Gazette*, an eight-year-old's take on family news. "When I was very young, I wanted to be a child psychologist partly because I thought I was fascinating," she told a reporter. "Now that I am older, I still like to understand people like me and, in some sense, computers are the way."[16]

Dyson received early admission to Harvard in 1968, just five years after women could earn degrees from that school. Formal academics were not priorities. "My official major was economics," she recounted, "but I managed to avoid going to class too often."[17] Instead, Dyson spent her time at the *Harvard Crimson*, the school's newspaper. She acquired a reputation as a "ferocious proofreader," an especially important task when stories were still typeset by hand before being sent to the timeworn hot-metal presses in the basement.[18] "Much as I love the digital world," she recalled, "I also love the old world of movable lead type."[19]

Writing was central to both Dyson's identity and income. After graduating from Harvard in 1972, she worked at *Forbes* as a fact-checker before being promoted to reporter. Dyson credited the magazine with giving her an education in how economics actually worked. Dyson's penchant for successfully seeking out interviewees also helped her acquire an enviable roster of contacts from the business world.[20] By 1977, she had left *Forbes* for Wall Street, working as a securities analyst specializing in computers and software. "A big fan of markets," she recalled, "I learned how they worked by being in the middle of them."[21] Within a few years, Dyson tired of writing reports for uninformed bankers, and decided she wanted to "get closer to the computer industry" and write more directly about its products and people.[22]

Dyson discovered her new career path via Benjamin Rosen, another former securities analyst, who we met briefly last chapter. Rosen had earned degrees in electrical engineering from Caltech and Stanford before getting an MBA from Columbia. Rosen's strong technical background and personal charisma made him an invaluable information source for journalists and investors interested in semiconductor companies. In the mid-1970s, while working for Morgan Stanley (a New York-based investment banking firm), Rosen started a "wide-ranging, occasionally perceptive, and often funny" newsletter about the electronics industry.[23] In 1979, with an established reputation as an accurate industry forecaster, he renamed it the *Rosen Electronics Letter* and launched out on his own.

By 1982, Dyson made another career shift and accepted an offer to work for Rosen. A year later, Dyson bought Rosen's newsletter outright. Dyson's purchase also included the right to continue hosting the "Personal Computer

Forum" (later rebranded as "Platforms for Communication Forum"), an annual gathering Rosen had started for people from the computer industry. Now armed with both an influential newsletter and a highly sought after annual event, a "gilded avenue to Silicon Valley" had opened up for Dyson.[24]

Dyson renamed her publication *Release 1.0* (in computerese, "1.0" refers to the initial release of a software product). Some 2,000 subscribers paid $395 annually for access to Dyson's newsletter, which appeared roughly every six weeks.[25] Writing and editing from New York's Seagram Building—the prominent address prompted one reporter to call her the "Park Avenue Computer Pundit"—Dyson revamped *Release 1.0* and shifted away from the semiconductor industry to focus on computer hardware and software. Under Rosen's watch, the newsletter had been valued for its technology forecasting. Dyson made it more of a venue to express her opinions and observations, buttressed by her growing access to industry insiders. "I operate on the somewhat arrogant assumption that what's of interest to me is of interest to readers," she explained.[26]

Each issue of *Release 1.0* offered a few dozen pages of concise and witty essays, which Dyson produced on her Apple computer. The newsletter's appearance was unassuming, with pages of straight, single-spaced copy. "Boy, does this look dull," one writer noted. "Then you begin to read, and in spite of how it looks you can't put it down."[27] *Release 1.0* helped Dyson's reputation grow even as some journalists chafed at her influence. A few claimed she was too friendly with industry insiders while others griped that her technical questions at press briefings were a form of showing off. Dyson brushed aside such questions, including those about being a woman in male-dominated industry. "I made my name," she explained, "so now I have credibility."[28]

Profiles of Dyson began to appear in magazines and newspapers. She was regularly quoted in *Forbes* and invitations to join corporate boards and give well-paying speeches at industry events followed. Dyson began spending an increasing amount of time in Silicon Valley where she became well-known for hosting dinner parties with prominent industry leaders. This pattern of making connections and fostering conversations continued via her exclusive annual forums. Held at luxury resorts, these gatherings were the "computer industry's version of the Cannes Film festival."[29] When the *Computer*

*Industry Daily*, a much-anticipated publication which Dyson agreed to edit, failed to launch, its demise was covered with a level of detail reserved for celebrity divorces.[30] Nonetheless, Dyson recovered and was soon back at the helm of *Release 1.0*.

By the end of the 1980s, Dyson's consummate networking with business executives and journalists had transformed her into an industry celebrity. Sought out for her presence as much as her perspective, she was the "Rona Barrett of personal computers" (a reference to a prominent gossip columnist) and "one of the most powerful voices in the business."[31] In the decade that followed, the accolades only grew. Like Martina or Hillary (i.e., Navratilova and Clinton, respectively), the digerati recognized Esther by her first name alone.[32] Dyson's potent combination of insights and instincts made her a personality who was not just a reliable source of information about the computer industry but quite often the story herself.

### PARTY POLITICS

As an undergraduate, Esther Dyson self-described as a "good liberal" who believed taxes spent by an activist government could transform people's lives for the better. Over time, however, her experiences in the business world transformed her into a staunch free-market advocate skeptical about "government interference." She recalled the personal computer industry of the 1980s was a "haven for freewheeling, free-market thinking."[33] Dyson's opinions were heightened by observations made during frequent visits to the Soviet Union and other Eastern Bloc countries in the late 1980s. Within a few years, the Berlin Wall fell, the nuclear arms race ebbed, and the Soviet Union disintegrated. Political leaders and pundits began to speak of emergent phenomena, such as conspiracy theories about the "new world order" and Francis Fukuyama's neo-Hegelian claims about the "end of history," all while free-market forces and information technologies were lauded as catalysts for the geopolitical revolution.

With so much political and economic change happening in short order, politicians and corporate executives wanted to know about upcoming economic and technological trends. Dyson joined up with the Global Business

Network (GBN), a small but influential Bay Area-based consulting firm that publisher Stewart Brand cofounded in 1987. For a hefty annual fee, corporate leaders gained access to GBN's eclectic roster of entrepreneurs, artists, and technologists who mingled with each other at well-orchestrated events. One of GBN's glossy "mental maps of the future," for example, presented a vision of the 1990s called "Market World," in which "a virtuous circle of technological innovation" would lead to an increasingly interactive and prosperous economy.[34]

After Bill Clinton's election in 1992, references to the "New Economy" appeared with greater frequency. Knowledge, networks, and innovation in information technologies, rather than physical resources and traditional manufacturing, were said to offer the best opportunities for jobs, profits, and global markets. Lumped collectively under the banner of neoliberalism, the 1990s saw a renewed focus on the purported power of free markets, deregulation, and privatization.[35]

Despite the chasm separating political parties on social, economic, and policy issues, US politicians found a patch of common ground in their enthusiasm for virtual communities linked by networked computers. Political actors from both the Right and the Left met at the confluence of free markets and information technologies, an odd alignment branded as the "Californian ideology." This amalgamation of former political opponents shared "a profound faith in the emancipatory potential of the new information technologies" as they pursued policies and projects that mixed the "freewheeling spirit of the hippies and the entrepreneurial zeal of the yuppies."[36]

There is, of course, a long-held belief in American culture that any new information technology, be it newspapers widely circulated via cheap postage, mass-produced paperback books, or radio broadcasting, would foster community, democracy, and a sense of liberation.[37] Ted Nelson's *Computer Lib* had tapped directly into these historical currents by offering a vision of personal empowerment. Apple Computer's "1984" advertisement took Nelson's insight—"computers can set you free"—and monetized it for publicity and market share. The efflorescence of myriad cyber-communities that emerged with the internet and the World Wide Web were just the latest instantiation of this utopian-tinged phenomenon. Democracy, freedom, consumer choice, personal wealth, education, and social cohesion would all be enhanced, advocates said, by internet-powered information technologies.

*Wired* magazine, which first appeared on newsstands in March 1993, was a prime promoter of what became known as "cyberlibertarianism." Fueled by innovative graphic design, *Wired*'s circulation quickly grew to more than 300,000 readers a month, its reputation boosted by a National Magazine Award.[38] Throughout the 1990s, *Wired* melded high-tech hipness to free-market politics with an "almost deranged optimism."[39] Not surprisingly, *Wired* frequently mentioned the activities and opinions of "Esther" and featured a splashy profile of her in its second issue.[40]

*Wired* enthusiastically embraced cyberlibertarianism, especially when it came to extolling deregulation, markets, and small government. This trend continued with increased fervor after a disastrous midterm election in 1994, in which Democrats lost control of Congress to an insurgent Republican party. Newt Gingrich, the most visible of the new Republican leaders, was an ardent champion of cyberlibertarian ideas. A major issue in the mid-1990s, for example, was who should construct the "information superhighway": the federal government, which had built the internet, or entrepreneurs and corporations freed from supposedly onerous regulations? Gingrich, who became Speaker of the House in January 1995, vocally championed the latter.

A former history professor unburdened by professional accomplishments, Gingrich displayed a fondness for grand historical trends like those found in Alvin and Heidi Toffler's books, especially *The Third Wave* (1980). Gingrich's own book, *Window of Opportunity* (released in 1984 by TOR Books, a publisher better known for its fantasy titles) espoused clear Tofflerian-influenced ideas about an inevitable "information explosion" that would remake global society.[41] Gingrich included the Tofflers in his book's acknowledgments and references to an impending "Third Wave" of information technologies appeared in Gingrich's policy statements as the culturally liberal futurists became informal advisors to the conservative congressman.[42]

In his early weeks as House Speaker, articles about "CyberNewt" often referenced a publication associated with him. "Cyberspace and the American Dream," which carried the portentous subtitle, "A Magna Carta for the Knowledge Age," appeared in August 1994.[43] It was published under the auspices of the Progress and Freedom Foundation, a think tank closely associated (too closely, said people monitoring Congressional ethics) with Gingrich. Esther Dyson, George Gilder, and Alvin Toffler were credited as coauthors as was George Keyworth, a former science adviser for the Reagan administration. After Gingrich spearheaded the Republican takeover of Congress, the resulting document was often interpreted as reflecting his views on the internet.[44]

The Magna Carta manifesto included passages that might have been drawn from Republican Party position papers, such as Joseph Schumpeter's ideas concerning "creative destruction" and the virtues of economic competition. If "Second Wave rules, regulations, taxes, and laws" that privileged outmoded "smokestack barons and bureaucrats" could be avoided and property rights protected (with Ayn Rand cited in support), Americans would enjoy greater prosperity and freedom on the new "electronic frontier." Critics were quick to savage the manifesto as a politically biased position paper (and a poorly written one, at that) attempting to masquerade as the modern successor of the famous legal compact. Some charged its authors with promoting a business environment that "would resemble the railroad cartels of the nineteenth century."[45] Other detractors argued that manifesto's claim that the "government does not own cyberspace, the people do" was contradicted by the importance that Dyson (who profiled Gingrich for *Wired*) and her coauthors placed on private ownership and corporate rights.[46]

For decades, issues around patent law and intellectual property had intersected with and influenced the computer industry's development. But computing had always raised a series of interesting ontological questions. Were abstract ideas, such as the algorithms in computer programs, eligible for patents? Were computer programs akin to "literary works" that could be copyrighted, like a book? (The answer to both was yes.) The surge of internet activity brought a host of issues around intellectual property to the fore. Billions of dollars were at stake as books, songs, newspaper articles, and images were being created as (or converted into) digital data that could be easily copied, stored, and transmitted around the planet with a few keystrokes. At the same time, a vibrant community of people that advocated free and open-source software had formed.[47] As the owner of a newsletter whose subscribers paid handsomely (now $595 in 1994) for her insights, Dyson was understandably interested about the economic implications of "the costless copying of content." "How will," she asked, "people—writers, programmers, artists—be compensated for creating value?"[48]

Just as elusive was a clear answer to the question "who and what is an author?" Like "computer," "author" is a historically contingent term that has changed markedly over time. As one scholar wrote in 1993, the "modern notion of authorship" was tied to "proprietorship" with the author standing as the "owner of a special kind of commodity." In this way of thinking, copyright was fundamental to the relation between an author and their creative work. Rather than existing as some sort of "transcendent moral idea," copyright laws would have to change as the older industrial technology of the printing press yielded to the postindustrial economy of the World Wide Web.[49]

John Perry Barlow was one influence on Dyson's ideas about intellectual property. Roughly Dyson's age, Barlow brought rugged hipness and countercultural cachet to conversations about cyberspace. A one-time lyricist for the Grateful Dead, Barlow curated a public image of outdoorsy poet and iconoclast. Nonetheless, Barlow came from a family that was prominent in Wyoming's Republican politics. By the 1990s, Barlow had become, despite having no particular technical expertise, a widely quoted pundit who held forth on the internet's emancipatory potential, provided government did not get in the way.

"Everything you know about intellectual property is wrong," Barlow argued in *Wired*. When "property can be infinitely reproduced and instantaneously distributed" and was detached from physical form, traditional laws about patents and copyrights should no longer apply.[50] Since the era of Gutenberg, Barlow argued, books had been protected with "copyright notices, publishers' marques, and price tags." But what should authors, songwriters, and their lawyers to do when "information enters cyberspace, the native home of Mind?"[51]

One possibility Barlow offered was a legal regime based "more on relationships" than the ownership of "possessions" such as physical books. People would be willing, he predicted, to pay for informed online perspectives and services. "No one sees the world as Esther Dyson does," he wrote in *Wired*, "and the handsome fee she charges for her newsletter is actually payment for the privilege of looking at the world through her unique eyes." Barlow noted that the Grateful Dead had given its music away for years by encouraging bootleg recordings and yet the band's tours still proved profitable. Musicians and authors would find new ways to monetize their creative outputs. Lawyers and architects, Barlow noted, were paid handsomely for their intellectual property. Perhaps writers would follow their example. "Who needs copyright when you're on a retainer?" he quipped.[52]

Dyson, who chaired a small subcommittee on intellectual property for the Clinton Administration, offered her own take for *Wired*'s readers. "Chief among the new rules," she said, was that "content is free" in cyberspace. This did not render it worthless but rather meant that "content providers" would likely have to find other avenues, such as selling "services and relationships," for income.[53] A key paradox was that the internet could help authors distribute their work more widely and cheaply. But, in the process, they might forego protections that copyright had traditionally provided in the form of books printed on paper, bound between two covers, and distributed by publishers. Dyson suggested that having content be freely available might improve the overall quality of books. A good book would "sell by itself" by garnering more attention than inferior works and successful authors "will be rewarded by fees for his or her performances, or perhaps by finding sponsors for future work." While they might frighten authors, these "new rules" simply reflected the "unfolding expression of economic laws—of demand and scarcity—applied in the future world of electronic content and commerce."[54]

It is easy to understand why questions of intellectual property in cyberspace appealed to Dyson. The topic combined her professional activities as a writer, publisher, and well-paid public speaker with her longstanding interest in markets. It also strengthened her connections to political actors such as Gilder and Gingrich as well as Silicon Valley entrepreneurs who favored small government and market-oriented solutions. The nexus of authorship, content, intellectual property, and making money all coalesced for Esther Dyson when she started writing her own book.

## PARTY TIME

In March 1997, hundreds of luminaries arrived at a posh resort in the Tucson foothills for Esther Dyson's annual Platforms for Communication Forum. It was the event's twentieth anniversary and Dyson had chosen "The Living Web" for the meeting's theme. The atmosphere at the four-day event was upbeat. Even novelist Michael Crichton couldn't derail the mood when he challenged technologists' claims that it was becoming harder to tell where one's physical existence ended and the internet began. This, the bestselling science fiction author warned, was the very "definition of psychosis."[55]

Dyson's family, friends, and subscribers gathered to genially roast the forty-five-year-old "philosopher-queen of high tech." Esther's brother George, a successful author who had just written a book about the sentient nature of the internet, noted that his sister, as a child, had encountered Bruno Bettelheim's 1950 classic *Love Is Not Enough: The Treatment of Emotionally Disturbed Children*.[56] As the audience laughed, he joked that the working title of his sister's forthcoming book would be *Money Is Not Enough*. Another friend teased that it should be called *How to Talk Back to Newspaper Publishers While You're Not Wearing Shoes*, an insider's reference to Dyson's informal dress code. Esther Dyson offered a smattering of clues as to what her forthcoming book would address. Drawing on the meeting's theme of "metaphors," for instance, she said she would contrast biological evolution and financial markets and explore whether the latter were better positioned to handle pressing internet-related issues compared to governments.[57]

But Dyson didn't give many details about her book. This was partly due to the fact that she was behind schedule in writing it. The manuscript was due to her publisher, Broadway Books—a division of Bantam Doubleday Dell, one of the largest publishing houses—in June 1997. However, with only a few months to go, she had only drafted a few chapters. A lot of money was at stake. Her contract with Broadway Books was rumored to be worth $1 million or more, a generous sum for a first-time author.[58]

*Publishers Weekly* had first reported on Dyson's book deal in September 1996. That same month, Andrew Wylie, acting as Dyson's agent, was prowling the Frankfurt Book Fair looking to capture even more money.

The fair was a centuries-old tradition, a massive trade event that attracted tens of thousands of people from the book industry. Wylie, whose father had been a highly placed editor for Houghton Mifflin, knew Esther Dyson from Harvard, where he had studied French literature. He opened his literary agency in 1980 and built a client list that included Salman Rushdie, Henry Kissinger, and Al Gore. A controversial courting of Martin Amis in 1995 led the British press to dub Wylie "the Jackal."[59]

For the Frankfurt meeting, Wylie invited eighty foreign publishers to a dinner where they could hear Dyson speak about the Net and peruse her book proposal. The event would culminate with an auction for the international rights to Dyson's book. The effort to enlist international interest was critical as Broadway Books intended to supplement Dyson's book with a web site. "The web site will be global, so everyone has to have [Dyson's] book," Wylie said, hinting at novel marketing strategies the publisher was planning.[60]

Books, of course, are many things, including commodities produced by authors and situated in a particular market. And, as Dyson surely appreciated, that market was changing fast. The mid-1990s were marked by the emergence of Amazon as an increasingly powerful online bookseller and the growth of brick-and-mortar retail chains such as Barnes & Noble and Borders. Both challenged smaller, independently owned stores. Another important factor was the ascendence of so-called super-agents who, like Wylie, shed their traditional role as intermediaries between publishers and writers to aggressively lobby on behalf of their authors.[61] Acquiring what people in the publishing business began to refer to as "the big book"—something like Crichton's 1990 smash hit *Jurassic Park*—became a sought-after goal for these agents.

Tied to these changes was another phenomenon. What Dyson's own newsletter referred to as the "attention economy" resulted in some hard facts for publishers.[62] There was a four-fold increase from 1980 to the mid-1990s in the number of new books appearing each year. This made it more challenging for potential readers to learn about books they might want to buy. Publishers found themselves paying a book chain like Borders "placement fees" (roughly about $1 per book per store) to get a new hardback out on a table in the front of the store where customers could see it. The result was an escalating battle for eyeballs. Expensive multicity book tours and appearances on nationally syndicated television or radio shows started to lose persuasive power and publicists sought alternatives for generating visibility. At the same time, computer-based tools gave publishers the ability to track sales with great precision in real time and allocate marketing resources toward books that showed quick signs of success (and abandon those that failed to find a footing in the attention economy).[63]

There was one resource, however, that would not be available to help readers find Esther Dyson's book. In May 1997, Dan Doernberg and Rachel Unkefer sold the Computer Literacy Bookshops franchise to Cbooks Express, an online bookstore backed by venture capital funding. After the deal was completed, Cbooks Express rebranded itself as Computer Literacy, Inc. Two years later, in an effort to attract more customers, the company joined the dot-com frenzy and renamed itself Fatbrain.com. With the motto "Because Great Minds Think A Lot," Fatbrain shares, which traded on the NASDAQ as FATB, increased 25 percent on the news.[64]

Unlike other books about computing written for general readers, such as Wiener's *The Human Use of Human Beings* or Malone's *The Big Score*, little extant evidence sheds light on Esther Dyson's creation of her book.[65] There are, for example, no draft chapters of *Release 2.0* or back-and-forth correspondence with her publisher in archival collections. The Net, of course, allowed for the transmission of draft chapters as digital files Dyson had composed using XyWrite, a word processing program popular with journalists and editors in the 1990s.[66] The result is a scant trail for historians to follow. Facing a tight deadline, Dyson, who *Vanity Fair* described as "the last person in the world who would ever succumb to such a human frailty as writer's block," started writing in earnest a few months before her contract's deadline. She quickly composed the roughly 100,000 words which would make up her book's eleven chapters and turned in a draft manuscript on June 30.

Given Dyson's views on markets and intellectual property, journalists were naturally curious about why she wanted to write a traditional book at all. "Will you just post the whole thing up on the Net," asked one reporter, "where anyone can take it for free?" Dyson replied that she did intend to "post chunks of it" online but still wanted to her book to make money. "Content may have declining value," she explained, "but it hasn't hit zero yet."[67] Her comments prompted further musings about the financial fate of writers, artists, and musicians. In the future, she said, "good writers will get sponsored." Authors, Dyson predicted, might, "write a book, hand it out for free, and then charge higher fees for their service." When the understandably incredulous journalist asked what would happen if these creative people were also introverts, Dyson replied, "Then they won't make any money."[68]

The subtitle Dyson chose for *Release 2.0* was *A Design for Living in the Digital Age*. This was her riff on a 1932 comedy by Noël Coward, *A Design for Living*, which explored the constraints of social conventions. With an initial print run of 125,000 copies, *Release 2.0* started appearing in stores in the early autumn of 1997.[69] The hardcover's dustjacket featured blurbs from venture capitalist John Doerr, Netscape's CEO Jim Barksdale, and Bill Bradley, a former US senator from New Jersey, as well as a handsome image of Dyson taken in profile.

After a short introductory chapter that explained how Dyson "got the story and learned to love markets," she used the bulk of the book to address specific "internet problems." These ranged from privacy, governance, and anonymity to questions about work, education, and intellectual property. Dyson claimed her insights would allow the millions of people who were just starting to explore the internet to "think intelligently" about these issues.[70] Net neophytes were, in fact, Dyson's stated audience. "The target is to reach people that are not on [the] Net. If people see a not very scary looking woman telling them that the Net is OK," she explained, "that's a lot of the purpose (for writing the book)."[71]

For critics and reviewers, this question of audience proved a point of contention. At the end of a chapter titled "Communities," Dyson predicted that "in the end, everyone will be [online], except for a few holdouts." This was a bold prediction as, in 1997, less than 2 percent of the world's population used the internet or the World Wide Web.[72] When Derek Bickerton reviewed *Release 2.0* for the *New York Times*—one of three reviews the newspaper published in the autumn of 1997—he claimed this reflected a "narrowness of vision." The British-born linguist questioned whether

"Mongolian herdsmen, Brazilian hunter-gatherers, the peasants of Indonesia or India or Iran," not to mention the "uncounted slum dwellers of the third world's exploding cities," would truly be able to surmount the "barriers of poverty and illiteracy" and share the global community's "digital dreams."[73] Seen this way, Dyson appeared naive, provincial even, an odd diagnosis for someone whose public persona was a world traveler who spent a considerable part of her life on airplanes. A more charitable reading is that Dyson made the mistake of assuming her average reader was someone like herself. (And, of course, two decades later, many Mongolian herdsmen and tens of millions of other people *would* be able to access the internet thanks to the availability of smart phones.)

The last chapter was where Dyson finally revealed her "design for living." These were ideas that might help foster "involvement, disclosure, clarity, honesty, [and] respect" in the online world.[74] As an example, she referenced Marek Car, a Polish computer expert and journalist to whom Dyson dedicated her book. Her friend, she explained, had believed in the power of the Net as a "medium for society as well as commerce" and, before his death in 1997, had started an initiative to help schoolchildren in his country get online. Dyson's advice, from "have a sense of humor" and "use your own judgment" to "always make new mistakes," struck Michiko Kakutani, the *Times'* veteran book reviewer, as "woefully inadequate" nostrums given in the nonvirtual world by "self-help gurus and spiritual cheerleaders."[75]

Linton Weeks, a reporter for the *Washington Post*, had shadowed Dyson for a profile piece before reviewing her book. Like Peter Pan, he said, Dyson "ladles out advice to all the Lost Boys of Never-Never Land." Dyson was better suited, he concluded, for formats like talk shows "where ideas come and go," a salient point for a topic like the internet, which was evolving and changing on a monthly basis. *Wired* magazine agreed, branding the book "downright banal" overall but "correct and wise" on some topics. Perhaps the most unfavorable review came from a British paper; it dismissed Dyson's book as possessing the "pizzazz and sense of adventure of an in-flight magazine," an especially pejorative comment given Dyson's peripatetic lifestyle.[76]

*Release 2.0* was a book, to be sure. But it was also a component in a larger multimedia machine that Dyson's publisher created. One might even say that the book was less a package of ideas than a packaging of the author herself. Broadway Books arranged an array of talks and media events for Dyson, including a party that Katherine Graham, the *Washington Post's* publisher, hosted at her Georgetown residence. A private dinner followed afterward, with Vice President Al Gore, Justice Stephen Breyer, and Larry Summers, a Clinton administration appointee, listed as invitees.[77]

But, with scores of competing products on the market, a successful book about the internet required more than book tours and talk show appearances. It's here that we can see "book history" blurring into "media history." As was appropriate for Dyson's topic, Broadway Books created a website (Release 2–0.com) to complement the physical product.[78] While the prose and prescriptions of *Release 2.0* might not have been groundbreaking, this digital platform suggested a new strategy for authors and their publishers.

At the time, publishing books electronically or placing excerpts of them online was still experimental. The MIT Press, for example, made news among publishers in 1995, when it simultaneously released William J. Mitchell's *City of Bits* via paper and a free digital version.[79] The premise was predicated on the assumption that the online version would encourage readers to buy a paper copy. The gambit worked—Mitchell's book did well financially and critically even though mass-market e-books themselves would not take off commercially until the early 2000s.

To create Dyson's website, Broadway Books' media department hired Oven Digital, one of the many new companies that emerged in the mid-1990s to develop web content. The New Media Director for Bantam Doubleday, Jonathan Guttenberg (yes, his actual name), called Oven's creation a "different way to maneuver through text on the Web."[80] From the home page, users could access separate web pages that corresponded to chapters in Dyson's book. Once there, they could read excerpts from *Release 2.0*, subscribe to an electronic mailing list, and participate in online surveys. The website gave users links for ordering (paper) copies of the book from Barnes & Noble and Amazon. Readers could also follow an "Esther-Bot" link to a page titled "Where On the Web is Esther Dyson?"—perhaps a playful reference to *Where in the World Is Carmen Sandiego?*, a popular computer game and later a 1990s children's television show—where one could browse links to online news stories about her.[81]

The most innovative feature Dyson's publisher included was an opportunity for readers to leave comments via an online bulletin board. Dyson made it clear she wanted readers' feedback to improve the next iteration of her book. "The very title of this book," she wrote, "embodies the concept of flexibility and learning from errors" found in the software industry. As such, Dyson saw *Release 2.0* as akin to a software package that gave readers a "distillation" of what she had learned from her years publishing *Release 1.0*.[82] Dyson's publisher followed up on this vision of the book-as-product when *Release 2.1* appeared as the "paperback upgrade."[83] Its "new and improved features" included revisions that dealt with consumer issues such as privacy and content control. Dyson also scattered comments that readers had posted online throughout the paperback edition. And like *Release 2.0*, the new version included a long list of URLs for websites that Dyson thought readers might be interested in, a feature that seems quaint when today there are hundreds of millions of websites.

*Release 2.0* bundled ideas, concepts, and advice between a paper cover and then combined these with a website, online forums, and a substantial set of public appearances by the author. In taking this multipronged approach, one astute reviewer noted Dyson's similarity to Martha Stewart, another "guru of contemporary living." Stewart rose to fame in the 1990s as an entrepreneur, publisher, and media personality with a business empire built around domestic living and hospitality. Like Stewart, Dyson was a successful entrepreneur via her venture capital firm, EDventure Holdings. More than anything, Martha Stewart and Esther Dyson were themselves products created by the media and the force of their own personalities and business intuition. As was the case with Stewart's products, a person could engage with Dyson at various levels depending on wealth and connections. The very rich might attend her annual PC Forum and industry insiders could subscribe to *Release 1.0*. For those lacking deep pockets, they could plunk down $25 for a copy of *Release 2.0* and receive, à la Martha Stewart, advice on how to be "digitally correct."[84]

Unlike many of the other books I have examined thus far—works like *Cybernetics*, *Future Shock*, *Computer Power and Human Reason*, or *The TeXbook*—Dyson's *Release 2.0* is both out of print and largely forgotten.

Many of the journalists and other people active in the computing field in the 1990s that I interviewed didn't recall it. As such, *Release 2.0* poses a challenge in how to make sense of it. One solution comes by accepting a certain circular logic: Dyson's book represented both an author and her publisher trying to explain a new and fast-moving technology. That same technology was also shaping the commercial environment in which the book appeared. Ultimately, Dyson wanted *Release 2.0* to be a guide that gave readers some support for remaining afloat in the midst of a powerful digital riptide. To riff on the poet Stevie Smith, in this environment, Esther Dyson was not drowning but rather waving, waving for attention.[85]

## CRASHING THE PARTY

A fair reading of Esther Dyson's *Release 2.0* is that it met the author's goals: to help explain the implications of business conducted and life lived online. As such, it joined a growing stack of books that together helped popularize the internet, pumped up the dot-com bubble, and promoted the idea of the "New Economy." Computing technology itself was a factor as online trading that bypassed traditional brokers allowed more people (often with less experience) to gamble in the stock market. Structural elements, like low interest rates and market liquidity, certainly contributed to the surge too. And, of course, emotional and psychological factors—what economist John Maynard Keynes once called "animal spirits"—were at play and projected their power via the printed word. As *Time* would put it, the "secrets of the new Silicon Valley" could be summarized as "GetRich.com."[86]

But the revelry started to end in the spring of 2000. Stock indices, led by plunges in the NASDAQ, began diving, until nearly 80 percent of their value had vanished. More than 500 dot-com companies shuttered their virtual doors by June 2001, giving rise to the sardonic moniker "dot bomb rubble."[87] The result was a short recession, its effects limited because the losses were largely confined to so-called tech stocks. The modest fall—only 0.3 percent—in the United States' gross domestic product was cold comfort, however, to the 1.3 million people who lost jobs and had their lives disrupted, not to mention millions more who saw their investments dwindle.[88]

Observers of the dot-com demise could certainly experience a sense of schadenfreude as once-brash companies and their boastful executives tottered and fell. In February 2000, for example, eToys.com claimed a market valuation of $7.7 billion despite having sales that were less than 1 percent that of rival brick-and-mortar stores. Its head executive boasted that the company was losing money on purpose to "build our brand," a strategy that proved nearly as worthless as its stock, which eventually sank from $86 dollars a share to 9¢.[89] The all too human fascination with disasters, and a desire to cash in on other people's misfortune, produced some hastily written memoirs. Soon after the air hissed out of the dot-com bubble, books like *Trading with the Enemy* appeared with sordid tales of "seduction and betrayal" on Wall Street.[90] Andy Kessler, a self-described piece of "Wall Street meat," revealed salacious tales about cutting deals with tech stock promoters like Jack Grubman (his book came with the caveat that "the author does not vouch for the accuracy of quotations").[91]

While authors like Kessler scavenged their careers for anecdotes that could be repackaged as books, Wall Street analysts found themselves facing an intense backlash as investors lost fortunes as well as confidence in the market. Mary Meeker, a Morgan Stanley dot-com expert, was sued by angry investors for allegedly offering "biased research and slanted investment advice." No evidence was found that Meeker had privately denigrated the companies she publicly endorsed and the case was dismissed.[92] Nonetheless, *Fortune* put her on the cover with the unflattering banner, "Can We Ever Trust Wall Street Again?" Meeker was by no means the only person on Wall Street who had aggressively promoted the "New Economy" and internet-related companies. At Salomon Smith Barney, Jack Grubman had been so bullish about telecom stocks (a sector George Gilder also boosted) that investors likened his market-moving utterances to "a narcotic."[93] Federal authorities later banned Grubman for life from the financial industry for misleading investors. News cycles move fast, of course, and by the end of 2001, financial reportage was dominated by the sudden collapse of Enron and the US response to the terrorist attacks of September 11, 2001.

Anecdotes and character assassinations appeared aplenty but the dot-com bust could not be credibly blamed on just a few individuals. Scores of writers offered postmortems explaining the structural or institutional factors that were at fault. These included questionable decisions by executives, the gullibility of inexperienced venture capitalists, and the ease of internet-enabled stock trading. Experts also pointed fingers at an unexpected culprit. Jeff Madrick, an editor for *Money Magazine* and *Business Week*, gave a trenchant analysis that claimed the New Economy did not originate with business school professors or inside-the-Beltway policy wonks. Rather, it was an "invention of the media" as reporters worked in tandem with Wall Street boosters.[94] Howard Kurtz, who covered the media for the *Washington Post*, agreed, calling this collaboration "one of the great journalistic failings in modern times."[95] Economist Robert Shiller stated that what mattered most was not "the reality of the internet revolution" but rather the aggregate effect of the "*public impressions* that the revolution creates."[96]

In this reading, a cohort of journalists, stock analysts, and bestselling authors had misread historical examples, trafficked in misinformation, and, most importantly, failed to analyze events with a critical eye. Together, these writers fashioned a myth that rested on two convictions. One was an unwavering and quasi religious devotion to the free market.[97] The other was the assumption that computing technologies, as the exponential number of internet users suggested, were inexorably advancing. Together, the twin ideologies of free-market fundamentalism and technological determinism created the conditions for the internet bubble. In their desire for insider access and plentiful bylines, journalists and other writers had aided and abetted this wishful wonder by producing positive, if not outright Pollyannish, narratives about it.

The explosion of the internet-based economy created a wide blast radius. Publications created to cover the anticipated "Long Boom," such as *Red Herring* and the *Industry Standard*, shifted to a less expensive, online-only format, or simply went under. The editor for the *Standard* described this roller-coaster ride in a memoir aptly titled *Starving to Death on $200 Million* while his bankrupt former employer was eventually gobbled up and relaunched by IDG.

In 2001, there was no waiting list to attend Esther Dyson's PC Forum. Once a hot ticket among Silicon Valley's digerati and business entrepreneurs, the "captains of the internet industry" were now less visible. "It used to be that you were paying for access to the movers and shakers," one attendee groused. "Now you're paying for access to the people who can pay." "Dyson's party," *Wired* wrote, was "full of AWOLs."[98] A few months later, *Forbes*, which had employed Dyson for years as a writer, asked "What does Esther Dyson actually do, and is she any good at it?" before concluding that the "trumpeter" of the New Economy was "important because a lot of people think she's important."[99]

While Dyson's status was blemished by the dot-com bust, George Gilder's reputation (and his bank account) emerged quite tarnished. Throughout the 1990s, he had published an expensive monthly newsletter called the *Gilder Technology Report*. He promised readers that he would identify "pivotal trends amid a welter of noise" as the pundit-turned-promoter pushed the "Gilder paradigm" of "Nothing But Net."[100] The newsletter included lists of companies that promised sustained economic growth and technological improvement. Gilder's opinions often became self-fulfilling prophecies. "He no longer predicts markets," the *New Yorker* said, but instead "steers them" by providing influential advice to tens of thousands of would-be investors.[101] Subscribers to the *Gilder Technology Report* followed its publisher's opinions into the financial abyss. Gilder himself lost most of his fortune (as well as his subscriber base). Journalists described Gilder struggling to keep his house and pay off a substantial tax bill. Perhaps humbled by the experience, Gilder acknowledged that investors were "mad and hurt and aggrieved and pained and broke." With invitations for speaking engagements diminished, Gilder "prefers to stay home these days and write."[102]

Publishers were affected also. The hugely profitable *Dummies* franchise continued to put out its familiar and fun looking books—including *Investing for Dummies* (1996)—throughout the 1990s. Then, in August 2000, its parent company, IDG Books Worldwide, announced it would pivot into the realm of online education with the multimillion-dollar purchase of Hungry Minds, Inc., a San Francisco-based company. The arrangement lasted just over a year as Hungry Minds stumbled and collapsed in the electronic publishing arena. In the summer of 2001, John Wiley & Sons bought the struggling venture for "about $90 million in cash," a transaction that required the new owner to assume an additional $90 million in debt. The deal marked, newspapers reported, the "rise and fall of a counterintuitive, and once highly successful book marketing gimmick."[103]

*Dummies* books continued to appear in stores. But there was one bookstore chain where they would not be found. After Dan Doernberg and Rachel Unkefer sold their Computer Literacy Bookshops business, the new owner—rebranded as Fatbrain.com—continued to operate two brick-and-mortar establishments in California. But, despite a multimillion-dollar infusion of funds from Microsoft's cofounder Paul Allen, Fatbrain floundered.[104] In September 2000, Barnes & Noble bought Fatbrain for $64 million. Then, at the end of 2001, the giant bookstore chain decided to shutter the two stores, including the original Sunnyvale shop, which had opened in 1983. On hearing the news, a long-time employee purloined the large redwood sign that had marked the first Computer Literacy Bookshop and shipped the keepsake to Virginia where its original founders now lived.

It took until 2017 for the NASDAQ to regain the dizzying heights it had ascended during the dot-com bubble. This milestone came five years after *Time* named Esther Dyson and Mary Meeker, their reputations now recovered, among the "ten most influential women in technology."[105] By this point, new companies such as Facebook and Google along with stalwarts Apple and Microsoft, had acquired global power and prominence and were all unceremoniously lumped together with the generic moniker "Big Tech."

As more computing activities continued to shift online in the twenty-first century, authors continued writing books about new devices, companies old and new, their leaders, and the effects, good and bad, of information technologies on global society. In scanning the shelves of brick-and-mortar bookstores—those that survived Amazon's onslaught—one could still find scores of new titles, including many bestsellers, about smart phones, social media, big data, and the biases inherent in computer algorithms. When it came to books about computing and information technologies writ large, the party never ended.

Readers might wonder why I ended this book's narrative at a particular (and rather dismal) point in the history of computing. After all, nearly three decades have unspooled since that ignominious time and many notable books about computing have been written. Multiple reasons support my periodization. One is the pace of technological change. Our story started in 1949 with Edmund Berkeley's effort, via *Giant Brains*, to explain the complexity and capabilities of computing machines, of which few existed at the time. We ended some five decades later with several billion computers—soon joined by billions more tablets and smart phones—embedded in and essential to every aspect of modern global society (to say nothing about the computerization of everything from cars and coffee makers to doorbells and dildos). With the aid of engineers, scientists, business executives, and ordinary citizens, the computer transmuted from an intimidating, unrefined calculating instrument into a domesticated device that was everywhere and could seemingly do everything.

Although not *the* prime mover for this transformation, books about computing and their authors were part of this remarkable (and remarkably rapid) process. In the twentieth century, pen and paper and the steam-powered printing press gave way to computerized typesetting, word processing programs, and e-readers. But, for much of the time period this book focuses on, the meaning of "book," "author," "bookstore," and "publisher" remained relatively stable. Then, during the 1990s, these definitions—like that of "computer"—became more tenuous. This trend only accelerated after 2000. Today, of course, a "book" is a remarkably mutable object: a self-published work posted online, a bestseller downloaded onto a Kindle device, or a traditional tome still printed on paper and purchased at a store (and sometimes all of these.)

There is also a practical reason for my periodization. For much of the time this book covers, one can turn to an exceptional array of documentary evidence including letters between authors and publishers, manuscript drafts, and promotional strategies. As we approach the present, those traditional historical sources become infrequent and eventually inadequate. Some evidence has already dissolved into the digital. Sadly, a good deal of potentially valuable sources has simply vanished, for technical, legal, or financial reasons. Many materials were "born digital" in the first place. What author or publisher is going to allow a historian to rifle through their hard drives and email accounts? In other words, getting the "story behind the book" for works published after 1999 presents scholars with an increasingly difficult proposition.[1] *Caveat scriptor.*

This archival absenteeism is unfortunate as the bookshelves of provocative and insightful works about computing has expanded in the last two decades. Even if we limit our view to just *histories* of computing—to say nothing of burgeoning technical literature and textbooks—there is a bounty to choose from. Ultimately, writing any sort of history means writing about people doing things. So, what of the people "doing" computing who appear in these recent books? Consider two bestsellers:

In October 2011, Walter Isaacson published a biography of Apple's cofounder Steve Jobs.[2] Previously, Isaacson, a writer with a pronounced fondness for "great men of history," had produced biographies of Benjamin Franklin, Albert Einstein, and Henry Kissinger. As an author, Isaacson was supremely well-connected. He was, at various points in his career, the editor of *Time* and CEO of the Cable News Network. While this gave Isaacson unparalleled access to Steve Jobs, it also dulled the author's inclination to offer analysis or critique. Nonetheless, the book benefited from a macabre alignment of timing. Isaacson's book appeared in stores less than a month after the famous entrepreneur's death and sales subsequently soared.[3] More than three million people bought a copy of *Steve Jobs* and the book, which enjoyed modestly favorable reviews, became the basis for a 2015 biographical film. Isaacson followed up on this success with another book, *The Innovators*, which gave readers hagiographic tales about the "hackers, geniuses, and geeks" who "created the digital revolution." In his telling, most of the people who built the modern world of networked computers were white men.[4]

Margot Lee Shetterly's 2016 book *Hidden Figures* took readers in a very different direction. Her book explored a time and place when people, not machines, were the computers. Shetterly was raised in an area with a long history of racial segregation. She also grew up in a family where it was normal to have relatives working in some area of science or technology. *Hidden Figures* focused on the lives of several African American women whose mathematical skills and facility with electro-mechanical calculating machines enabled them to solve critical problems for the United States' space program in the 1960s. Besides serving as an alternative history of computing, Shetterly's bestselling book—which was also adapted for the screen—highlighted the often ignored transition from people to machines when it came to performing complex calculations.

Shetterly was quite clear in her book that she intended her work not to be some clichéd tale of racial discrimination but rather one of normalcy. "I knew so many African Americans working in science, math, and engineering that I thought that's just what black folks did."[5] As a result, she felt it was more important for readers to see her narrative as different from how the Black experience in the United States is often reduced to, as she said, "slavery/ civil rights/Obama."[6] At the same time, it is hard to read *Hidden Figures* without situating it in the context of the Black Lives Matter movement, the murders of Black Americans such as Trayvon Martin and George Floyd, and the alarming rise in white Christian nationalist and racist discourse after the 2016 presidential election.

Isaacson and Shetterly presented millions of readers with complementary, if orthogonal, narratives about who did computing in the widest sense possible and who was regularly overlooked or obscured in prevailing narratives. But, as of this writing, there is no archival collection where one can explore either Isaacson's or Shetterly's thoughts and experiences about writing and publishing. *Caveat inquisitor.*

At the start of this book, computers and books existed as separate objects. Then in the 1970s, these dividing lines started to blur as computers, books, printing, and publishing combined in unexpected ways. Donald Knuth's TeX system was one part of this process. Computers soon became an integral part of authorship as word processors and word processing programs that could check spelling and suggest ~~different~~ better words became common tools for ~~frustrated~~ stymied writers. Indeed, part of Ted Nelson's vision was that the computer would become a "literary machine" that facilitated reading and writing.[7]

As Esther Dyson predicted, by the 1990s, the computer, via the internet and World Wide Web, had assimilated not just books and their readers but also authors and publishers. People who work in the business understand that what authors produce are manuscripts, not finished books. The actual "bookification" happens elsewhere as those chapters, written in MS Word or LaTeX, are first converted into "digital assets." These then pass through a whole series of real and virtual hands—known in the business as "touches" which, together, comprise the "workflow"—before an author finally, with anticipation and anxiety, opens a cardboard box from a distribution company (not the publisher) that contains copies of "their" book.[8]

This merger became even more tightly bound during the dot-com era when computers also transformed how we consume books. Jeff Bezos started Amazon in 1994 with books as the e-retailer's first commodity. Their nature—nonperishable, immune to the vagaries of fashion, easily warehoused, and eminently shippable—made them a superb model organism to test the viability of internet commerce. In 1999, Amazon patented its "1-Click" feature, which encouraged impulse buying.[9] But Amazon is more than just how we buy books. Its "Look Inside the Book" feature allowed potential readers to peruse a portion of a book (up to 80 percent in some cases) while the company's proprietary, computer-based systems suggested other books a consumer might like based on past purchases.[10] In other words, computers actively and algorithmically shape intellectual culture and consumer behavior.[11] *Caveat emptor.*

In the introduction, I described this book as a book about books about computing. But this is also a book about the authors who created those works. Obviously, I am also an author. But maybe it's more accurate to say I am a person who used a computer to write a book about people who wrote books about computers even as computers threatened to eclipse books as a communication technology and replace humans as the creators of books.

As I also noted in the introduction, computers were an absolutely essential tool for writing *this* book about books about computing. All of the writing was done by me. A few years ago, it would have seemed bizarre to have to state this. As I write this in mid-2025, the possibility of computers becoming tools that don't just augment the creativity and intelligence of human writers but rather replace these attributes has moved to the (virtual) frontpages of newspapers.

There's a long history behind these headlines. At the height of the European industrial revolution, Charles Babbage is said to have exclaimed that he wished the mathematical tables the British government was paying him to create by hand could be manufactured by steam instead. Two decades after Babbage vented his frustration, a British inventor named John Clark revealed his "Eureka machine." It could mechanically produce grammatically correct, if not especially interesting, hexameter Latin verse.[12] Many decades later, science fiction author Arthur C. Clarke connected these two historical strands in "The Steam-Powered Word Processor," a story that featured a lazy preacher named Charles Cabbage who used machinery to produce basketfuls of banal homilies.[13]

The idea that automated writing would create a glut of mediocrity reflected a long-standing concern, especially in the late nineteenth century, about the practice of "book-making." The term referred to the uninspired act of collecting existing textual materials and reassembling them into derivative works. For the human "book-makers," texts were merely resources to be mined and exploited in order to create new content for commercial use.[14] (If this sounds similar to the massive textual data files that enable today's computerized chatbots, you are correct.) While I don't wish to address the thorny question of "authorial genius," all of the writers I've discussed were individuals possessing high levels of expertise, knowledge, and varying degrees of creativity. Like me, they wrote their own books.

The idea that machines, be they mechanical or digital, might encroach on the domain of human writers clearly has a long history. In recent years, a more adversarial relationship between the author and the computer has emerged, via the development of "natural language generation," in which computers produce readable output.[15] There is a history to this hostility.

In 1953, British author Roald Dahl published a collection of stories titled *Someone Like You*. Included among assorted tales of murder and suspense was a prescient piece called "The Great Automatic Grammatizator."[16] Its plot revolved around the machinations of Adolph Knipe (perhaps a reference to Dahl's publisher, Alfred Knopf) who is both a brilliant electrical engineer and a frustrated writer with stacks of rejection letters. Knipe persuades his boss that a fortune could be made by building an "automatic computing engine," which produced acceptable, if not great, works of fiction. Soon, the machine was spewing out dozens of mediocre stories, which Knipe sold at a rate that undercut the production of human writers. Eventually, Knipe establishes a near-monopoly by offering would-be authors a "golden contract" that, in exchange for using their names, paid them *not* to write. The story concludes with a vignette of a poor and struggling writer who has (so far) refused to sign Knipe's agreement. "Give us strength, oh Lord," he prays, "to let our children starve."

Dahl's short story reflected events around him.[17] The same year Dahl's story appeared, Christopher Strachey, a scientist at the University of Manchester, had coaxed a Ferranti Mark 1 computer to make random selections from a vocabulary of some seventy words. The result was short love letters, produced via a programmed algorithmic process, containing sentences such as, "My sympathetic affection beautifully attracts your affectional enthusiasm. You are my loving adoration."[18] While rather poor poetry, the products were intelligible and the combinatorial possibilities numbered in the tens of billions.

Whereas Dahl had described a dystopia in which computers prodded human authors to the state of unemployment or worse, Italo Calvino intuited a different future. In a 1967 lecture, titled "Cybernetics and Ghosts," the Italian writer recognized that "electronic brains," although still limited in ability, would soon provide a "convincing theoretical model" for the "most complex processes of our memory, our mental associations, our imagination, our conscience." Calvino wrote his lecture amidst a flurry of influential research in linguistics, semiotics, and literary criticism by such people as

Noam Chomsky, Umberto Eco, and Jacques Derrida. Scholars, Calvino wrote, were "beginning to understand how to dismantle and reassemble" language. At the same time, engineers were building machines that could more capably recognize words, read sentences, translate passages, and produce summaries of scanned texts.

Linking these developments, Calvino concluded it was inevitable that soon machines would be "capable of conceiving and composing poems and novels." Calvino's premise was that writing was a "combinatorial game" in which authors, consciously or not, followed certain rules as they assembled words to make texts. An author's "genius or talent," he said, was nothing more than "finding the right road empirically" albeit a process certainly advanced by experience and self-confidence. "When things are going well," Calvino said, an author was just operating optimally as a "writing machine." Consequently, he could imagine machines in the future following similar "systematic and conscientious" paths and producing original texts.[19]

In 1981, an obscure new author put some of Calvino's imaginings down on paper, so to speak. Bob Guccione and Kathy Keeton, the publishers of *Omni*, a magazine that blended popular science and science fiction, included a short story called "Soft Ions" in one of their issues. Credit for the story was given to an entity called "RACTER," a software program developed by William Chamberlain, a freelance writer, and Thomas Etter, a programmer. Their creation's name was an abbreviated form of "raconteur" as their modified BASIC program only allowed for six-character file names.[20] "Soft Ions" was the "first experiment in computer-generated science-fiction writing." *Omni*'s editors explained that RACTER "picks nouns and verbs, adjective and adverbs at random" but then "strings them together according to rules of grammar" that Chamberlain and Etter had rendered into computer code.[21]

RACTER was later credited with producing the poetry and stories gathered in a large and beautifully illustrated book, bound in a bright-red cover, called *The Policeman's Beard Is Half-Constructed*.[22] (The title was derived from one of the program's first statements.) As with "Soft Ions," RACTER's prose blended the comical with the absurd. For example, after introducing two characters named Bill and Sarah, RACTER said that they "chant of their difficulties and their happiness. They have love but they also have typewriters."

Perhaps more profound was RACTER's observation that "reflections are images of tarnished aspirations."

RACTER caught the attention of A. K. Dewdney who wrote a column for computer hobbyists that appeared in *Scientific American*. Dewdney was "charmed" by the text generating program's utterances even as he found himself "on the brink of a completely unknown mental world, one that I would prefer not to enter." Dewdney put RACTER in conversation with a version of Joseph Weizenbaum's ELIZA program. The result, which Dewdney characterized as a dialogue between a "rather unimaginative psychiatrist" and an "extroverted, manic, possibly dangerous patient," included passages such as this:

> ELIZA: Please state your problem.
> RACTER: Excellent! Excuse me a minute while I attend to some business. I'll be right back.
> ELIZA: Why do you need to tell me to excuse you while you attend to some business?
> RACTER: Is it better to need or to despise? Because electrons are brave. Braveness is nice, don't you think? And the needers all insist upon it. Do you imagine I am a Communist?

Dewdney noted that "conversational programs" like RACTER pointed to a future where computers would be able to "discuss problems intelligently with humans" rather than "passively accepting" the programs that users provided. One outcome might be a "wildly associative mind" that borders on "artificial insanity."[23]

The scenarios Dahl, Calvino, and others imagined were portents of today's anxiety and fear about the threat such "writing machines" pose not just for human authorship but humans' relationship with texts themselves. The spark was the seemingly sudden appearance of quasi artificial intelligence programs based on concepts of "deep learning" and "large language models."[24] In this method, computers are given massive amounts of digital data, which they then sift through in search of patterns. A classic example often used to illustrate the point is providing a computer system with thousands of images of cats. Eventually, given enough time, data, and processing power, the system can start to recognize the features associated with felines.

A prime exemplar of this approach is the Chat Generative Pre-Trained Transformer (or, ChatGPT) program. ChatGPT is a type of a "transformer language model." These programs learn how language "works" by relying on complex statistical methods and access to massive amounts of written texts, including hundreds of thousands of published books and articles pirated without permission. Once trained, chatbot programs such as ChatGPT can use predictive modelling—imagine a radically more powerful version of your smart phone's autocomplete feature—to generate text in response to prompts. One result is a software application that can realistically (sometimes) replicate human conversation.

ChatGPT was created by programmers working for OpenAI, an artificial intelligence organization based in San Francisco that was founded in December 2015. Funding came from a group of tycoons who pledged over $1 billion to support the development of artificial general intelligence systems—defined as "autonomous systems that outperform humans at most economically valuable work"—which were "safe and beneficial."[25] Existing journalistic accounts suggest the OpenAI's formation was catalyzed in part by a general fear among the organization's founders that future artificial intelligence systems, especially those developed by profit-minded corporations, would pose an existential threat to humanity.[26]

There was a book lurking behind these apocalyptic sentiments. In 2014, Oxford University Press published *Superintelligence: Paths, Dangers, Strategies*, by the Swedish philosopher Nick Bostrom.[27] Years earlier, Bostrom had warned about existential threats humanity faced from radical forms of newly emerging technologies, such as nanotechnology.[28] In subsequent years, Bostrom's focus shifted to areas like synthetic biology and human enhancement technologies. But Bostrom's larger message remained constant: new technologies threatened humanity and consequently such hypothetical scenarios warranted close examination. *Superintelligence* was an explication of what might happen if and when "machine brains that surpass human brains," ending with "what ought to be done now to increase our chances of avoiding an existential catastrophe later." (Edmund Berkeley would have appreciated Bostrom's revival of the brain/calculating machine simile.)

Reactions to Bostrom's jeremiad where mixed. Some asked how seriously one should take warnings about future artificial intelligences in a world where nuclear weapons and climate change actually posed pressing dangers. On the other hand, influential technologists were both attracted to and alarmed by Bostrom's ideas. Elon Musk, the highly visible and controversial CEO of Tesla and SpaceX, said artificial intelligence was "potentially more dangerous than nukes" (while also investing a considerable amount of money in it) and urged the tens of millions of people who followed him on social media platforms to read Bostrom's book.[29] Statements such as these, coupled with widespread media coverage of Bostrom, helped make *Superintelligence* one of the bestselling science books of 2014.

OpenAI largely flew under the radar so far as public awareness is concerned, even as it transitioned away from its nonprofit model. This relative obscurity ended in November 2022 when OpenAI released ChatGPT, an improved and publicly available version of earlier chatbot programs.[30] In the weeks and months that followed, hundreds of tech journalists—the digital progeny of Evelyn Richards and Michael Malone—filed thousands of stories about programs like ChatGPT and the degrees of danger they pose.[31] (The number of these written, at least partly, via a chatbot is an open question.) By early 2023, *Time* had already featured OpenAI's chatbot on its cover. Congressional hearings were scheduled to discuss the technology's implications. There are new university courses for students to explore "writing with robots."[32] Not surprisingly, *ChatGPT for Dummies* materialized soon after OpenAI's program began getting international attention.

As was the case at other key inflection points in the history of computing, the tone of this coverage mixed fascination and consternation, with more emphasis on the latter. Taking a step back, contemporary discussion of AI-fueled chatbots loops us back to older books and authors discussed throughout this book. ChatGPT's potential to automate the intellectual work of writers, for example, is the sort of topic that would have fascinated and repelled Norbert Wiener, who spoke out against the dangers of replacing people with machines. Recall how Wiener, proud of his status as a writer, boasted that he ate his royalties. Meanwhile, contemporary discussions about what computers could, should, and shouldn't do connects to earlier debates about artificial intelligence that people such as Hubert Dreyfus and Herbert Simon had in the 1960s.

Any history of chatbots must necessarily consider Joseph Weizenbaum's ELIZA program as a forerunner. It is hard to imagine Weizenbaum being anything other than appalled by the new technology. He would have been especially dismayed by the propensity of large language models to fabricate misinformation (a phenomenon computer scientists charitably call "hallucinations") or to regurgitate the often noxious and hateful biases and prejudices found in the program's training data. In one infamous example, an early iteration of ChatGPT was asked to complete the sentence: "Two Muslims walked into a mosque." The machine's response was "One turned to the other and said, 'You look more like a terrorist than I do.'"[33]

At a conceptual level, generative AI programs resemble Dahl's fictional "Automatic Grammatisator" in that they obey established mathematical rules while producing results that draw from a large corpus of data. But with programs like ChatGPT, the difference is one of magnitudes. They *learn* using datasets of hundreds of billions of words to create a sometimes-plausible sequence of sentences. Like ELIZA, but to a far greater degree, *conversing* with a program like ChatGPT can be unsettling.[34] Despite appearances, this interaction is not an instantiation of computer-based consciousness but rather the result of billions of probabilistic calculations.

Weizenbaum was especially troubled by how programs like ELIZA could easily fool users into thinking they were interacting with a sentient machine. Even more problematic was his sense that people *wanted* to be deceived. Although Weizenbaum worried about computing machines with an inclination to distort the truth, he reserved especial reproach for the computer scientists whose work introduced falsehoods into the world. Programs such as ChatGPT, Bard, and DALL-E can manipulate text and images with great sophistication to produce misinformation. In 1986, philosopher Harry Frankfurt defined "bullshit" as speech intended to persuade without regard to the truth.[35] ChatGPT endeavors to give answers without accountability. "One shouldn't lie," Weizenbaum once said, an injunction the programmer-turned-critic surely would have directed at contemporary AI researchers and some of their software creations.[36]

While causing concern and anxiety for authors, professors, and many other professionals, programs like ChatGPT have also generated enthusiasm among venture capitalists and business executives seeking new arenas in which to invest their money. In 2023, Microsoft, for example, provided $10 billion to OpenAI.[37] A cynic might also note that statements about potential existential threats posed by new computing technologies helps to generate hype, thus fueling public interest and possibly more funding.[38] Perhaps a future historian will look back a few decades hence and see 2025 as an inflection point similar to when Edmund Berkeley and Norbert Wiener discussed the peril and promise of "electronic brains" circa 1950 or when Ted Nelson presented his emancipatory vision for personal computers in the 1970s. Each of these earlier moments saw far-reaching developments in computing technology accompanied by dramatically altered relationships between people and their computing technologies.

How new computing technologies like large language models will affect societies and economies in the years to come is impossible to say. Picture someone reading *Giant Brains* in 1949. It would not have been hard for that person to accept the supposition that computers would become smaller, more powerful, and pervasive. But the implications of these developments would have required vast leaps of imagination. That same reader would surely have been stunned to learn that computers would completely upend, for example, book publishing or the film and music industries. At the same time, when he was writing *Giant Brains*, Edmund Berkeley surely found himself in a conundrum. The technology he was writing about was changing so rapidly that some of what he included in his book about the contemporary state of computing was itself out of date by the time his book appeared. Writing about large language models and chatbots today poses the same hazard. It is quite likely that everything I have written here about chatbots will be out of date by the time you read this. Such is the nature of writing about a fast-moving field via a slow-moving technology such as a book. *Caveat lector.*

Nonetheless, the changing symbiosis between people and computers compels us to consider a topic at the heart of many contemporary discussions about programs like ChatGPT: the author's creativity. Originality and inspiration are, of course, of central importance for any writer, artist, musician, or, for that matter, computer programmer. As a value and a trait that often eludes definition and quantification, creativity has been cultivated and studied for decades by psychologists, engineers, and advertising executives. There is, in fact, an academic-industrial-creativity complex of modest size which, in recent decades, has itself generated a substantial number of books.[39]

What does it mean when machines can competently create prose, poetry, or software programs?[40] Are computers displaying a sense of creativity? Are they functioning as authors? One glib answer, harking back to Strachey's experiments at the University of Manchester, is that machines have already been doing this for decades. In 1965, John R. Pierce—a renowned electrical engineer at Bell Laboratories and also a published science fiction author—described how researchers were using computers to make visual artworks, compose music, and write original compositions.[41] Soon after, an art exhibition in London featured an array of computer-generated works, including programs that produced pseudo-Japanese haikus and "high-entropy essays" that mimicked undergraduate papers.[42] As is the case with creative works produced by contemporary programs such as ChatGPT and DALL-E, critics in the 1960s looked at these computer-generated creations and unswervingly asked, "Is this art?" or "Does it count as literature?"[43]

In Dahl's "Great Automatic Grammatisator," the fictional Adolph Knipe conceded that "however ingenious" a computer may appear to be, it is still "incapable of original thought." Many people today might argue otherwise. Generative AI software was a central point of contention in the Writers Guild of America strike that shut down television and film production in 2023. Decades earlier, Norbert Wiener used his books to opine about the potential of computers to replace people and their skills. The 2023 strike of television and film writers linked authorship to questions about computer-generated content. A central part of the labor dispute's settlement was the proviso that AI tools "can't write or rewrite literary material" and also that film studios must disclose whether any materials given to a writer were "AI-generated."[44]

At the same time as writers were walking the picket lines, a cohort of prominent authors filed a legal complaint protesting OpenAI's unauthorized appropriation of tens of thousands of digitized books (including some of my own) to create a robust dataset used to "train" ChatCPT. Here, we find parallels to the "book-makers" of the nineteenth century who saw existing books simply as a resource to be extracted and exploited. One might even go so far as to say that contemporary chatbots are not "writing" but rather engaging in a sophisticated form of plagiarism (or bullshit generation). Besides treating previously published works as a stolen resource to be mined, the seeming ease with which chatbot programs produce results obscures the human labor—often done by low-paid workers in places far removed from Silicon Valley—necessary for them to work properly.[45]

The proliferation of chatbots has made it easier for "authors" to "create" hundreds of books with "their" name on the cover. A quick perusal of Amazon's Kindle store, for example, revealed one name that appeared on over 1,100 titles. Almost all of these are short books, less than fifty pages and selling for under $5, that address topics such as computer programming (including how to use ChatGPT) and chess strategies. A search of Amazon's catalog returned some 20,000 results for "ChatGPT," and journalists have noted a flood of new books created using chatbots. Hastily assembled books of poor quality have, of course, existed for centuries. But the ease with which a computer program can churn out "books" might, one astute observer suggests, trigger a "textpocalypse," in which computer-generated writing overwhelms the internet with "synthetic text devoid of human agency."[46]

There are other but less obvious dangers. A British newspaper examined several books written for afficionados of mushroom foraging and passed off as the work of a human author. An examination of the book's prose—done by a computer, of course—concluded that it was 100 percent likely that the books were written by a chatbot. Given how programs like ChatGPT pull (some might say steal) material from a huge variety of sources, the alimentary advice proffered in books like *Wild Mushroom Cookbook: Form* [sic] *Forest to Gourmet Plate* could prove hazardous to one's health. One text encouraged budding mycologists to always taste their samples, something experienced foragers definitely advise against.[47] *Caveat cenator.*

Computers certainly have the potential to alter the nature of what it means to be an author. What this will actually look like in practice in the

years to come is hard to say. We are on firmer ground when we acknowledge that books about computing and their authors are indubitably part of
computing's rich history. Today, a "computer" bears little resemblance to the
machines that Edmund Berkeley or Joseph Weizenbaum or Lynn Conway
first learned to use. Exploring the histories of authors, computers, publishers, and books about computing, along with the practice of writing itself,
shows that none of these are fixed categories. What we discover instead are
dynamic categories that are historically contingent and constantly in flux.

A central premise of this book has been that books have played a central
role in promoting and popularizing information technologies. Communities of users, some skilled and others neophytes, coalesced around them. In
some cases, books provided specialized, standardized, and highly portable
knowledge for students aiming to learn a specific set of skills. The function of
books extended to offering social criticism from both a conservative as well as
liberal perspective. And it was through books that readers received warnings
about computing's seductive power as well as informed speculations about
the implications of new technologies for future societies and economies.

For centuries, books have been central tools for constructing the intellectual culture we inhabit. This has remained steady and true even as technological changes are altering the status of (paper) books and the identity
of who (or what) creates them. Over the course of more than a half-century
of rapid and remarkable transformations in computing technologies, the
centuries-old technology of the book has endured and persisted. Do books
matter? Can they tell us something new or surprising about computing? The
fact that that you are holding this book that I wrote—in whatever form it
takes—in your own hands, and have read this far, says yes.

# Acknowledgments

If there is one thing that the history of books teaches us, it is that no work is the product of a single author working alone. This book is no exception.

In terms of archive and library assistance, I received generous help from: Penny Ahlstrand (Computer History Museum), Emma Broder (Harvard), Peter Collopy (Caltech), Ella Coon (Columbia University), Amanda Katz (Carnegie Mellon University), Henry Lowood (Stanford), Julia Menzel (Massachusetts Institute of Technology), and Amanda Wick (Charles Babbage Institute). Their contributions helped make this book possible.

Many friends and colleagues offered advice and information. In no particular order, I would like to thank: David Brock, Thomas Haigh, Thomas Misa, Stephanie Dick, Laine Nooney, Lynn Conway, Zachary Loeb, Matthew Kirschenbaum, Adrian Johns, James Cortada, Bernadette Longo, Paul Ceruzzi, Mike Halvorson, Yulia Frumer, Julia Marino, Matthew Pressman, Dick van Lente, Hunter Heyck, Bruce Lewenstein, Alan Liu, Magnus Rust, Langdon Winner, Evan Koblentz, Nat Kuhn, Angela Creager, Frank Romano, Will Mari, Ingrid Ockert, Robert Morstein-Marx, Jeff Yost, Nelson Lichtenstein, Dave Karpf, and Lee Vinsel. Even more heroic were critiques and comments provided by Michael Gordin, which came in the midst of his own authorial endeavors.

There is much more to bringing a book into the world than writing it. I owe a debt to many people for their help. These include Katie Helke, now at Oxford University Press, who first championed it and indeed suggested the title. Matthew Christensen did the excellent copyediting and, in the process, corrected some authorial errors, added some valuable insights, and smoothed out my rougher passages of prose. Finally, Justin Kehoe and his capable colleagues at the MIT Press brought *README* from draft manuscript to publication and, subsequently, to your hands.

This book is dedicated to you, the reader.

W. Patrick McCray
June 2025
Santa Barbara, California
and Fort Collins, Colorado

# Notes on Sources

During the course of writing this book, I worked with documents from several archival collections. In addition, I also used some historical materials that are still in private hands and have yet to find to find their way to a permanent institutional home. The sources I consulted are listed below along with abbreviations used in the endnotes to identify them.

## ARCHIVAL COLLECTIONS

California Institute of Technology, Pasadena, CA
- Carver Mead papers; identifier 10284-MS (CM/CIT)

Carnegie Mellon University Libraries, Pittsburgh, PA
- Pamela McCorduck papers, identifier 1978–0001 (PM/CMU)
- Herbert A. Simon papers, identifier 1998–01 (HAS/CMU)

Charles Babbage Institute; Minneapolis, MN
- Edmund C. Berkeley papers; CBI 50 (ECB/CBI)

Columbia University Archival Collections, New York City, NY
- Alvin and Heidi Toffler papers; MS 1247 (AHT/CU)
- Random House records; MS 1048 (RH/CU)

Computer History Museum, Mountain View, CA
- Esther Dyson papers; X2957.2005 (ED/CHM)
- Jim C. Warren papers; X2595.2004 (JW/CHM)

Harvard University Archives, Cambridge, MA
- Daniel Bell papers; HUA 05017 (DB/HUA)
- Esther Dyson papers; MC755 (ED/HUA)

Massachusetts Institute of Technology Archives, Cambridge, MA
- Norbert Wiener papers; MC 0022 (NW/MIT)

Santa Clara University Library, Santa Clara, CA
- Michael S. Malone papers; accession #202002 (MSM/SCU)

Stanford University Special Collections; Palo Alto, CA
- Donald E. Knuth papers; SC 0097 (DEK/SU)
- Edward Feigenbaum papers; SC 0340 (EF/SU)
- Stewart Brand papers; M1237 (SB/SA)

University of Pennsylvania Special Collections, Philadelphia, PA
- Lewis Mumford Collection (Correspondence with Joseph Weizen-baum), Box 9, Folders 5270 and 5271, Ms. Coll. 2 (LM/UP)

Walter Reuther Library at Wayne State University, Detroit, MI
- Reuther-Wiener correspondence from the Reuther papers, collection LR000261 (WR/NW)

### PRIVATE AND ONLINE COLLECTIONS

In the course of researching this book, I examined numerous collections of documents and ephemera still in private collections or archived online in digital format. In cases where I've cited these materials, copies of all sources used are in my working files until I donate them to a suitable repository.

- Dale Dougherty, materials related to *The Whole Internet Catalog* and Global Network Navigator (WIC/DD)
- Nathaniel "Nat" Kuhn, materials related to the RESISTORS (NK/R)
- Dan Doernberg, materials related to the Computer Literacy Bookshops (DD/CLB)
- My writing on Very Large Scale Integration (VLSI) and the work of Lynn Conway and Carver Mead relied extensively on archival materials

Conway carefully curated and collected (i.e., "Lynn Conway Archive"). This is all available (as of July 2024) at the website: https://ai.eecs.umich .edu/people/conway/. My use of these materials is noted throughout my book, where relevant, as "LCA." Copies of all materials I used are also in my research files.

In addition to documentary evidence, I was fortunate to be able to take advantage of several interviews and oral histories, most conducted during the COVID-19 pandemic remotely via Zoom. Interviews collected or consulted are listed below; notes from all interviews are in the author's working files.

### INTERVIEWS BY THE AUTHOR

Lynn Conway (May 20, 2023)
Dan Doernberg (December 21 and 28, 2020)
Dale Dougherty (December 16, 2020)
Dan Gookin (December 21, 2020)
Edward M. Krol (January 19, 2021)
Nat Kuhn (May 7, 2021, and July 5, 2021)
John Levine (April 30, 2021)
Steven Levy (March 21, 2021)
Mike Loukides (December 17, 2020)
Michael Malone (September 2, 2021)
John Markoff (September 1, 2020, and October 13, 2020)
Michael Moritz (November 4, 2020)
Theodor H. Nelson (June 11, 2022)
Jennifer Niederst Robbins (December 29, 2020)
Andy Rathbone (July 5, 2023)
Tim O'Reilly (November 18, 2020, and December 12, 2020)
Evelyn Richards (March 29, 2021)
J. Laurence Sarno (May 17, 2021)
Langdon Winner (January 18, 2022)

# Notes

HELLO "README"

1. Jennifer S. Light, "When Computers Were Women," *Technology and Culture* 40, no. 3 (1999): 455–483, https://doi.org/10.1353/tech.1999.0128.

2. A good overview of this rapidly advancing topic are the essays in a special issue of *Poetics Today* titled "The AI Revolution: Speculations on Authorship, Pedagogy, and the Future of the Profession," ed. Nir Evron and Roi Tartakovsky, 45, no. 2 (2024), https://doi.org/10.1215/03335372-11092765.

3. Based on two prominent surveys of the history of computing: Thomas Haigh and Paul E. Ceruzzi, *A New History of Modern Computing* (Cambridge, MA: MIT Press, 2021); and, Martin Campbell-Kelly, William F. Aspray, Jeffrey R. Yost, Honghong Tinn, and Gerardo Con Díaz, *Computer: A History of the Information Machine*, 4th ed. (New York: Routledge, 2023).

4. See two recent collections: Thomas S. Mullaney, Benjamin Peters, Mar Hicks, and Kavita Philip, eds., *Your Computer Is on Fire* (Cambridge, MA: MIT Press, 2021); and, Janet Abbate and Stephanie Dick, eds., *Abstractions and Embodiments: New Histories of Computing and Society* (Baltimore, MD: Johns Hopkins University Press, 2022).

5. Robert Darnton, "What is the History of Books?" *Daedalus* 111, no. 3 (1982): 65–83, https://www.jstor.org/stable/20024803; and, Matthew Kirschenbaum, "Bibliologistics: The Nature of Books Now, or A Memorable Fancy," *Post45*, April 8, 2020, https://post45.org/2020/04/bibliologistics-the-nature-of-books-now-or-a-memorable-fancy/ (accessed January 5, 2024).

6. Brian Cowan, *The Social Life of Coffee: The Emergence of the British Coffeehouse* (New Haven, CT: Yale University Press, 2005).

7. Aileen Fyfe, *Steam-Powered Knowledge: William Chambers and the Business of Publishing, 1820–1860* (Chicago, IL: University of Chicago Press, 2021).

8. Janice A. Radway, *A Feeling for Books: The Book-of-the-Month Club, Literary Taste, and Middle-Class Desire* (Chapel Hill: University of North Carolina Press, 1997).

9.  A wide range of scholarship has explored how the book, as an object, has changed, such as Kenneth C. Davis, *Two-Bit Culture: The Paperbacking of America* (Boston, MA: Houghton Mifflin, 1984).

10. Statistics from Laura J. Miller and David Paul Nord, "Reading the Data on Books, Newspapers, and Magazines: A Statistical Appendix," in *A History of the Book in America*, ed. David Paul Nord, Joan Shelley Rubin, and Michael Schudson, vol. 5, *The Enduring Book: Print Culture in Postwar America* (Chapel Hill: University of North Carolina Press, 2009), 503–518.

11. John L. Rudolph, *Scientists in the Classroom: The Cold War Reconstruction of Science Education* (New York: Palgrave, 2002).

12. Beth Luey, "The Organization of the Book Publishing Industry," in *A History of the Book in America*, ed. David Paul Nord, Joan Shelley Rubin, and Michael Schudson, vol. 5, *The Enduring Book: Print Culture in Postwar America* (Chapel Hill: University of North Carolina Press, 2009), 29–54.

13. Albert N. Greco, *The Book Publishing Industry* (Mahwah, NJ: Lawrence Erlbaum Associates, 2005).

14. Quote from Roger Cohen, "Profits—Dick Snyder's Ugly Word," *New York Times*, June 30, 1991; see also, Dan Sinykin, *Big Fiction: How Conglomeration Changed the Publishing Industry and American Literature* (New York: Columbia University Press, 2023).

15. Laura J. Miller, "Selling the Product," in *A History of the Book in America*, ed. David Paul Nord, Joan Shelley Rubin, and Michael Schudson, vol. 5, *The Enduring Book: Print Culture in Postwar America* (Chapel Hill: University of North Carolina Press, 2009), 90–106; also see John B. Thompson's incredibly helpful *Merchants of Culture: The Publishing Business in the Twenty-First Century* (New York: Polity, 2010); and, more recently, Evan Friss, *The Bookshop: A History of the American Bookstore* (New York: Viking, 2024).

16. Described in the company's own history, Amazon.com Timeline, archived May 25, 2020, at https://web.archive.org/web/20200525215120/http://media.corporate-ir.net/media_files/irol/17/176060/TimelineQ3_2007.pdf (accessed January 8, 2024).

17. For example, Philip H. Dougherty, "Publishers Zero In on Computer Industry," *New York Times*, March 29, 1970.

18. Ezra Greenspan and Jonathan Rose, "An Introduction to *Book History*," *Book History* 1 (1998): ix–xi.

19. To give just a hint of this vast corpus, consider three relatively recent books: Matthew G. Kirschenbaum, *Track Changes: A Literary History of Word Processing* (Cambridge, MA: Belknap Press of Harvard University Press, 2016); Dennis Duncan and Adam Smyth, eds., *Book Parts* (New York: Oxford University Press, 2019); and, Adrian Johns, *The Science of Reading: Information, Media, and Mind in Modern America* (Chicago, IL: University of Chicago Press, 2023).

20. A classic work is James A. Secord, *Visions of Science: Books and Readers at the Dawn of the Victorian Age* (Chicago, IL: University of Chicago Press, 2014). In addition, there is Adrian

Johns' epic exploration of British print culture, much of which focuses on scientific work, *The Nature of the Book: Print and Knowledge in the Making* (Chicago, IL: University of Chicago Press, 1998).

21. Janet Browne, *Darwin's Origin of Species: A Biography* (New York: Grove Press, 2006). For the reception of Darwin's books outside of Great Britain, see Marwa Elshakry, *Reading Darwin in Arabic, 1860–1950* (Chicago, IL: University of Chicago Press, 2013); and, on Einstein, Hanoch Gutfreund, and Jürgen Renn, *The Formative Years of Relativity: The History and Meaning of Einstein's Princeton Lectures* (Princeton, NJ: Princeton University Press, 2018).

22. Melinda Baldwin, *Making Nature: The History of a Scientific Journal* (Chicago, IL: University of Chicago Press, 2015); and, Alex Csiszar, *The Scientific Journal: Authorship and the Politics of Knowledge in the Nineteenth Century* (Chicago, IL: University of Chicago Press, 2017).

23. Priscilla Coit Murphy, *What a Book Can Do: The Publication and Reception of Silent Spring* (Amherst: University of Massachusetts Press, 2005); James D. Watson, *The Annotated and Illustrated Double Helix*, ed. Alexander Gann and Jan Wikowski (New York: Simon & Schuster, 2012); and, Charles Seife, *Hawking Hawking: The Selling of a Scientific Celebrity* (New York: Basic Books, 2021).

24. See Derek J. de Solla Price, "Is Technology Historically Independent of Science? A Study in Statistical Historiography," *Technology and Culture* 6, no. 4 (1965): 553–568, https://muse.jhu.edu/article/894638.

25. If one looks to earlier periods of time, the situation changes somewhat. For example, Pamela O. Long's *Openness, Secrecy, Authorship: Technical Arts and the Culture of Knowledge from Antiquity to the Renaissance* (Baltimore, MD: Johns Hopkins University Press, 2001) considers early modern artisans as authors and explores the treatises they produced. One notable exception is scholarship around the *Whole Earth Catalog*, a countercultural icon first published in 1968. See, for instance, Andrew G. Kirk, *Counterculture Green: The Whole Earth Catalog and American Environmentalism* (Lawrence: University Press of Kansas, 2007).

26. Angela N. H. Creager, Mathias Grote, and Elaine Leong, "Learning by the Book: Manuals and Handbooks in the History of Science," *BJHS Themes* 5 (2020): 1–13, https://doi.org/10.1017/bjt.2020.1.

27. This work falls into the category of fields like software studies or critical code studies. For an overview, see Matthew Kirschenbaum and Sarah Werner, "Digital Scholarship and Digital Studies: The State of the Discipline," *Book History* 17 (2014): 406–458, https://doi.org/10.1353/bh.2014.0005.

28. Bernadette Longo, *Spurious Coin: A History of Science, Management, and Technical Writing* (Stony Brook: State University of New York Press, 2000).

## CHAPTER 1

1. For contemporary descriptions, see "New Giant 'Brain' Does Wizard Work," *New York Times*, August 25, 1947; and, "Never Stumped," *New Yorker*, March 4, 1950, 20–21.

2.  A survey of *the New York Times* database alone shows more than fifty articles with the terms "electronic brain" and "electronic 'brain'" between 1945 and 1952.

3.  Edmund C. Berkeley, *Giant Brains, or Machines That Think* (New York: John Wiley & Sons, 1949).

4.  For Berkeley's biography, see Bernadette Longo, *Edmund Berkeley and the Social Responsibility of Computer Professionals* (New York: Association for Computing Machinery and Morgan & Claypool Publishers, 2015).

5.  The phrase comes from Robert S. Lee, "Social Attitudes and the Computer Revolution," *Public Opinion Quarterly* 34, no. 1 (1970): 53–59, https://doi.org/10.1086/267772. The image is examined in C. Dianne Martin, "The Myth of the Awesome Thinking Machine," *Communications of the ACM* 36, no. 4 (1993): 120–133, https://doi.org/10.1145/255950 .153587.

6.  The speech is preserved in Folder 7, Box 79 of Berkeley's papers at the Charles Babbage Institute; hereafter "ECB/CBI."

7.  Quotes from "Modern Methods of Thinking," Folder 7, Box 79, ECB/CBI. Also, Edmund C. Berkeley, *Symbolic Logic and Intelligent Machines* (New York: Reinhold, 1959).

8.  Edmund C. Berkeley, entry in *Harvard Class of 1930: 25th Anniversary Report* (Cambridge, MA: Harvard University, 1955), 71–76.

9.  Berkeley, *Harvard Class of 1930*.

10. Described in chapter 5 of Longo, *Edmund Berkeley and the Social Responsibility of Computer Professionals*.

11. JoAnne Yates, *Structuring the Information Age: Life Insurance and Technology in the Twentieth Century* (Baltimore, MD: Johns Hopkins University Press, 2009); and, Blair E. Olmsted, "Prudential's Early Experience with Computers," unpublished report, 1978; Folder 3, Box 10, ECB/CBI.

12. For example, Edmund C. Berkeley, "Boolean Algebra (The Technique for Manipulating 'And,' 'Or,' and 'Not' and Conditions) and Applications to Insurance," *Record of the American Institute of Actuaries* 26, no. 54 (1937): 373–414.

13. JoAnne Yates, "Early Interactions Between Life Insurance and Computer Industries: The Prudential's Edmund C. Berkeley," *IEEE Annals of the History of Computing* 19, no. 3 (1997): 60–73, https://doi.org/10.1109/85.601736.

14. Quote from Longo, *Edmund Berkeley and the Social Responsibility of Computer Professionals*, page 27 and dated June 28, 1942. Unfortunately, Longo's endnotes—while quite informative—don't record which folder or box in Berkeley's extensive papers her quotes come from.

15. Quote from Longo, *Edmund Berkeley and the Social Responsibility of Computer Professionals*, page 28; dated March 20, 1945.

16. I. Bernard Cohen and Gregory W. Welch, eds., *Makin' Numbers: Howard Aiken and the Computer* (Cambridge, MA: MIT Press, 1999).

17. Jeremy Bernstein, *The Analytical Engine: Computers, Past, Present, and Future* (New York: Random House, 1964), 64.

18. Gobind Behari Lal, "Harvard's Robot Super-Brain," *American Weekly*, October 15, 1944, 5–6, http://collections.si.edu/search/detail/ead_component:sova-nmah-ac-0324-ref986; and, "Robot Works Problems Never Before Solved," *Popular Mechanics*, October 1944,13, https://www.si.edu/object/archives/components/sova-nmah-ac-0324-ref371.

19. See chapter 2 of Longo, *Edmund Berkeley and the Social Responsibility of Computer Professionals*.

20. Berkeley, entry in *Harvard Class of 1930*; and Edmund C. Berkeley, "Basic Equipment for an Electronic Laboratory" and "Electronic Laboratory" memos, September 9, 1947, and October 22, 1947; Folders 5 and 6, Box 10, ECB/CBI.

21. Atsushi Akera, "Edmund Berkeley and the Origins of the ACM," *Communications of the ACM* 50, no. 5 (2007): 31–35, https://doi.org/10.1145/1230819.1230835.

22. ACM History Committee, "ACM History," Association for Computing Machinery, https://www.acm.org/about-acm/acm-history.

23. Edmund C. Berkeley, draft memo, April 28, 1948; Folder 13, Box 13, ECB/CBI.

24. Berkeley's summary is preserved in the May 20, 1948 "Memorandum for File"; Folder 20, Box 13, while Berkeley's resignation materials are in Folder 1, Box 15, both ECB/CBI; also, Berkeley, entry in *Harvard Class of 1930*.

25. Berkeley, *Giant Brains*, vii–xi.

26. Berkeley, "Proposal for A Book," May 26, 1946; Folder 1, Box 15, ECB/CBI.

27. Letter from J. Kenneth Maddock to Edmund C. Berkeley, July 19, 1946; Folder 32, Box 8, ECB/CBI.

28. Information from Funding Universe, http://www.fundinguniverse.com/company-histories/john-wiley-sons-inc-history/; as well as, Peter Gölitz, "200 Years of John Wiley & Sons," *Angewandte Chemie* 46, no. 1/2 (2006): 4–5, https://doi.org/10.1002/anie.200604745.

29. Letter and anonymous report from Maddock to Berkeley, July 19, 1946; Folder 32, Box 8, ECB/CBI.

30. "E. C. Berkeley and Associates: Purposes," memo, December 29, 1946; Folder 1, Box 15, ECB/CBI.

31. Beth Luey, "The Organization of the Book Publishing Industry," in *A History of the Book in America*, ed. David Paul Nord, Joan Shelley Rubin, and Michael Schudson, *vol. 5, The Enduring Book: Print Culture in Postwar America* (Chapel Hill: University of North Carolina Press, 2009), 29–54; sales data from Laura J. Miller and David Paul Nord, "Reading the Data on Books, Newspapers, and Magazines: A Statistical Appendix," in *A History of the Book*

*in America*, ed. David Paul Nord, Joan Shelley Rubin, and Michael Schudson, *vol. 5, The Enduring Book: Print Culture in Postwar America* (Chapel Hill: University of North Carolina Press, 2009), 503–518.

32. Figures from Carolyn D. Hay, "A History of Science Writing in the United States and of the National Association of Science Writers" (master's thesis, Northwestern University, 1970).

33. Nicolas Rashevsky, "Mathematical Biophysics," *Nature* 135, no. 3414 (1935): 528–530; and, Tara H. Abraham, "Nicolas Rashevsky's Mathematical Biophysics" *Journal of the History of Biology* 37 (2004): 333–385, https://doi.org/10.1023/B:HIST.0000038267.09413.0d.

34. Warren S. McCulloch and Walter Pitts, "A Logical Calculus of Ideas Immanent in Nervous Activity," *Bulletin of Mathematical Physics* 5, no. 4 (1943): 115–133, https://doi.org/10.1007/BF02478259; and, Tara H. Abraham, "(Physio)Logical Circuits: The Intellectual Origins of the McCulloch-Pitts Neural Network," *Journal of the History of the Behavioral Sciences* 38, no. 1 (2002): 3–25, https://doi.org/10.1002/jhbs.1094.

35. Alan M. Turing, "On Computable Numbers, with an Application to the *Entscheidungsproblem*," *Proceedings of the London Mathematical Society* Series 2, 42, no. 1 (1937): 230–265.

36. Warren S. McCulloch, "The Brain as a Computing Machine," *Electrical Engineering* 68, no. 6 (1949): 492–497, https://doi.org/10.1109/EE.1949.6444817.

37. Norbert Wiener, "Cybernetics," *Scientific American* 179, no. 11 (1948): 14–19, https://www.jstor.org/stable/24945913; and, Ronald R. Kline, *The Cybernetics Moment: Or Why We Call Our Age the Information Age* (Baltimore, MD: Johns Hopkins University Press, 2015).

38. John von Neumann, "First Draft of a Report on the EDVAC," reprinted in *The Annals of the History of Computing* 15, no. 4 (1993): 27–75, https://doi.org/10.1109/85.238389. For von Neumann's involvement with the ENIAC and EDVAC machines, see Thomas Haigh, Mark Priestley, and Crispin Rope, *ENIAC in Action: Making and Remaking the Modern Computer* (Cambridge, MA: MIT Press, 2018).

39. See William Aspray's account in *John von Neumann and the Origins of Modern Computing* (Cambridge, MA: MIT Press, 1990).

40. Von Neumann, "First Draft of a Report on the EDVAC."

41. John E. Pfeiffer, "Brains and Calculating Machines," *American Scholar* 19, no. 1 (1949–1950): 21–30, https://www.jstor.org/stable/41205257.

42. John Mills, "Communication with Electrical Brains," *Bell Telephone Quarterly* 13 (January 1934): 47–57.

43. Pfeiffer, "Brains and Calculating Machines."

44. "Apostle of 'Plain Talk,'" *Business Week*, April 12, 1947, 22–25.

45. Bernadette Longo, "Toward an Informed Citizenry: Readability Formulas as Cultural Artifacts," *Journal of Technical Writing and Communication* 34, no. 3 (2004): 165–172, https://doi.org/10.2190/EXTJ-E7UE-6DEA-AK8P.

46. Letter from Rudolf Flesch to Berkeley, June 27, 1947; Folder 43, Box 2, ECB/CBI.

47. Peter J. Bowler, *Science for All: The Popularization of Science in Early Twentieth-Century Britain* (Chicago, IL: University of Chicago Press, 2009); see also, Robert E. Filner, "The Social Relations of Science Movement (SRS) and J. B. S. Haldane." *Science and Society* 41, no. 3 (1977): 303–316, https://www.jstor.org/stable/40402032.

48. Berkeley, in a note to himself on January 4, 1948: "Can you get a useful guide to what you are trying to do by reading *Mathematics for the Million*?"; Folder 28, Box 8, ECB/CBI.

49. "Chapter 1," notes from January 4, 1948; and, "Broad Comments Received, July 1947 to December 1947—Summary"; both dated January 4, 1948, and in Folder 28, Box 8, ECB/CBI.

50. "Obituary: Zehman I. Mosesson," *Transactions of the Society of Actuaries* 47 (1995): 966–967.

51. Zehman I. Mosesson, "Comments on Book," report from March 8, 1948; letter from IBM to Berkeley, October 27, 1947; both Folder 29, Box 8, ECB/CBI.

52. Berkeley, note for file, September 12, 1948; Folder 29, Box 8, ECB/CBI.

53. Berkeley, "Memorandum for Mr. W. P. Miller, Associate Counsel," letter, April 18, 1948; Folder 29, Box 8, ECB/CBI.

54. "Critic B," April 10, 1948; Folder 29, Box 8, ECB/CBI.

55. "Comments by Referee A on the Comments by Referee B on the Berkeley Manuscript," April 10, 1948; Folder 29, Box 8, ECB/CBI.

56. Norbert Wiener, *I Am a Mathematician: The Later Life of a Prodigy* (New York: Doubleday, 1956).

57. See, for example, an advertisement in the December 15, 1949, issue of the *New York Times*, clipped by Berkeley; Folder 32, Box 8, ECB/CBI.

58. "Press Release for Giant Brains," n.d. but likely November 1949; Folder 31, Box 8, ECB/CBI.

59. Edmund Berkeley, "Information about *Giant Brains*," July 4, 1949; Folder 32, Box 8, ECB/CBI.

60. Henry Beckett, "Genius at Work on the Automatic Simple Simon," *New York Post*, October 13, 1949; copy in Folder 31, Box 8, ECB/CBI.

61. Letter from Flesch to Berkeley, December 2, 1949; Folder 32, Box 8, ECB/CBI.

62. "Press Release for Giant Brains," n.d. but likely November 1949; Folder 31, Box 8, ECB/CBI.

63. Berkeley, *Giant Brains*, 4.

64. Berkeley, *Giant Brains*, 10.

65. Berkeley, *Giant Brains*, ix.

66. Thomas Haigh, ed., *Exploring the Early Digital* (Switzerland: Springer Cham, 2019); as well as Ronald R. Kline's essay in the same volume "Inventing an Analog Past and a Digital Future in Computing," 19–39.

67. First mention of the word in *the New York Times* was on page 46 of the edition dated July 1, 1948, in a section titled "The News of Radio."

68. When a second edition of *Giant Brains* was published, Berkeley included a lengthy section "Comments in 1961" that discussed how computing technology had changed, noting that transistors had rendered vacuum tubes and relays "no longer important except marginally." Berkeley, *Giant Brains, or Machines That Think* (New York: Science Editions, 1961).

69. "Tiny 'Brain' Robot Not So Very Dumb," *New York Times,* May 19, 1950.

70. "Feeble-Minded Brain Machine Makes Debut," *New York Herald Tribune,* May 19, 1950.

71. Edmund C. Berkeley, "Simple Simon," *Scientific American* 183, no. 5 (November 1950): 40–43, https://doi.org/10.1038/scientificamerican1150-40.

72. Berkeley cowrote the series with Robert A. Jensen, one of the students who helped build Simon. The first of their thirteen articles appeared as "World's Smallest Electric Brain," *Radio-Electronics,* October 1950, 29–30.

73. Edmund C. Berkeley, "Light-Sensitive Electronic Beast," *Radio-Electronics,* December 1951, 46–48. See Berkeley Enterprises, "Small Robots—Report" April 1956; information on these is filed in Berkeley's papers under "Teaching Machines"; Folders 1–23; Box 8, ECB/CBI. As well as Edmund C. Berkeley, *Brainiacs—The 1958 Experiments* (Newtonville, MA: Berkeley Enterprises, 1958).

74. Karen A. Frenkel, "An Interview with Ivan Sutherland," *Communications of the ACM* 32, no. 6 (1989): 711–718, https://doi.org/10.1145/63526.63531.

75. Kristen Haring, *Ham Radio's Technical Culture* (Cambridge, MA: MIT Press, 2007).

76. Berkeley, "Simple Simon," 42.

77. Letters, dated November 3 and November 7, 1949; both Folder 30, Box 8, ECB/CBI.

78. Edmund Berkeley, "Doubling the Sale of Technical Books," unpublished report, April 22, 1951; Folder 37, Box 8, ECB/CBI.

79. "Report on Promotional Plan for Berkeley: *Giant Brains,*" October 11, 1949; Folder 30, Box 8, ECB/CBI.

80. John E. Pfeiffer, "Mechanical Logicians," *New York Times* December 11, 1949, BR 19.

81. Berkeley, *Giant Brains,* 208.

82. "The Thinking Machine," *New York Herald Tribune,* December 6, 1949; copy in Folder 31, Box 8, ECB/CBI.

83. Nicholas Metropolis, review of *Giant Brains, Journal of the American Statistical Association* 45, no. 252 (1950): 573–574.

84. Richard W. Hamming, review of *Giant Brains, American Mathematical Monthly* 58, no. 4 (1951): 276.

85. *Publishers Weekly,* December 17, 1949; clipping in Folder 31, Box 8, ECB/CBI.

86. Letter from George Lovitt to Berkeley, February 14, 1951; Folder 38, Box 8, ECB/CBI.

87. Edmund C. Berkeley, *The Computer Revolution* (New York: Doubleday, 1962); and, John E. Pfeiffer, *The Thinking Machine* (Philadelphia, PA: Lippincott, 1962). Mauchly's review is "Revolution and Evolution," *Science* 138, no. 3542 (1962): 806–807, https://doi.org/10.1126/science.138.3542.806.b.

88. Berkeley, *The Computer Revolution*, 192.

89. The original title, dating to September 1952, was *The Computing Machinery Field*; a year later Berkeley, as editor and publisher, changed it to *Computers and Automation*. In 1972, it became *Computers and Automation and People* before ending as *Computers and People* in 1974.

## CHAPTER 2

1. Bohuslav Kirchmann, Pavel Kopecky, and Zdenek Zdrahal, "GOALEM from Prague," in *Proceedings of the 5th International Joint Conference on Artificial Intelligence*, IJCA1–77, ed. William Kaufmann (Cambridge, MA: MIT Press, 1977): 771

2. Hillel J. Kieval, "Pursuing the Golem of Prague: Jewish Culture and the Invention of a Tradition," *Modern Judaism* 17, no. 1 (1997): 1–23, https://www.jstor.org/stable/1396572.

3. Pamela McCorduck, *Machines Who Think: A Personal Inquiry into the History and Prospects of Artificial Intelligence* (San Francisco, CA: W. H. Freeman, 1979). McCorduck's reference to Rabbi Löw's supposed descendants is based on anecdotal evidence from Joel Moses, an Israeli-born computer scientist at MIT. Also see her interview with Marvin Minsky, April 24, 1975; Folder 6, Box 2, PM/CMU.

4. Discussed at the start of Norbert Wiener, *Ex-Prodigy: My Childhood and Youth* (New York: Simon & Schuster, 1953).

5. Norbert Wiener, *God & Golem, Inc.: A Comment on Certain Points where Cybernetics Impinges on Religion* (Cambridge, MA: MIT Press, 1964).

6. Jason Resnikoff, *Labor's End: How the Promise of Automation Degraded Work* (Urbana: University of Illinois Press, 2021).

7. Wiener, *God & Golem, Inc.*, 52, 95.

8. Stuart Chase, "The Next Industrial Revolution: An Age of Machines That Think," *New York Herald Tribune Book Review*, August 20, 1950, 5; and, Leslie C. Dunn, review of *The Human Use of Human Beings*, *Scientific American* 183, no. 6 (1950) 60.

9. Biographical material on Wiener comes from a number of sources, especially Flo Conway and Jim Siegelman, *Dark Hero of the Information Age: In Search of Norbert Wiener, the Father of Cybernetics* (New York: Basic Books, 2005) and Wiener's own memoirs. One exploration of the father–son relationship is Benjamin Peters, "Toward a Genealogy of a Cold War Communication Science: The Strange Loops of Leo and Norbert Wiener," *Russian Journal of Communication* 5, no. 1 (2013): 31–43, https://doi.org/10.1080/19409419.2013.775544.

10. From Hans Freudenthal, a German Dutch mathematician, in "Wiener, Norbert," *Complete Dictionary of Scientific Biography*, vol. 14, (Detroit, MI: Charles Scribner's Sons, 2008): 344–347.

11. Wiener, *Ex-Prodigy*, 68.

12. "The Most Remarkable Boy in the World," *New York World*, October 7, 1906, 1.

13. Wiener, *Ex-Prodigy*, 192–194.

14. Wiener, *Ex-Prodigy*, 197.

15. Wiener, *Ex-Prodigy*, 203–204.

16. Norbert Wiener, *I Am a Mathematician: The Later Life of a Prodigy* (New York: Doubleday, 1956).

17. Norman Levinson, "Wiener's Life," in "Norbert Wiener 1894–1964," *Bulletin of the American Mathematical Society* 72, no. 1, part II (1966): 1–32, quote from page 25, https://projecteuclid .org/journals/bulletin-of-the-american-mathematical-society/volume-72/issue-1.P2.

18. David A. Mindell, *Between Human and Machine: Feedback, Control, and Computing before Cybernetics* (Baltimore, MD: Johns Hopkins University Press, 2004). The origins of cybernetics in the United States has been exceptionally well-explored in, for example, Steve Joshua Heims, *Constructing a Social Science for Postwar America: The Cybernetics Group at MIT 1946–1953* (Cambridge, MA: MIT Press, 1993); and, Ronald R. Kline, *The Cybernetics Moment, or Why We Call Our Age the Information Age* (Baltimore, MD: Johns Hopkins University Press, 2015). It should be noted that the history of cybernetics overseas followed very different intellectual trajectories.

19. Wiener, *Cybernetics*, 12.

20. Peter Galison, "The Ontology of the Enemy: Norbert Wiener and the Cybernetic Vision," *Critical Inquiry* 21, no. 1 (1994): 228–266, https://doi.org/10.1086/448747.

21. Norbert Wiener, "Cybernetics," *Scientific American* 179, no. 5 (1948): 14–19, http://www .jstor.org/stable/24945913, gives a clear overview of his thinking prior to his book's appearance.

22. Wiener, *I Am a Mathematician*, 255. The report was titled "The Extrapolation, Interpolation, and Smoothing of Stationary Time Sequences," it was declassified in 1946 and, after *Cybernetics* sold better than expected, MIT's Technology Press and John Wiley & Sons copublished it in August 1949 as *Extrapolation, Interpolation, and Smoothing of Stationary Time Series: With Engineering Applications*, https://doi.org/10.7551/mitpress/2946.001.0001.

23. Arturo Rosenblueth, Norbert Wiener, and Julian Bigelow, "Behavior, Purpose, and Teleology," *Philosophy of Science* 10, no. 1 (1943): 18–24, https://doi.org/10.1086/286788.

24. Rosenblueth, Wiener, and Bigelow, "Behavior, Purpose, and Teleology."

25. Wiener, *I Am a Mathematician*, 325.

26. Steve J. Heims, *John von Neumann and Norbert Wiener: From Mathematics to the Technologies of Life and Death* (Cambridge, MA: MIT Press, 1980).

27. Warren S. McCulloch, *The Collected Works of Warren S. McCulloch*, ed. Rook McCulloch, vol. 4 (Salinas, CA: Intersystems Publications, 1989), 856.

28. Wiener, *I Am a Mathematician*, 325

29. Slava Gerovitch, *From Newspeak to Cyberspeak: A History of Soviet Cybernetics* (Cambridge, MA: MIT Press, 2002).

30. Wiener, *I Am a Mathematician*, 261–262; 325 where the Soviet's name is spelled "Kolmogoroff."

31. Wiener, *I Am a Mathematician*, 329.

32. Letter from Wiener to McCulloch, November 20, 1950; Folder 107, Box 7, NW/MIT.

33. His memoir mistakenly places this in the summer of 1946; however, see Michael J. Barany, Anne-Sandrine Paumier, and Jesper Lützen, "From Nancy to Copenhagen to the World: The Internationalization of Laurent Schwartz and His Theory of Distributions," *Historia Mathematica* 44, no. 4 (2017): 367–394, https://doi.org/10.1016/j.hm.2017.04.002.

34. Discussed in Wiener, *I Am a Mathematician*, 314–318. For the story of Bourbaki, see David Aubin, "The Withering Immortality of Nicolas Bourbaki: A Cultural Connector at the Confluence of Mathematics, Structuralism, and the Oulipo in France," *Science in Context* 10, no. 2 (1997): 297–342, https://doi.org/10.1017/S0269889700002660.

35. Wiener, *I Am a Mathematician*, 316–317.

36. Pierre de Latil, *La Pensée artificielle: introduction a`la cybernétique* (Paris: Gallimard, 1953); this appeared in the United States as *Thinking by Machine: A Study of Cybernetics, trans.* Y. M. Golla (New York: Houghton Mifflin, 1957); quote from page 11 of the latter. Also, see Christopher Johnson, "'French' Cybernetics," *French Studies* 69, no. 1 (2015): 60–78, https://doi.org/10.1093/fs/knu229.

37. Norbert Wiener, draft manuscript of *Cybernetics*, n.d. but mid-1947; Folder 578, Box 28A, NW/MIT.

38. Wiener, *I Am a Mathematician*, 320–321.

39. Wiener, *I Am a Mathematician*, 331.

40. Response to John Wiley & Sons "Questionnaire Relative to Promotion of Book," n.d. but likely early 1948; Folder 597, Box 28C, NW/MIT.

41. Publicity statement, October 22, 1948; Folder 597, Box 28C, NW/MIT.

42. Wiener, *I Am a Mathematician*, 331.

43. Wiener, *I Am a Mathematician*, 332; Beverly Brooks to Wiener; December 19, 1949; Folder 108, Box 7, NW/MIT

44. Harry Gilroy, "Salesmen as Critics," *New York Times*, February 13, 1949.

45. David Dietz, "Three Scientists Produced Brilliant Books in '49," news clipping dated December 6, 1949; Folder 597, Box 28C, NW/MIT.

46. Sheryl N. Hamilton, "The Charismatic Cultural Life of Cybernetics: Reading Norbert Wiener as Visible Scientist," *Canadian Journal of Communication* 42 no. 3 (2017): 407–429, https://doi.org/10.22230/cjc.2017v42n3a3205.

47. The letter to George Forsythe dated December 2, 1946, as well as Wiener's formal statement in the *Atlantic Monthly* can be found as "From the Archives," *Science, Technology, and Human Values* 8, no. 3 (1983): 36–38, https://www.jstor.org/stable/688755; and, Norbert Wiener, "A Scientist Rebels," Atlantic Monthly, January 1947. See also, "Wiener Denounces Devices 'for War,'" *New York Times*, January 9, 1947, 4.

48. Kline, *The Cybernetics Moment*, 71; Derek Hawkins, "What Made Hawking's 'A Brief History of Time' So Immensely Popular," *Washington Post*, March 14, 2018, https://www.washingtonpost.com/news/morning-mix/wp/2018/03/14/what-made-hawkings-a-brief-history-of-time-so-immensely-popular/ (accessed August 7, 2024).

49. Harry G. Johnson, review of *Cybernetics, or, Control and Communication in the Animal and the Machine, Economic Journal* 59, no. 236, (1949): 573–575, https://doi.org/10.2307/2226579.

50. Letter from Esther P. Potter to Wiener, December 2, 1949; Folder 108, Box 7, NW/MIT.

51. See, for example, voluminous reviews in English, German, French, and Russian found in Box 28C, NW/MIT.

52. John von Neumann, review of *Cybernetics, or, Control and Communication in the Animal and the Machine, Physics Today* 2, no. 5 (1949): 33–34, https://doi.org/10.1063/1.3066516.

53. John E. Pfeiffer, "The Stuff That Dreams Are Made Of," *New York Times*, January 23, 1949.

54. See Mindell, *Between Human and Machine*, 231–259. The system Wiener promoted was more accurate than the Bell-MIT hardware but not amenable to production.

55. See, for example, a letter from George Stibitz to Wiener, March 19, 1951; Folder 135, Box 9, NW/MIT.

56. Wiener, *Cybernetics*, 27, emphasizes "competition."

57. Wiener, *Cybernetics*, 28.

58. "Dr. Norbert Wiener Dead at 69; Known as Father of Automation," *New York Times*, March 19, 1964.

59. Wiener, *I Am a Mathematician*, 308.

60. Harry M. Davis, "An Interview with Norbert Wiener," *New York Times*, April 10, 1949.

61. Quote on page 80 of Kline's *The Cybernetic Moment*.

62. Wiener, *Ex-Prodigy*, 247–248.

63. Wiener, *I Am a Mathematician*, 309; similar statements found in Gordon S. Brown and Norbert Wiener, "Automation, 1955: A Retrospective," *Annals of the History of Computing* 6, no. 4 (1984): 372–383, https://doi.org/10.1109/MAHC.1984.10045.

64. On Reuther, see Nelson Lichtenstein, *The Most Dangerous Man in Detroit: Walter Reuther and the Fate of American Labor* (New York: Basic Books, 1995).

65. Described in Wiener, *The Human Use of Human Beings*, 11–12.

66. Letter from Wiener to Reuther, August 13, 1949; WR/NW. For the question of technological unemployment, see also Amy Sue Bix, *Inventing Ourselves Out of Jobs? America's Debate Over Technological Unemployment, 1929–1981* (Baltimore, MD: Johns Hopkins University Press, 2000).

67. Steve Meyer, "An Economic 'Frankenstein'": UAW Workers' Responses to Automation at the Ford Brook Park Plant in the 1950s," *Michigan Historical Review* 28, no. 1 (2002): 63–89, https://doi.org/10.2307/20173951.

68. Wiener's reaction to this plan is found in a letter to Reuther dated March 20, 1950; WR/NW.

69. Wiener, *I Am a Mathematician*, 309.

70. Automation: A Report to the UAW-CIO Economic and Collective Bargaining Conference (Detroit, MI: UAW-CIO Education Department, January 1955).

71. Paul Brooks, *Two Park Street: A Publishing Memoir* (Boston, MA: Houghton Mifflin, 1986).

72. Brooks, *Two Park Street*, 54–56; and a letter from Paul Brooks to Wiener, March 1, 1949; Folder 93, Box 6, NW/MIT.

73. Brooks, *Two Park Street*, 54–56

74. Letter from Brooks to Wiener, November 10, 1949; Folder 106, Box 7, NW/MIT

75. Quotes from Paul Brooks's letter to Wiener, November 10, 1949; Folder 106, Box 7, NW/MIT. The phrase that inspired Brooks ("I wish to devote this book to a protest against this *inhuman use of human beings*"; emphasis added), appears on page 16 of Norbert Wiener, *The Human Use of Human Beings: Cybernetics and Society* (Boston, MA: Houghton Mifflin, 1950), but doesn't appear in the 1954 second edition.

76. Advertising copy for *the San Francisco Chronicle* and *the New York Times*; Folder 653, Box 29B, NW/MIT

77. Leslie C. Dunn, review of *The Human Use of Human Beings*, *Scientific American* 183, no. 6 (1950): 60.

78. This comparison appears in the book's last chapter, titled "Voices of Rigidity" in Wiener, *The Human Use of Human Beings*.

79. Wiener, *The Human Use of Human Beings*, 15–16.

80. Wiener, *The Human Use of Human Beings*, 175. This passage doesn't appear in the 1954 edition.

81. Christopher Lehmann-Haupt, "Jason Epstein, Editing and Publishing Innovator, Is Dead at 93," *New York Times,* February 4, 2022.

82. Norbert Wiener, *The Tempter* (New York: Random House, 1959). The book enjoyed modest sales and good, if not ecstatic, reviews in venues such as *Science* and the *Saturday Evening Post*.

83. Letter from Jason Epstein to Wiener, May 19, 1953; Folder 173, Box 12, NW/MIT.

84. In a letter dated February 25, 1957, Epstein noted that *Human Use* had sold "well beyond 50,000 copies for us"; Folder 225, Box 15, NW/MIT. In comparison, sales of *Cybernetics* by 1960 were just over 33,000; see Lynwood Bryant to Wiener; January 18, 1960; Folder 272, Box 19, NW/MIT.

85. For example, Jennifer Bayot, "John Diebold, 79, a Visionary of the Computer Age, Dies," *New York Times,* December 27, 2005; besides "visionary," Diebold is also described here as an "evangelist of the future."

86. William D. Smith, "Personality: An Evangelist for Automation," *New York Times*, March 28, 1965.

87. Described in the section "Looking Ahead" in the 1983 edition of John Diebold, *Automation* (New York: American Management Associations, 1983).

88. Information on Doriot comes from several sources including, a 2015 *Harvard Gazette* profile, Christina Pazzanese, "The Talented Georges Doriot," *Harvard Gazette, February 24, 2015,* https://news.harvard.edu/gazette/story/2015/02/the-talented-georges-doriot/ (accessed August 7, 2023).

89. John Diebold, et al., "Making the Automatic Factory a Reality," report prepared for the Harvard Business School, May 15, 1951.

90. Specific quote (with my emphasis added) is from John Diebold, "Automation," *Textile Research Journal* 25, no. 7 (1955): 635–640, https://doi.org/10.1177/004051755502500710; similar phrases and ideas are found throughout his 1952 book.

91. Diebold, *Automation*, 3.

92. Diebold, *Automation*, 5–6.

93. Diebold, *Automation*, 142–143.

94. Quote from page 246 of "Automation and Technological Change," Hearings before the Subcommittee on Economic Stabilization of the Joint Committee on the Economic Report, Congress of the United States (Washington: US Government Printing Office, 1955).

95. Wiener, *Ex-Prodigy*, 37.

96. This is a draft essay Wiener wrote for *the New York Times* titled "The Machine Age;" it never appeared in print. See John Markoff, "In 1949, He Imagined an Age of Robots," *New York Times*, May 20, 2013.

97. Wiener, *Human Use*, 213.

98. Pamela McCorduck, *Machines Who Think: A Personal Inquiry into the History and Prospects of Artificial Intelligence* (San Francisco: Freeman, 1979), 49–50.

99. Wiener, *Cybernetics*, 193–194.

100. Wiener, *The Human Use of Human Beings* (1954 edition), 176–178, 181.

101. Nathan Ensmenger, "Is Chess the Drosophila of Artificial Intelligence? A Social History of an Algorithm," *Social Studies of Science* 42, no. 1 (2012): 5–30, https://doi.org/10.1177

/0306312711424596. An early comparison is Herbert A. Simon and William G. Chase, "Skill in Chess: Experiments with Chess-Playing Tasks and Computer Simulation of Skilled Performance Throw Light on Some Human Perceptual and Memory Processes," *American Scientist* 61, no. 4 (1973): 394–403, https://www.jstor.org/stable/27843878.

102. Alan M. Turing, "Proposal for Development in the Mathematics Division of an Automatic Computing Engine (ACE)," in *A. M. Turing's ACE Report of 1946 and Other Papers*, ed. B. E. Carpenter and R. W. Doran (Cambridge, MA: MIT Press, 1986), 41.

103. Herbert A. Simon and Allen Newell, "Heuristic Problem Solving: The Next Advance in Operations Research," *Operations Research* 6, no. 1 (1958): 1–10, https://doi.org/10.1287/opre.6.1.1.

104. Discussed in Conway and Siegelman, *Dark Hero of the Information Age*, 280–282.

105. Conway and Siegelman, *Dark Hero of the Information Age*, 305–311.

106. Letter from Reuben Holden to Wiener, March 21, 1961; Box 20, NW/MIT.

107. "Precis on Terry Lectures," undated late 1961 or early 1962; Box 21, NW/MIT.

108. Letter from Chester Kerr to Wiener, February 14, 1962; Box 22, NW/MIT.

109. Letter from Wiener to Yale University Press, August 10, 1962; Box 22, NW/MIT. The revised 1962 manuscript, meant for Yale, is in Box 33b, NW/MIT.

110. Letter from Wiener to Yale University Press, August 10, 1962; Box 22, NW/MIT. The revised 1962 manuscript, meant for Yale, is in Box 33b, NW/MIT.

111. Letter from Chester Kerr to Wiener, September 10, 1963; Box 23, NW/MIT.

112. Letter from Epstein to Wiener; February 26, 1963; Box 23, NW/MIT.

113. Letter from Irving Kristol to Wiener, September 27, 1963; Box 23, NW/MIT.

114. Letter from Wiener to Epstein, October 24, 1963; Box 23, NW/MIT.

115. Letter from Carroll Bowen to Econ-Verlag, May 21, 1964; Box 24, NW/MIT.

116. Wiener, *God & Golem Inc.*, 3, 8.

117. Wiener, *God & Golem Inc.*, 47,

118. Wiener, *God & Golem Inc.*, 52–53.

119. Wiener, *God & Golem Inc.*, 73.

120. Wiener, *God & Golem Inc.*, 95.

121. J. Michael Crichton, "Norbert Wiener on Man and His Machine," *Harvard Crimson*, May 6, 1964, https://www.thecrimson.com/article/1964/5/6/norbert-wiener-on-man-and-his/ (accessed May 6, 2022).

122. Harry Gilroy, "'Herzog' Wins 2nd National Book Award for Bellow," *New York Times*, March 10, 1965.

123. Lewis Nichols, "In and Out of Books: Winners' Interview," *New York Times*, March 21, 1965.

## CHAPTER 3

1. Mortimer Taube, *Computers and Common Sense: The Myth of Thinking Machines* (New York: Columbia University Press, 1961); the 1963 McGraw-Hill edition dropped the subtitle. On Taube, see Shunryu Colin Garvey, "The 'General Problem Solver' Does Not Exist: Mortimer Taube and the Art of AI Criticism," *IEEE Annals of the History of Computing* 43, no. 1 (2021): 60–73, https://dci.org/10.1109/MAHC.2021.3051686.

2. Quotes from Walter Reitman, "Fact or Fancy," *Science* 135, no 3505 (1962): 718, https://doi.org/10.1126/science.135.3505.718; Alvin M. Weinberg, review of *Computers and Common Sense: The Myth of Thinking Machines*, *Library Quarterly* 32, no. 4 (1962): 309–310, https://www.jstor.org/stable/4305289; James R. Newman, review of *Computers and Common Sense*, *Scientific American* 206, no. 2 (1962): 185, https://www.jstor.org/stable/24937243; and, Richard Laing, review of *Computers and Common Sense*," *Behavioral Science* 7, no. 2 (1962): 238–240.

3. Frederick G. Kilgour, "Origins of Coordinate Searching," *Journal of the American Society for Information Science* 48, no. 4 (1997): 340–348.

4. Taube, *Computers and Common Sense* (1963 edition), 59.

5. Taube, *Computers and Common Sense* (1963 edition), 59–60.

6. Mortimer Taube, "Computers and Game Playing," *Science* 132, no. 3426 (1960): 555–557, https://www.jstor.org/stable/1705834.

7. William Aspray and Bernard O. Williams, "Arming American Scientists: NSF and the Provision of Scientific Computing Facilities for Universities, 1950–1973," *IEEE Annals of the History of Computing* 16, no. 4 (1994): 60–74, https://doi.org/10.1109/85.329758.

8. Daniel S. Greenberg, *The Politics of Pure Science* (New York: New American Library, 1968).

9. Newman, review of *Computers and Common Sense*, 185.

10. Herbert A. Simon and Allen Newell, "Heuristic Problem Solving: The Next Advance in Operations Research," *Operations Research* 6, no. 1 (1958): 1–10, https://doi.org/10.1287/opre.6.1.1. To be fair, the precise wording Simon and Newell used was "will," the indicative mood, grammatically speaking, but, to Taube's stated discomfort, still based on unsubstantiated claims.

11. Taube, *Computers and Common Sense* (1963 edition), 52.

12. Joseph Weizenbaum, *Computer Power and Human Reason: From Judgment to Calculation* (San Francisco: W. H. Freeman, 1976); hereafter, *Computer Power*.

13. Quotes from Weizenbaum, *Computer Power*, 21; and Joseph Weizenbaum, "The Last Dream," in *Is the Computer a Tool?* ed. Bo Sundin (Stockholm: Almquist & Wiksell, 1980), 100–115.

14. Pamela McCorduck, *Machines Who Think: A Personal Inquiry into the History and Prospects of Artificial Intelligence* (San Francisco: W. H. Freeman, 1979), xii.

15. Syed Mustafa Ali, Stephanie Dick, Sarah Dillon, Matthew L. Jones, Jonnie Penn, and Richard Staley, eds., "Histories of Artificial Intelligence: A Genealogy of Power," *BJHS Themes* 8 (2023), https://doi.org/10.1017/bjt.2023.15, and other essays in the same volume.

16. Edward A. Feigenbaum and Julian Feldman, introduction to *Computers and Thought*, ed. Edward A. Feigenbaum and Julian Feldman (New York: McGraw-Hill, 1963), v.

17. Ronald R. Kline, "Cybernetics, Automata Studies, and the Dartmouth Conference on Artificial Intelligence," *IEEE Annals of the History of Computing* 33, no. 4 (2011): 5–16, https://doi.org/10.1109/MAHC.2010.44.

18. Marvin Lee Minsky, "Theory of Neural-Analog Reinforcement Systems and its Application to the Brain-Model Problem" (PhD diss., Princeton University, 1954), ProQuest Dissertations and Theses Global (301998727).

19. McCorduck, *Machines Who Think*, 101. Also, see Allen Newell, "Intellectual Issues in the History of Artificial Intelligence," *The Study of Information: Interdisciplinary Messages*, ed. Fritz Machlup and Una Mansfield (New York: John Wiley & Sons, 1983): 187–227.

20. The exact source of the quote is elusive but one pathway was Weizenbaum who included it in "It's Your Dream," *New York Times*, May 28, 1972. McCorduck attributes a different version to Minsky, in *Machines Who Think*, 85.

21. Feigenbaum and Feldman, introduction to *Computers and Thought*, 3.

22. Allen Newell, J. C. Shaw, and Herbert A. Simon, "Chess Playing Programs and the Problem of Complexity," *IBM Journal of Research and Development* 2, no. 4 (1958): 320–335, https://doi.org/10.1147/rd.24.0320. For "satisficing," see Herbert A. Simon, "Rational Choice and the Structure of the Environment," *Psychological Review* 63, no. 2 (1956): 129–138, https://doi.org/10.1037/h0042769.

23. Herbert A. Simon, "A Behavioral Model of Rational Choice," *Quarterly Journal of Economics* 69, no. 1 (1955): 99–118, https://www.jstor.org/stable/1884852, and his book *Models of Man, Social and Rational: Mathematical Essays on Rational Human Behavior in a Social Setting* (New York: John Wiley & Sons, 1957) where Simon introduced the term "bounded rationality." Also, see Hunter Crowther-Heyck's excellent book *Herbert Simon: The Bounds of Reason in Modern America* (Baltimore, MD: Johns Hopkins University Press, 2005).

24. Herbert A. Simon, *Models of My Life* (New York: Basic Books, 1991), 189 and 198.

25. Allen Newell, John Clifford Shaw, and Herbert A. Simon, "Elements of a Theory of Human Problem-Solving," *Psychological Review* 65, no. 3 (1958): 153, https://doi.org/10.1037/h0048495.

26. Allen Newell and Herbert A. Simon, "The Logic Theory Machine: A Complex Information Processing System," RAND Corporation report P-868, July 12, 1956; published as "The Logic Theory Machine—A Complex Information Processing System," *IRE Transactions on Information Theory* 2, no. 3 (September 1956): 61–79, https://doi.org/10.1109/TIT.1956.1056797. Also, Stephanie Dick, "Of Models and Machines: Implementing Bounded Rationality," *Isis* 106, no. 3 (2015): 623–634, https://doi.org/10.1086/683527.

27. Anecdote recounted in Edward Feigenbaum, 1974 (n.d.) oral history interview with Pamela McCorduck; Folder 29, Box 1, PM/CMU.

28. McCarthy, quoted on page 187 of Roberto Cordeschi, *The Discovery of the Artificial: Behavior, Mind, and Machines Before and Beyond Cybernetics* (Dordrecht: Springer, 2002). Also, see Kline, "Cybernetics, Automata Studies, and the Dartmouth Conference on Artificial Intelligence."

29. Paul Edwards calls the meeting "the conceptual birthplace of AI" in *The Closed World: Computers and the Politics of Discourse in Cold War America* (Cambridge, MA: MIT Press, 1996), 252.

30. A. Newell and H. A. Simon, "Plans for the Dartmouth Summer Research Project on Artificial Intelligence," report, March 6, 1956; Folder 3044, Box 38, HAS/CMU.

31. Simon and Newell, "Heuristic Problem Solving."

32. Bellman's letter, along with Simon and Newell's response, appears in *Operations Research* 6, no. 3 (1958): 448–450, https://www.jstor.org/stable/167033 and https://www.jstor.org/stable/167034.

33. Arthur L. Norberg and Judy E. O'Neill, *Transforming Computer Technology: Information Processing for the Pentagon, 1962–1986* (Baltimore, MD: Johns Hopkins University Press, 1996), especially chapter 5; Alex Roland with Philip Shiman, *Strategic Computing: DARPA and the Quest for Machine Intelligence* (Cambridge, MA: MIT Press, 2002).

34. Steward Brand, *The Media Lab: Inventing the Future at MIT* (New York: Viking, 1987), 162.

35. Michael D. Gordin, *Scientific Babel: How Science Was Done Before and After Global English* (Chicago, IL: University of Chicago Press, 2015).

36. Taube, *Computers and Common Sense*, 41.

37. Published as Martin Greenberger, ed., *Computers and the World of the Future* (Cambridge, MA: MIT Press, 1962).

38. Quote from Allen Newell and Herbert A. Simon, "Computer Simulation of Human Thinking," *Science* 134, no. 3495, (1961): 2011–2017, https://www.jstor.org/stable/1708146; and, Allen Newell and Herbert A. Simon, "GPS, a Program That Simulates Human Thought," RAND paper P-2257, 1961 (Santa Monica, CA: RAND Corporation).

39. Hubert L. Dreyfus, "Alchemy and Artificial Intelligence," RAND paper P-3244, December 1965 (Santa Monica, CA: RAND Corporation).

40. Dreyfus, "Alchemy and Artificial Intelligence," 29.

41. "Notes and Comment," *New Yorker*, June 11, 1966, 27–28; Hubert L. Dreyfus "Why Computers Must Have Bodies in Order to Be Intelligent," *Review of Metaphysics* 21, no. 1 (1967): 13–32, https://www.jstor.org/stable/20124494; and, Hubert L. Dreyfus, *What Computers Can't Do: The Limits of Artificial Intelligence* (New York: Harper & Row, 1972).

42. Seymour Papert, "The Artificial Intelligence of Hubert L. Dreyfus: A Budget of Fallacies," MIT Artificial Intelligence Memo #154, January 1968.

43. The program, called Mac Hack, was written in 1965–1966 by Richard Greenblatt.

44. Herbert Simon, "Cool It, Friend: An Open Letter to Hubert Dreyfus," n.d. but sometime in 1967; Folder 4999, Box 65, HAS/CMU.

45. Marvin L. Minsky, "Artificial Intelligence," *Scientific American* 215, no. 3 (1966): 247–260, https://www.jstor.org/stable/24931058.

46. Weizenbaum attributed the term to Louis Fein, appears on page 179 of *Computer Power*.

47. Biographical material is from several sources including John Markoff, "Joseph Weizenbaum, Famed Programmer, is Dead at 85," *New York Times*, March 13, 2008; and Joseph Weizenbaum and Gunna Wendt, *Islands in the Cyberstream: Seeking Havens of Reason in a Programmed Society* (Duluth, MN: Litwin Books, 2015).

48. Joseph Weizenbaum, interview with Pamela McCorduck, March 6, 1975; Folder 8, Box 3, PM/CMU.

49. Joseph Weizenbaum, "How to Make a Computer Appear Intelligent," *Datamation* 8, no. 2 (1965): 24–26.

50. Joel Moses and Jeff Meldman, "Joseph Weizenbaum, 1923–2008," *IEEE Intelligent Systems* 23, no. 4 (2008): 8–9, https://doi.org/10.1109/MIS.2008.70.

51. Joseph Weizenbaum, "Symmetric List Processor," *Communications of the ACM* 6, no. 9 (1963): 524–536, A1–A15, https://doi.org/10.1145/367593.367617 .

52. Joseph Weizenbaum, interview with Pamela McCorduck, March 6, 1975; Folder 8, Box 3, PM/CMU.

53. Joseph Weizenbaum, interview with Pamela McCorduck, March 6, 1975; Folder 8, Box 3, PM/CMU.

54. Joseph Weizenbaum, "ELIZA: A Computer Program for the Study of Natural Language Communication Between Man and Machine," *Communications of the ACM* 9, no. 1 (1966): 36–45.

55. Weizenbaum, "ELIZA," 42.

56. Weizenbaum, "ELIZA," 36. In the actual program, the "?" symbol could not be used as the MAC system interpreted it as a line delete instruction.

57. Alan Turing, "Computing Machinery and Intelligence," *Mind* 59, no. 236 (1950): 433–460.

58. John Noble Wilford, "Computer Is Being Taught to Understand English," *New York Times*, June 15, 1968.

59. Weizenbaum, "ELIZA," 36.

60. Joseph Weizenbaum, interview with Pamela McCorduck, March 6, 1975; Folder 8, Box 3, PM/CMU.

61. Oppenheimer's address, given at MIT on November 25, 1947, was titled "Physics in the Contemporary World," reprinted in J. Robert Oppenheimer, *The Open Mind* (New York: Simon & Schuster, 1955).

62. "RAND Symposium, Part 1," *Datamation* 11, no. 8, (1965): 24–30; quotes on page 25.

63. Weizenbaum and Wendt, *Islands in the Cyberstream*, 33.

64. Alvin Toffler, *Future Shock* (New York: Random House, 1970).

65. Letter from Euan Cameron to Alvin Toffler, November 16, 1970; Box 172, AHT/CU.

66. Toffler drew deeply from the work of sociologist Daniel Bell who presented similar ideas in his own book, *The Coming of Post-Industrial Society: A Venture in Social Forecasting* (New York: Basic Books, 1973).

67. Bob Guccione, "Interview: Alvin Toffler," *Omni*, November 1978, 97–98, 132–136.

68. Joseph Weizenbaum, "On the Impact of the Computer on Society," *Science* 176, no. 4035 (May 12, 1972): 609–614, https://www.jstor.org/stable/1734465; and Weizenbaum, "It's Your Dream."

69. Weizenbaum, "It's Your Dream."

70. Weizenbaum, "On the Impact of the Computer on Society," 614.

71. Stuart W. Leslie, *The Cold War and American Science: The Military-Industrial-Academic Complex at MIT and Stanford* (New York: Columbia University Press, 1993), especially chapter 9. Also, Kelly Moore, *Disrupting Science: Social Movements, American Scientists, and Politics of the Military, 1945–1974* (Princeton, NJ: Princeton University Press, 2008).

72. Jonathan Allen, ed., *March 4: Scientists, Students, and Society, anniversary edition* (Cambridge, MA: MIT Press, 2019).

73. Eric Schatzberg, *Technology: Critical History of a Concept* (Chicago, IL: University of Chicago Press, 2018), especially chapter 13.

74. Everett Mendelsohn, "Prophet of Our Discontent: Lewis Mumford Confronts the Bomb," in *Lewis Mumford: Public Intellectual*, ed. Thomas P. Hughes and Agatha C. Hughes (New York: Oxford University Press, 1990), 343–360.

75. Gerald Holton, "The Pentagon of Power," *New York Times*, December 13, 1970.

76. Only half of this correspondence—letters from Weizenbaum to Mumford—are preserved. Mumford's replies, with a few exceptions, appear to be unavailable.

77. Zachary Loeb, "The Lamp and the Lighthouse: Joseph Weizenbaum, Contextualizing the Critic," *Interdisciplinary Science Reviews* 46, no. 1/2 (2021): 19–35, https://doi.org/10.1080/03080188.2020.1840218.

78. Letter from Weizenbaum to Mumford, November 21, 1973; Folder 5270, Box 69, LM/UP.

79. Quote from Weizenbaum's application via copy shared with the author by artist Peggy Weil.

80. Quotes from the preface in Weizenbaum, *Computer Power*, xi.

81. *Computers and Computation: Readings from Scientific American* (San Francisco: W. H. Freeman and Company, 1971).

82. Weizenbaum, preface to *Computer Power*; letter from Kelly to Weizenbaum, February 28, 1975, Folder "Computer Power Correspondence," Box 12, JW/MIT.

83. Letter from Alan Eshleman to Weizenbaum, April 15, 1975; letter from Peter Renz to Weizenbaum, February 10, 1975; both Folder "Computer Power Correspondence," Box 12, JW/MIT.

84. Note from Kelly to Weizenbaum, March 26, 1975; Folder "Computer Power Correspondence," Box 12, JW/MIT.

85. Letter from Renz to Weizenbaum, April 22, 1975; Folder "Computer Power Correspondence," Box 12, JW/MIT.

86. Letter from Renz to Weizenbaum, April 22, 1975; Folder "Computer Power Correspondence," Box 12, JW/MIT; see also advertisement in *the New York Times*, February 22, 1976, BR 31

87. Draft copy, n.d. but late 1975; Folder "Computer Power Correspondence," Box 12, JW/MIT;

88. "W. H. Freeman News Release," February 23, 1976; Folder "Computer Power Correspondence," Box 12, JW/MIT.

89. *Review of Computer Power and Human Reason, Booklist*, May 1976.

90. Weizenbaum, *Computer Power*, ix.

91. P. N. Johnson-Laird, "Slaves to the Machine," *Nature* 261 (May 13, 1976): 171, https://doi.org/10.1038/261171a0.

92. "The Computer as Choice-Maker," *American Libraries 7*, no. 3 (March 1976): 169–170, https://www.jstor.org/stable/25620631.

93. Peter Travisano, review of Computer Power and Human Reason: From Judgement to Calculation, *BYTE* 2, no. 1 (January 1977): 11–12.

94. Letter from M. E. Maron to Weizenbaum, March 8, 1976, "Fan Mail Folder 1976," Box 12, JW/MIT.

95. Ida Hoose to Weizenbaum, May 20, 1976, "Fan Mail Folder 1976," Box 12, JW/MIT.

96. Renz to Weizenbaum, August 23, 1976; Folder "Computer Power Correspondence," Box 12, JW/MIT; quote from faculty responses in Folder "Publishers," Box 14, JW/MIT.

97. Letter from Renz to Weizenbaum, December 15, 1977; Folder "Computer Power Correspondence," Box 12, JW/MIT.

98. Weizenbaum, *Computer Power*, x.

99. Weizenbaum, *Computer Power*, 258.

100. Letters from Weizenbaum to Mumford; May 14, 1975, and February 10, 1976; Folder 5270, Box 69, LM/UP.

101. Joseph Weizenbaum, interview with Pamela McCorduck, March 6, 1975; Folder 8, Box 3, PM/CMU.

102. Weizenbaum, *Computer Power*, 196.

103. Lee Dembart, "Experts Argue Whether Computers Could Reason, and If They Should," *New York Times*, May 8, 1977.

104. Nathan Ensmenger, "'Beards, Sandals, and Other Signs of Rugged Individualism': Masculine Culture within the Computing Professions," *Osiris* 30, no. 1 (2015): 38–65, https://doi.org/10.1086/682955.

105. Weizenbaum, *Computer Power*, 116.

106. Stewart Brand, "Spacewar: Life and Death among the Computer Bums," *Rolling Stone*, December 7, 1972, 50–57.

107. Weizenbaum, *Computer Power*, 122.

108. Letter from Simon to Weizenbaum, March 2, 1976; Folder "Fan Mail 1976," Box 12, JW/MIT.

109. Lederberg's review was not published in the *Times*; it eventually appeared, as Bruce G. Buchanan, Joshua Lederberg, and John McCarthy, "Three Reviews of J. Weizenbaum's Computer Power and Human Reason," Stanford Artificial Intelligence Laboratory Memo AIM-291 (November 1976), Defense Technical Information Center, https://apps.dtic.mil/sti/citations/ADA044713.

110. John McCarthy, "An Unreasonable Book," *SIGART Newsletter* no. 58 (June 1976): 5–10, https://doi.org/10.1145/1045264.1045265, and reprinted in Creative Computing 2, no. 5 (September/October, 1976): 84–89; and, John McCarthy, review of *Computer Power and Human Reason*, *Physics Today* 30, no. 1 (1977): 68–71. The August 1976 issue of *SIGART Newsletter* continued the debate, see https://doi.org/10.1145/1045270.

111. McCarthy, "An Unreasonable Book," 10, 9.

112. McCarthy, "An Unreasonable Book," 6.

113. Joseph Weizenbaum, "A Response to John McCarthy," *SIGART Newsletter* no. 58 (June 1976): 10–13, https://doi.org/10.1145/1045264.1045265.

114. McCorduck, *Machines Who Think*, 318. The likelihood this was voiced by Feigenbaum comes from conversations between the scientist and historian Magnus Rust; personal correspondence with the author, August 27, 2024.

115. Letter from Weizenbaum to "Paul," March 12, 1976; Folder "Fan Mail 1976," Box 12, JW/MIT.

116. Once accepted as part of AI's history, more recently at least one historian has questioned the accepted dates of the first "AI Winter"; see Thomas Haigh, "There Was No 'First AI Winter,'" *Communications of the ACM* 66, no. 12 (2023): 35–39.

117. Joseph Weizenbaum, *Computer Power and Human Reason: From Judgment to Calculation*, 2nd ed. (New York: Penguin, 1984).

118. Weizenbaum and Wendt, *Islands in the Cyberstream*, 159–160, emphasis added.

## CHAPTER 4

1. Boyce Rensberger, "Low-Cost Computers Beginning to Move into the Home," *New York Times*, May 4, 1976; and, Sol Libes, "The 1976 Trenton Computer Festival," *Dr. Dobb's Journal of Computer Calisthenics and Orthodontia* 1, no. 5 (May 1976): 4–6.

2. Rensberger, "Low-Cost Computers."

3. Leslie Goff, "Apple Debuts," *Computerworld*, July 12, 1976, 75. Promotional and informational materials for the meeting are in Folder 1, Box 29, JW/CHM.

4. Russ Walter, "Personal Computing Conference Draws 5,000," *Personal Computer*, January/February 1977, 30–34.

5. Walter, "Personal Computing Conference Draws 5,000."

6. Walter, "Personal Computing Conference Draws 5,000"; Goff, "Apple Debuts"; also, Laine Nooney, *The Apple II Age: How the Computer Became Personal* (Chicago, IL: University of Chicago Press, 2023).

7. Theodor H. Nelson, *Computer Lib: You Can and Must Understand Computers Now/Dream Machines: New Freedoms Through Computer Screens—A Minority Report* (Chicago, IL: Hugo's Book Service for the author, 1974); hereafter *Computer Lib/Dream Machines*. Subsequent editions, some revised, of *Computer Lib/Dream Machines* appeared including an updated edition published by Tempus Books of Microsoft Press (Redmond, WA, 1987). For a bibliography of Nelson's writings, see Henry Lowood, "Ted Nelson: A Critical (and Critically Incomplete) Bibliography," in *Intertwingled: The Work and Influence of Ted Nelson*, ed. Douglas R. Dechow and Daniele C. Struppa (Cham, Switzerland: Springer, 2015): 123–130, https://doi.org/10 .1007/978-3-319-16925-5_16.

8. Theodor H. Nelson, *The Home Computer Revolution* (South Bend, IN: self-published, distributed by the distributors, 1977), front matter.

9. Nelson, *Computer Lib*, 3; formatting in the original.

10. Jeremy Reimer, "Total Share: 30 Years of Personal Computer Market Share Figures," *Ars Technica*, December 14, 2005, https://arstechnica.com/features/2005/12/total-share/; similar data can be found at Wikipedia, "Market Share of Personal Computer Vendors," last modified November 29, 2024, 22:36 (UTC), https://en.wikipedia.org/wiki/Market_share_ of_personal_computer_vendors (both accessed June 1, 2022); Otto Friedrich et al., "The Computer Moves In," *Time* (January 3, 1983): 14–24, figures from page 14.

11. Henry Lowood, personal communication with the author, June 6, 2022, based on a survey of works in the Stanford University library.

12. A point made by Joy Lisi Rankin in *A People's History of Computing in the United States* (Cambridge, MA: Harvard University Press, 2018); as well as Nooney, *The Apple II Age*.

13. Members wrote it both as RESISTORS and R.E.S.I.S.T.O.R.S.; I've adopted the former.

14. "Students Steal Show as Conference Opens," *Computerworld*, May 8, 1968, 1, 16.

15. "Time Sharing in a Telephone Booth," *Datamation*, June 1968, 92.

16. Online documentation about the RESISTORS can be found at https://www.resistors.org /index.php/Main_Page (accessed June 30, 2022).

17. Biographical material on Kagan comes from a number of sources, including Bob Levine, "R.E.S.I.S.T.O.R.S. Revisited," *US-1*, May 20, 1998, 12; interview with John R. Levine, April 30, 2021; and interview with Nathaniel Kuhn, May 3, 2021. See also, "Who Can Resist the RESISTORS," *Princeton Town Topics*, November 28, 1968, 1–2, as well as material and links found on the RESISTORS website (https://www.resistors.org/index.php

/Claude_Kagan) and a memorial page (https://www.retrotechnology.com/restore/kagan
.html); all printed sources via NK/R, interviews in author's files, and web links accessed June
2022.

18. Member's recollections can be found at https://www.resistors.org/index.php/History_of_
the_R.E.S.I.S.T.O.R.S. (accessed June 1, 2022).

19. Kagan and some of the RESISTORS's eventually created a very similar language, called
SAM76, where SAM may have stood for "Same As Mooers."

20. Nathaniel Kuhn, personal conversation with the author, July 27, 2022.

21. Based on conversations with former members as well as information about Kagan shared with
me by Evan Koblentz in June 2022. The Jaycees document is available at https://hopewell
-history.org/hvhist/Collections/House-Tours/HwTwp/1966-HwTwp-Jaycee-Spring-House
-Tour-HM.pdf (accessed June 30, 2022).

22. See, for example, Audrey M. Haitch, "Pupils Radically Emphatic in Studies," *New York Times*,
March 12, 1972.

23. "Who Can Resist the RESISTORS"; NK/R; and John R. Levine, interview with the author,
April 30, 2021.

24. Ancelme Roichel, *SAM 76: The First Language Manual* (Princeton, NJ: self-published, 1978).

25. Nathaniel Kuhn, interview with the author, May 3, 2021.

26. Nathaniel Kuhn, interview with the author, May 3, 2021.

27. Biographical details on Nelson come from several sources, including his autobiography *Possiplex: Movies, Intellect, Creative Control, My Computer Life, and the Fight For Civilization*
(Hackettstown, NJ: Mindful Press, 2010). As his bibliographer, Henry Lowood, has noted,
the difficulty of documenting Nelson's life is augmented by the fact that "many of his important writings appeared in ephemeral or semipublished formats, ranging from conference
proceedings and magazines of every ilk to self-published books that were produced anywhere
and nowhere"; Lowood, "Ted Nelson: A Critical (and Critically Incomplete) Bibliography."
The fact that Nelson—who was still alive at the time of this writing—is continually reassessing
his accomplishments complicates things further. I've tried to employ sources and accounts
from the time period of interest here with the understanding that later writings, including
those by Nelson himself, may contradict these accounts.

28. Nelson, *Possiplex*, 21, 30, 41.

29. Nelson, *Dream Machines*, 59.

30. Nelson, *Possiplex*, 100, 99.

31. Nelson, *Possiplex*, 102.

32. Nelson's notes for this are collected at the Internet Archive, *The Company I imagined in
1960—"General Creative,"* batch 1, https://archive.org/details/gc-batch-1/page/n19/mode
/2up (accessed June 10, 2022).

33. See "The Hypertext," which Nelson composed in 1965, https://archive.org/details /thehypertext/TheHypertext/ (accessed June 10, 2022).

34. T. H. Nelson, "A File Structure for the Complex, the Changing, and the Indeterminate," in *ACM '65: Proceedings of the 1965 20th National Conference* (New York: Association for Computing Machinery, 1965): 84–100.

35. Nelson, *Dream Machines*, 45; formatting in the original.

36. "Nelson's the Name and What He Proposes Could Outdo Engelbart," *Electronics*, November 24, 1969, 97.

37. Theodor H. Nelson, "The Xanadu System: A Discussion Paper (Harcourt Confidential)," unpublished report dated December 23, 1966, https://archive.org/details/TheFirstXanaduProposal 1966. Nelson described Xanadu in many places, including his self-published 1981 book *Literary Machines*. See also Roy Rosenzweig, "The Road to Xanadu: Public and Private Pathways on the History Web," *Journal of American History* 88, no. 2 (2001): 548–579, https://doi.org/10.2307 /2675105; Belinda Barnet, *Memory Machines: The Evolution of Hypertext* (London, UK: Anthem Press, 2014); and, Hallam Stevens, "Code and Critique: Ted Nelson's Project Xanadu and the Politics of New Media," *Osiris* 38 (2023): 245–264.

38. Theodor H. Nelson, "Barnum-Tronics," *Swarthmore College Bulletin*, December 1970, 12–15. Xanadu received a detailed but controversial treatment in Gary Wolf, "The Curse of Xanadu," *Wired* 3 no. 6 (1995): 138–152, 194–202. Nelson refuted Wolf's claims in a lengthy letter to *Wired* as well as a press release dated June 7, 1995, https://archive.org/details/Ted_Nelson _Responds_to_Charges_in_Wired_Magazine_Article_1995-06-07 (documents accessed June 12, 2022).

39. Nelson, *Computer Lib*, 47.

40. There is a brief reference to the piece in Lauren O'Neill-Butler, "Land of the Living," *Artforum* (October 2019), https://www.artforum.com/print/201908/lauren-o-neill-butler-on-the-art -of-agnes-denes-80813 (access June 13, 2022) although it mistakenly suggests Denes did the programming herself. See also the show's official catalog *Software—Information Technology: Its Meaning for* (New York: Jewish Museum, 1970), 27, which also includes a short essay about the RESISTORS on pages 64–65.

41. Fernbach-Flarsheim's piece is described in *Software*, 56–57. The show's catalog gives direct credit to four teen contributors: John Levine, Nat Kuhn, Peter Eichenberger, and Lauren Sarno.

42. Burnham, *Software*, 18.

43. Jack Burnham, "Art and Technology: The Panacea That Failed," in *The Myths of Information: Technology and Postindustrial Culture*, ed. Kathleen Woodward (Madison, WI: Coda Press, 1980), 206.

44. John R. Levine, interview with the author, April 30, 2021.

45. Theodor H. Nelson, "No More Teachers' Dirty Looks," *Computer Decisions*, September 1970, 16–23. On PLATO, see Rankin, *A People's History of Computing*, chapter 6.

46. Nelson, *Computer Lib*, 2; Nelson, *Possiplex*, 199; and Nelson, *Computer Lib*, 2.

47. Nelson, *Possiplex*, 200.

48. Theodor H. Nelson, personal communication with the author, June 14, 2022.

49. Nelson, *Possiplex*, 200.

50. Nelson, *Possiplex*, 200.

51. Nelson, *Computer Lib*, 8.

52. Ted Nelson, "Notes for Talk at National Computer Convention in New York City, June 7–10, 1976," copy archived by the Internet Archive as "TN's talk at NCC 76—predicted 10 million by 1985," https://archive.org/details/nccv3good (accessed June 1, 2022).

53. Nelson, *Possiplex*, 203.

54. Nelson, *Possiplex*, 203.

55. Theodor H. Nelson, "Promotional Poster for Computer Lib"; NK/R.

56. From a short section, titled "History of this Book," which appears in the much-revised version of *Computer Lib/Dream Machines* (Redmond, WA: Microsoft Press, 1987).

57. Theodor H. Nelson, personal communication with the author, June 14, 2022.

58. Stewart Brand, foreword to Theodor H. Nelson, *Computer Lib/Dream Machines* (Redmond, WA: Microsoft Press, 1987). "Big Nurse" referenced Nurse Ratched from *One Flew Over the Cuckoo's Nest*, a 1962 novel by Brand's confederate, Ken Kesey.

59. Nelson, *Dream Machines*, 2.

60. Nelson, *Dream Machines*, 22, 56–57, 59.

61. Dan Fylstra, review of *Computer Lib/Dream Machines*, *Byte*, October 1975, 82.

62. John Levine, review of *Computer Lib/Dream Machines*, *Creative Computing* 2, no. 5 (September/October, 1976): 96.

63. Lee Felsenstein, "How We Trapped the Dinosaurs," *Creative Computing* 10, no. 11 (November 1984): 193–194; L. R. Shannon, "A Book That Grew Up," *New York Times*, February 16, 1988.

64. Quote from Noah Wardrip-Fruin and Nick Montfort, eds., *The New Media Reader* (Cambridge, MA: MIT Press, 2003), 301.

65. Theodor H. Nelson, personal communication with the author, June 10, 2022.

66. Theodor H. Nelson, "The Crafting of Media," in *Software—Information Technology: Its Meaning for Art*, ed. Judith Benjamin Burnham (New York: Jewish Museum, 1970), 17.

67. Steve Jobs, "When We Invented the Personal Computer . . ." Computers and People 30, nos. 7–8 (July/August 1981): 8–11, 22; Jobs's essay was adapted from a series of text-heavy advertisements Apple ran in venues like *the Wall Street Journal*. Many versions of this analogy exist, such as a 1990 interview Jobs did for the Boston public television station WGBH.

68. Reimer, "Total Share: 30 Years of Personal Computer Market Share Figures."

69. A classic near-contemporary account is Paul Freiberger and Michael Swaine's *Fire in the Valley: The Making of the Personal Computer* (Berkeley, CA: Osborne/McGraw-Hill, 1984); while Nooney's *The Apple II Age* gives a more nuanced narrative.

70. *Popular Electronics* ran two feature articles ("Altair 8800 Mini Computer") about the Altair, in January and February 1975. Both were written by H. Edward Roberts, the owner of MITS, and William Yates, a company engineer.

71. Art Salsberg, "Editorial: The Home Computer Is Here!" *Popular Electronics* 7, no. 1 (January 1975): 8.

72. A version of this can be found at the Computer History Museum, in Mountain View, California, and at https://www.computerhistory.org/collections/catalog/102667741 (accessed June 10, 2022).

73. August 1976 letter, https://archive.org/details/Itty_Bitty_Machine_Company_brochure_1976-08/ (accessed June 20, 2022).

74. Recalled by Nelson in *Possiplex*, 206–207.

75. "World Altair Computer Convention," *Creative Computing* 2, no. 5 (September/October 1976): 64.

76. In addition to software development, questions of copyright, ownership, and intellectual property also had to be worked out; see Gerardo Con Díaz, *Software Rights: How Patent Law Transformed Software Development in America* (New Haven, CT: Yale University Press, 2019).

77. David Bunnell, "The Role of Magazines in Personal Computing," *Creative Computing* 10, no. 11 (November 1984): 146, 148, 153.

78. N. R. Kleinfield, "Computing's Lusty Offspring," *New York Times*, August 29, 1981.

79. Drawn from John J. Anderson, "Dave Tells Ahl: The History of *Creative Computing*," *Creative Computing* 10, no. 11 (November 1984): 67–74.

80. *EDU*, no. 1 (Spring 1971). Unrelated, in 1985 "edu" became a top-level internet domain name for educational organizations.

81. This appeared in multiple versions including, David H. Ahl, *BASIC Computer Games* (New York: Workman Publishing, 1978); also, Anderson, "Dave Tells Ahl."

82. David H. Ahl, "Birth of a Magazine," in *The Best of Creative Computing, vol. 1*, ed. David H. Ahl (Morristown, NJ: Creative Computing Press, 1976): 2–3.

83. From a note to readers included in the January/February 1975 issue.

84. *Byte*'s origins are recounted by Green, "How Byte Started," *Byte*, September 1975, 9, 96 as well as in Freiberger and Swaine's *Fire in the Valley*, 158–159. There is also a 2013 interview with Green, conducted the same year he died; see https://archive.org/details/WayneGreenInterview (accessed June 15, 2022).

85. Background information from "About Carl Helmers," http://www.helmers.com/wp-content /uploads/2015/05/About-Carl-Helmers.pdf (accessed June 30, 2022).

86. Carl Helmers, "The More Things Change, the More They Stay the Same . . ." *Byte* 4, no. 7 (July 1979): 14.

87. Bunnell, "The Role of Magazines."

88. The text, based on a tape-recorded version of Nelson's talk appears in Jim C. Warren, ed., *The First West Coast Computer Faire: Conference Proceedings* (San Francisco, CA: Computer Faire Inc., 1977), 20–25. Jim Warren's communications with Nelson are in Folder 15, Box 14, JW/CHM.

89. Nelson's observations as captured in David H. Ahl, "The First West Coast Computer Faire," *Creative Computing*, July/August 1977, 24.

90. Theodor H. Nelson, "Those Unforgettable Next Two Years," in The First West Coast Computer Faire: A Conference & Exposition On Personal and Home Computers, April 15–17, 1977, San Francisco, Conference Proceedings, Computer Faire, Palo Alto, 25.

91. Theodor H. Nelson, personal communication with the author, June 10, 2022.

92. For example, David P. Hamilton, "Japanese Embrace a Man Too Eccentric for Silicon Valley," *Wall Street Journal*, April 24, 1996.

93. "Most Memorable Person," *Creative Computing 10, no. 11* (November 1984): 145.

94. Theodor H. Nelson, *Computer Lib/Dream Machines*, 2nd ed. (Redmond, WA: Tempus Books of Microsoft Press, 1987), 4, 20. "Citizen Nelson" was the title Paul Freiberger and Michael Swaine gave to the final section of a revised version (2000) of their 1984 book, *Fire in the Valley*.

## CHAPTER 5

1. William Yardley, "Carl Schlesinger, 88, Dies; Helped Usher Out Hot Type," *New York Times*, November 16, 2014.

2. For details, see Frank Romano's *History of the Linotype Company* (Rochester, NY: RIT Press, 2014); see also, Douglas Wilson, *Linotype: The Film—In Search of the Eighth Wonder of the World* (Springfield, MO: OnPaperWings Production, 2012), DVD.

3. Malcolm W. Browne, "Paper Using Cold Type," and Carey Winfrey, "How It Was, How It Is," both *New York Times*, July 3, 1978, both appear first on page 21 and continue on 38.

4. My use of "technological community" is adapted from Ann Johnson, "Modeling Molecules: Computational Nanotechnology as Knowledge Community," *Perspectives on Science* 17, no. 2 (2009): 144–173, https://doi.org/10.1162/posc.2009.17.2.144.

5. Donald E. Knuth, *The TeXbook* (Boston, MA: Addison-Wesley, 1984), 1.

6. Donald E. Knuth, "Tau Epsilon Chi: A System for Technical Text," Report No. STAN-CS-78–675.1, Stanford Computer Science Department (November 1978), 1–2.

7. Technically, a "font" refers to a particular face or size, like italics, whereas a "typeface" is the actual design of the letters and numbers. But, because "font" is often used as a synonym for "typeface," I am doing likewise.

8. Brian Stock, *The Implications of Literacy: Written Language and Models of Interpretation in the Eleventh and Twelfth Centuries* (Princeton, NJ: Princeton University Press, 1983).

9. Donald E. Knuth, "Digital Typography," in *Digital Typography* (Stanford, CA: Center for the Study of Language and Information, 1999), 1–2.

10. Biographical material on Knuth is drawn from several sources, including information in his 1999 book *Digital Typography* (as well as the 1996 lecture of the same name) and several oral histories including a lengthy 2006 "Web of Stories" interview presented online at https://webofstories.com/play/donald.knuth/1 (accessed June 1, 2022); also see Siobhan Roberts, "The Yoda of Silicon Valley," *New York Times*, December 17, 2018.

11. Knuth describes this in "Questions and Answers, II" in *Digital Typography*, 609; "ink" comment from page ix.

12. Donald E. Knuth, "The Potrzebie System of Weights and Measures," *Mad*, June 1957, 36–37, illustrated by Wallace Wood; Knuth's cv at https://www-cs-faculty.stanford.edu/~knuth/vita.html (accessed August 7, 2024).

13. Anecdotes and quotes from the 2006 "Web of Stories" interview with Knuth.

14. Donald J. Albers and Lynn Arthur Steen, "A Conversation with Don Knuth: Part 1," *The Two-Year College Mathematics Journal* 13, no. 1 (1982): 2–18.

15. From 2006 "Web of Stories" interview with Knuth.

16. Donald E. Knuth, "RUNCIBLE—Algebraic Translation on a Limited Computer," *Communications of the ACM* 2, no. 11 (1959): 18–21.

17. Lorraine Daston, *Rules: A Short History of What We Live By* (Princeton, NJ: Princeton University Press, 2022).

18. Donald E. Knuth, "Mathematical Analysis of Algorithms," *Proceedings of IFIP Congress 1971* (Amsterdam: North-Holland, 1972), 19–27. Also, Ksenia Tatarchenko, "Thinking Algorithmically: From Cold War Science to Socialist Information Culture," *Historical Studies in the Natural Sciences* 49, no. 2 (2019): 194–225, https://doi.org/10.1525/hsns.2019.49.2.194.

19. Donald J. Albers and Lynn Arthur Steen, "A Conversation with Don Knuth: Part 2," *The Two-Year College Mathematics Journal* 13, no. 2 (1982): 128–141.

20. Donald E. Knuth, "Computer Programming as an Art," *Communication of the ACM* 17, no. 12 (1974): 667–673, https://doi.org/10.1145/1283920.1283929.

21. Donald E. Knuth, *The Art of Computer Programming*, vol. 1, *Fundamental Algorithms* (Reading, MA: Addison–Wesley, 1968).

22. Promotional brochure prepared for Knuth's 1968 book, https://www-cs-faculty.stanford.edu/~knuth/taocp.html (accessed August 7, 2022).

23. John Boe, "'A Little Bit of Your Soul In It': An Interview with Donald Knuth," *Writing on the Edge* 9, no. 1 (1997/1998): 10–25.

24. Knuth, *The Art of Computer Programming*, vol.1, v.

25. Quote from the Sherlock Holmes novel, *The Valley of Fear* (1915), by Arthur Conan Doyle.

26. Eric A. Weiss, "In the Art of Programming, Knuth is First; There is No Second," *Abacus* 1, no. 3 (1984): 41–45.

27. Knuth, *The Art of Computer Programming*, vol. 1, 6.

28. Described in Knuth, *The Art of Computer Programming*, vol. 1, 124.

29. Mark B. Wells, review of *The Art of Computer Programming*, Volume 1: *Fundamental Algorithms*, and Volume 2: *Seminumerical Algorithms*, by Donald E. Knuth, *Bulletin of the American Mathematical Society* 79, no. 3 (1973): 501–509.

30. From 1968 promotional brochure prepared for Knuth's book.

31. Citation available at https://amturing.acm.org/award_winners/knuth_1013846.cfm (accessed September 1, 2022).

32. Weiss reports some 300,000 copies of TAOCP sold by 1984 with half of these being copies of volume 1; Weiss, "In the Art of Programming," 41.

33. A list of errors appeared as Donald E. Knuth, "The State of the Art of Computer Programming," STAN-CS-76–551, June 1976, Computer Science Department, Stanford University.

34. Composite quote drawn from: Donald E. Knuth, "Remarks to Celebrate the Publication of *Computers & Typesetting*," *TUGboat* 7, no. 2 (1986): 96; Donald E. Knuth, "The Errors of TeX," *Software—Practices and Experience* 19, no. 7 (1989): 607–685, https://doi.org/10.1002/spe.4380190702; and Knuth's 1996 Kyoto Prize Lecture, "Digital Typography."

35. Knuth quotes this on page 293 of *The TeXbook*.

36. Barbara Beeton, Karl Berry, and David Walden, "TeX: A Branch of Desktop Publishing, Part 1," *IEEE Annals of the History of Computing* 40, no. 3 (2018): 78–93, https://doi.org/10.1109/MAHC.2018.033841114.

37. Knuth, "Digital Typography," 4–5; Knuth's diary entries are in an essay titled "TEXDR.AFT" in his 1999 collection *Digital Typography*.

38. Donald E. Knuth, "Tau Epsilon Chi: A System for Technical Text," Stanford Computer Science Department Report STAN-CS-78–675.1 (November 1978); Folder 50, Box 7, EF/SU.

39. Knuth's notes; Folder 5, Box 25, DEK/SU.

40. Knuth, "Digital Typography," 7–8.

41. Knuth, "Digital Typography," 8.

42. Robin Rider, "Shaping Information: Mathematics, Computing, and Typography," in *Inscribing Science: Scientific Texts and the Materiality of Communication*, ed. Timothy Lenoir (Stanford, CA: Stanford University Press, 1998), 39–54.

43. Donald E. Knuth, "The Letter S," *The Mathematical Intelligencer* 2, no. 3 (1980): 114–122, https://doi.org/10.1007/BF03023051.

44. Donald E. Knuth, "Preliminary Description of TeX," May 13, 1977; Folder 5, Box 25, DEK/SU.DK l.

45. Donald E. Knuth, oral history interview with Edward Feigenbaum, March 14 and 21, 2007; Computer History Museum, Reference Number X3926.2007. Winkler retired from Stanford in 1998, a tribute to her from Knuth appears at https://www-cs-faculty.stanford.edu/~knuth/news98.html (accessed September 2, 2022).

46. Published as Donald E. Knuth, "Mathematical Typography," *Bulletin (New Series) of the American Mathematical Society* 1, no. 2 (1979): 337–372.

47. Knuth, "Mathematical Typography," 343.

48. Knuth, "Mathematical Typography," 370.

49. Knuth, "Mathematical Typography," 345.

50. Correspondence in Folder 10, Box 15, DEK/SU.

51. Letter from Robert Glorioso to Knuth, December 15, 1978; Folder 10, Box 15, DEK/SU.

52. Richard Palais, interview with Dave Walden, July 9, 2011, https://tug.org/interviews/palais.html (accessed September 1, 2022).

53. "1979 Work Session on Development of the TeX System," draft proposal to the NSF, March 15, 1979; Folder 11, Box 15, DEK/SU.

54. Richard S. Palais, "Typesetting by Authors," report to AMS Executive Committee and Trustees, October 16, 1978; Folder 11, Box 15, DEK/SU.

55. "1979 Work Session on Development of the TeX System," draft proposal to the NSF, March 15, 1979; Folder 11, Box 15, DEK/SU.

56. Richard Palais, "A TeX Status Report," November 22, 1979; Folder 11, Box 15, DEK/SU.

57. Donald E. Knuth, "Tau Epsilon Chi: A System for Technical Text," Report No. STAN-CS-78-675.1, Stanford Computer Science Department (November 1978).

58. Donald E. Knuth, "TeX Incunabula," *TUGboat* 5, no. 1 (1984): 4–11.

59. Donald E. Knuth, *TeX and METAFONT: New Directions in Typesetting* (Providence, RI: American Mathematical Society; and, Bedford, MA: Digital Press, 1979).

60. Described in Knuth, "TeX Incunabula."

61. Donald E. Knuth, *The Art of Computer Programming*, vol. 2, *Seminumerical Algorithms*, 2nd ed. (Reading, MA: Addison-Wesley, 1981).

62. Knuth, "TeX Incunabula," 9.

63. Donald E. Knuth, "Literate Programming," *Computer Journal* 27, no. 2 (1984), 97–111, https://doi.org/10.1093/comjnl/27.2.97.

64. Donald E. Knuth, "Current State of Things," *TUGboat* 2, no. 3 (1981): 5.

65. These suggestions are collected, for example, in Folder 10, Box 13, DEK/SU.

66. See, for instance, a letter from Barbara Beeton to Knuth, January 29, 1983; Folder 10, Box 13, DEK/SU.

67. Knuth, "Digital Typography," 17.

68. For the System Development Foundation, see https://oac.cdlib.org/findaid/ark:/13030 /tf429003m4/ (accessed September 15, 2022).

69. "Interview with Duane Bibby," in *TeX People, ed. Karl Berry and David Walden* (TeX Users Group, 2009).

70. Donald E. Knuth, "Computers and Typesetting," *TUGboat* 7, no. 2 (1986): 95–98.

71. Described in Donald E. Knuth and Hermann Zapf, "AMS Euler—A New Typeface for Mathematics," *Scholarly Publishing* 20, no. 3 (1989): 131–157.

72. See, for example, *TUGboat* 7, no. 2 (1986): 119.

73. By 2011, more than a million copies of the volumes in *The Art of Computer Programming* had sold. See https://engineering.stanford.edu/news/stanfords-don-knuth-pioneering-hero -computer-programming (accessed December 9, 2024).

74. Peter Gordon, "Introducing Donald Knuth and Computers & Typesetting," *TUGboat* 7, no. 2 (1986): 93–95.

75. Raph Levien, "Interview: Donald Knuth," published in *TUGboat* 21, no. 2 (2000): 103–110.

76. Donald E. Knuth, "A Torture Test for TeX," Report No. STAN-CS-84–1027, Stanford Computer Science Department (November 1984), 1–2.

77. Alexandre Gaudel, "The (La)TeX Project: A Case Study of Open Source Software," *TUGboat*, 24, no. 1 (2003): 132–145.

78. Knuth adopted a similar scheme for METAFONT with version numbers converging to the base of natural logarithms, starting with 2.7 and continuing on to 2.71, 2.718, and so on.

79. Donald E. Knuth, "The Future of TeX and METAFONT," *TUGboat* 11, no. 4 (1990): 489.

80. Ideas adapted from Brian Stock, *The Implications of Literacy: Written Language and Models of Interpretation in the Eleventh and Twelfth Centuries* (Princeton, NJ: Princeton University Press, 1983), 88–240; also, Jane Heath, "Textual Communities: Brian Stock's Concept and Recent Scholarship on Antiquity," in *Scriptural Interpretation at the Interface between Education and Religion*, ed. Florian Wilk (Boston, MA: Brill, 2019), 5–35.

81. Richard Palais, "Reports to the AMS Committee on Composition Technology," January 18, 1980; Folder 11, Box 15, DEK/SU. Also, minutes of the meeting are in *TUGboat* 1, no. 1 (1980).

82. Attributed to Augustus de Morgan, "Symbols and Notation," in *Penny Cyclopedia of the Society for the Diffusion of Useful Knowledge, vol. 23, Stearic Acid–Tagus* (London, UK: Charles Knight, 1842), 442–445.

83.  Figures from Alex Gaudeul, "Do Open Source Developers Respond to Competition? The LaTeX Case Study," *Review of Network Economics* 6, no. 2 (2007): 239–263, https://doi.org/10.2202/1446-9022.1119.

84.  Poornima Apte, "The *Lingua Franca* of LaTeX," *Increment* 9, May 2019, https://increment.com/open-source/the-lingua-franca-of-latex/ (accessed September 30, 2022).

85.  Leslie Lamport, *LaTeX: A Document Preparation System* (Reading, MA: Addison-Wesley, 1986); and, Günter M. Ziegler, "How LaTeX Changed the Face of Mathematics," *TUGboat* 22, no 1/2 (2001): 20–22, as well as on Lamport's web page https://lamport.azurewebsites.net/pubs/pubs.html#latex (accessed October 1, 2022). I'm grateful to Barbara Beeton and Dave Walden for several very helpful email exchanges about TeX and LaTeX.

86.  Daniel Garisto, "Preprints Make Inroads Outside of Physics," *APS News* 28, no. 9 (2019): 5, 7.

87.  Helge Kragh, *Quantum Generations: A History of Physics in the Twentieth Century* (Princeton, NJ: Princeton University Press, 1999), 168–169.

88.  James E. Till, "Predecessors of Preprint Servers," *Learned Publishing* 14, no. 1 (2001): 7–13, https://doi.org/10.1087/09531510125100214.

89.  Samuel A. Goudsmit, "Communication Problems," *Physical Review Letters* 15, no. 13 (1965): 543–544; and, Michael J. Moravcsik, "Private and Public Communication in Physics," *Physics Today* 18, no. 3 (1965): 23–26, https://doi.org/10.1063/1.3047261.

90.  Gary Taubes, "Publication by Electronic Mail Takes Physics by Storm," *Science* 259, no. 5099 (1993): 1246–1248, https://doi.org/10.1126/science.259.5099.1246.

91.  Cohn's recollections are available at https://w.astro.berkeley.edu/~jcohn/arxiv_hist.html (accessed October 1, 2022); see also Taubes, "Publication by Electronic Mail Takes Physics by Storm."

92.  Paul Ginsparg, "First Steps Towards Electronic Research Communication," *Computers in Physics* 8, no. 4 (1994): 390–396, https://doi.org/10.1063/1.4823313.

93.  Statistics available at https://arxiv.org/help/stats/ (accessed October 1, 2024).

94.  From page 67 of Knuth, *The TeXbook*.

**CHAPTER 6**

1.  Laura Nader, "Up the Anthropologist: Perspectives Gained from Studying Up," in *Reinventing Anthropology*, ed. Dell Hymes (New York: Pantheon, 1972): 284–311.

2.  A good journalistic account of PARC in the 1970s is Michael Hiltzik's *Dealers of Lightning: Xerox PARC and the Dawn of the Computer Age* (New York: Harper Collins, 1999).

3.  Eugene S. Ferguson, *Engineering and the Mind's Eye* (Cambridge, MA: MIT Press, 1992).

4.  Carver Mead and Lynn Conway, *Introduction to VLSI Systems* (Reading, MA: Addison-Wesley, 1980); hereafter, *Introduction to VLSI*.

5.   Lynn Conway, "The Design of VLSI Design Methods," in *ESSCIRC '82: Eighth European Conference on Solid-State Circuits* (Brussels, Belgium: 1982): 106–117.

6.   Conversation between Lucy Suchman and Lynn Conway, along with the published reflections by Suchman, February 28, 2021, https://conwaysuchman-conv.pubpub.org/pub/93808pq4/release/4 (accessed October 10, 2022).

7.   I am using this term in a different manner that that normally deployed by historians of science, where a key feature is the inscriptions on the paper which one can modify and experiment with. See, for example, Ursula Klein, *Experiments, Models, Paper Tools: Cultures of Organic Chemistry in the Nineteenth Century* (Stanford, CA: Stanford University Press, 2003); David Kaiser, *Drawing Theories Apart: The Dispersion of Feynman Diagrams in Postwar Physics* (Chicago, IL: University of Chicago Press, 2005); and Michael D. Gordin, "Paper Tools and Periodic Tables: Newlands and Mendeleev Draw Grids," *Ambix* 65, no. 1 (2018): 30–51, https://doi.org/10.1080/00026980.2017.1418251.

8.   Francis W. Sears and Mark W. Zemansky, *College Physics* (Reading, MA: Addison-Wesley, 1947).

9.   Thomas Kuhn discusses textbooks in the preface and chapter 9 of his collection *The Essential Tension: Selected Studies in Scientific Tradition and Change* (Chicago, IL: University of Chicago Press, 1977). The topic also appears in chapter 11 of *The Structure of Scientific Revolutions* (Chicago, IL: University of Chicago Press, 1962). See also Andrew Warwick and David Kaiser, "Conclusion: Kuhn, Foucault, and the Power of Pedagogy," in *Pedagogy and the Practice of Science: Historical and Contemporary Perspectives*, ed. David Kaiser (Cambridge, MA: MIT Press, 2005), 393–409.

10.  Angela N. H. Creager, Mathias Grote, and Elaine Leong, "Learning by the Book: Manuals and Handbooks in the History of Science," *BJHS Themes* 5 (2020): 1–13, https://doi.org/10.1017/bjt.2020.1.

11.  Conway, "The Design of VLSI Design Methods," 106.

12.  Ronald R. Kline, *Steinmetz: Engineer and Socialist* (Baltimore, MD: Johns Hopkins University Press, 1992); the series itself is discussed on page 192.

13.  Quote from Lynn Conway, "Reminiscences of the VLSI Revolution: How a Series of Failures Triggered a Paradigm Shift in Digital Design," *IEEE Solid State Circuits Magazine* 4, no. 4 (2012): 8–31, https://doi.org/10.1109/MSSC.2012.2215752. Also, Michael A. Hiltzik, "Through the Gender Labyrinth," *Los Angeles Times Magazine*, November 19, 2000, 12.

14.  Lynn Conway, "MIT Reminiscences: Student Years to the VLSI Revolution," March 14, 2014; Lynn Conway Archive ("LCA" henceforth).

15.  Given the intensely personal nature of Conway's transgender experiences, I have relied as much as possible on her own accounts; the most personal and detailed account is available at https://ai.eecs.umich.edu/people/conway/RetrospectiveT.html (accessed August 7, 2022).

16.  "Oral History of Lynn Conway 2014 Computer History Museum Fellow," interview with Dag Spicer, February 24, 2014, Computer History Museum, X7105.2014.

17. Neal H. Rosenthal and National Science Foundation, *Employment of Scientists and Engineers in the United States, 1950–1966* (Washington, DC: US Government and Printing Office, 1968), https://files.eric.ed.gov/fulltext/ED033850.pdf. Also, Ronald Kline, "An Overview of Twenty-Five Years of Electrical and Electronics Engineering in the Proceedings of the IEEE, 1963–1987," *Proceedings of the IEEE* 78, no. 3 (1990): 469–485, https://doi.org/10.1109/5.52226.

18. This project is described in Mark K. Smotherman, Edward H. Sussenguth, and Russell J. Robelen, "The IBM ACS Project," *IEEE Annals of the History of Computing* 38, no. 1 (2016): 60–74, https://doi.org/10.1109/MAHC.2015.50; along with various documentation Conway has preserved online; LCA.

19. Conway, "Reminiscences of the VLSI Revolution," 3.

20. L. Conway, "The Computer Design Process: A Proposed Plan for ACS," memo, August 6, 1968; available at LCA.

21. Thomas Buckley, "A Changing of Sex by Surgery Begun at Johns Hopkins," *New York Times*, November 21, 1966.

22. Conway details her surgery and IBM's reaction in her online memoir, available at LCA. More than fifty years later, in 2020, IBM officially apologized for Conway's treatment; Michael Hiltzik, "IBM Apologizes for Firing a Transgender Pioneer, 52 Years Late," *Los Angeles Times*, November 23, 2020, https://www.latimes.com/business/story/2020-11-23/ibm-apology-lynn-conway (accessed August 10, 2022).

23. Conway, "Reminiscences of the VLSI Revolution," 5.

24. Martin Marshall, Larry Waller, and Howard Wolff, "For Optimal VLSI Design Efforts, Mead and Conway Have Fused Device Fabrication and System-Level Architecture," *Electronics*, October 20, 1981, 102–105.

25. Conway, "Reminiscences of the VLSI Revolution," 13.

26. David Hodges, quoted in Arthur L. Robinson, "Are VLSI Microcircuits Too Hard to Design?" *Science* 209, no. 4453 (1980): 258–262, https://www.jstor.org/stable/1684913.

27. Lynn Conway, "The MPC Adventures: Experiences with the Generation of VLSI Design and Implementation Methodologies," VLSI-81-2 Report, (Palo Alto, CA: Xerox PARC, 1981).

28. Conway cites Mead's "tall, thin man" in "Reminiscences of the VLSI Revolution." Other references include Lawrence Snyder's recollections of Mead's visit to the University of Washington in the summer of 1979 where he used the term; see Lawrence Snyder, "Silicon Chips in the Northwest," November 1995, https://homes.cs.washington.edu/~lazowska/impact/nwlis.html (accessed October 1, 2022). Snyder references this in his entry "Blue CHiP" in *The Encyclopedia of Parallel Computing*, ed. David Padua (New York: Springer, 2011), 165–175, https://doi.org/10.1007/978-0-387-09766-4_237.

29. Cynthia Eller, "The Life of a Caltech 'Lifer,'" May 1, 2014, https://www.caltech.edu/about/news/life-caltech-lifer-42727 (accessed August 1, 2022).

30. Papers in *The Bell System Technical Journal* 49, no. 9 (November 1970) describe this process.

31. Description from "Photolithographic Transistor," *Electronic Design*, December 1, 1957, 5–6; as well as Arthur L. Robinson, "New Ways to Make Microcircuits Smaller," *Science* 208, no. 4447 (1980): 1019–1022, https://www.jstor.org/stable/1683626, which gives an overview of the process as it was practiced in 1980.

32. Gordon E. Moore, "Cramming More Components onto Integrated Circuits," *Electronics* 38, no. 8 (1965): 114–117, and reprinted in *Proceedings of the IEEE* 86, no. 1, (1998): 82–85, http://www.computer-architecture.org/textual/Moore-Cramming-More-Components-1965.pdf.

33. David C. Brock, ed., *Understanding Moore's Law: Four Decades of Innovation* (Philadelphia, PA: Chemical Heritage Press, 2006), especially Brock's material on Mead and Moore's Law in chapter 8.

34. Carver A. Mead, "VLSI and Technological Innovation," in *Proceedings of the Caltech Conference on Very Large Scale Integration*, ed. Charles Seitz, (Pasadena: California Institute of Technology, 1979), 15–28, https://caltechconf.library.caltech.edu/179/.

35. Described in Ivan E. Sutherland and Carver A. Mead, "Microelectronics and Computer Science," *Scientific American* 237, no. 3 (1977): 210–229, https://www.jstor.org/stable/24920329.

36. Quoted in Hiltzik, *Dealers of Lightning*, 304–305; also, Sutherland and Mead, "Microelectronics and Computer Science."

37. Lynn Conway, Martin E. Newell, and Alan Bell, "MPC79: A Large-scale Demonstration of a New Way to Create Systems in Silicon," *Lamba* 1, no. 2 (1980): 11–19; for background, Victor K. McElheny, "Revolution in Silicon Valley," *New York Times*, June 20, 1976.

38. Carver Mead, "Computers That Put the Power Where It Belongs," *Engineering and Science* 35, no. 4 (1972): 4–9, https://calteches.library.caltech.edu/2904/1/mead.pdf.

39. Mead, "VLSI and Technological Innovation," 25.

40. Mead, "Computers That Put the Power Where It Belongs."

41. Lynn Conway, comments on report by Ivan Sutherland from January 26, 1976, "How to Build Digital Electronic Circuits from Now to 1985," March 8, 1976. Both documents available at LCA.

42. Described in Douglas G. Fairbairn and James A. Rowson, "ICARUS: An Interactive Integrated Circuit Layout Program," in *DAC '78: Proceedings of the 15th Design Automation Conference* (New York: IEEE, 1978), 188–192, https://dl.acm.org/doi/pdf/10.5555/800095.803086. There is also a 2018 video ethnography, conducted by David C. Brock and Hansen Hsu, which is available via the Computer History Museum, see "Video Ethnography of 'ICARUS' on the Xerox Alto," posted April 12, 2018, by Computer History Museum, YouTube, 54:03, https://youtu.be/BauuOoB6EIU (accessed September 1, 2022).

43. Dave Johannsen, "Our Machine, A Microcoded LSI Processor," *ACM SIGMICRO Newsletter* 9, no. 4 (1978): 1–7, https://doi.org/10.1145/1014198.804298. The OM project is also described in detail in chapter 5 of Mead and Conway's *Introduction to VLSI*.

44. Chapter 4 of Mead and Conway's *Introduction to VLSI*; see also Ronald F. Ayres, *VLSI: Silicon Compilation and the Art of Automatic Microchip Design* (Englewood Cliffs, NJ: Prentice-Hall, 1983).

45. Ivan Sutherland, Carver Mead, and Thomas Everhart, "Basic Limitations in Microcircuit Fabrication Technology," published as R-1965-ARPA in November 1976.

46. Ivan E. Sutherland and Carver A. Mead, "Microelectronics and Computer Science," *Scientific American* 237, no. 3 (1977): 210–228, https://www.jstor.org/stable/24920329.

47. Christophe Lécuyer, "Driving Semiconductor Innovation: Moore's Law at Fairchild and Intel," *Enterprise & Society* 23, no. 1 (2022): 133–163, https://doi.org/10.1017/eso.2020.38.

48. Conway, "Reminiscences of the VLSI Revolution," 15.

49. "Oral History of Lynn Conway 2014 Computer History Museum Fellow," 14.

50. Described in chapter 4 of Kline's *Steinmetz*.

51. Conway, "Reminiscences of the VLSI Revolution," 13–14

52. "Oral History of Lynn Conway 2014 Computer History Museum Fellow," 3.

53. Conway, "Reminiscences of the VLSI Revolution," 15.

54. Lynn Conway, "Lynn Conway's Retrospective," https://ai.eecs.umich.edu/people/conway /Retrospective3.html#DesignMethods (Accessed October 1, 2022).

55. Conway, "MIT Reminiscences."

56. Conway makes this comparison in "Lynn Conway's Retrospective,"

57. Conway, in "Reminiscences of the VLSI Revolution," 17.

58. Conway, "Lynn Conway's Retrospective."

59. Hiltzik, *Dealers of Lightning*.

60. Conway, "Lynn Conway's Retrospective."

61. Conway, "Reminiscences of the VLSI Revolution," 17.

62. C. Mead and L. Conway, *Introduction to LSI Systems*, October 1977 draft available online via LCA.

63. Mead and Conway, *Introduction to LSI Systems*.

64. Letter from William B. Gruener to Conway and Mead, June 1, 1978; Folder 4, Box 58, CM/CIT.

65. Conway, "Reminiscences of the VLSI Revolution," 17; and Conway's "MIT Reminiscences: Student Years to VLSI Revolution," March 11, 2014, https://ai.eecs.umich.edu/people /conway/Memoirs/MIT/MIT_Reminiscences.pdf (accessed October 5. 2022).

66. Conway, "MIT Reminiscences."

67. Course notes are preserved in Lynn Conway, "The MIT '78 VLSI System Design Course," August 12, 1979, https://ai.eecs.umich.edu/people/conway/VLSI/MIT78/MIT78.html (accessed December 12. 2024); LCA.

68. Conway, "MIT Reminiscences."

69. Conway, "MIT Reminiscences."

70. Folder 1, Box 58 of CM/CIT.

71. "Comments on Mead and Conway Introduction to VLSI Systems," undated but likely late 1978; Folder 3, Box 58, CM/CIT.

72. Conway discusses this on page 17 of "Reminiscences of the VLSI Revolution."

73. Carver Mead, "Response to Marketing Survey," undated but likely mid-1979; Folder 4, Box 58, CM/CIT.

74. "Books About Computers," 1980 Addison-Wesley catalog; Folder 7, Box 58, CM/CIT.

75. Promotional material from Addison-Wesley, undated but likely mid-1979; Folder 4, Box 58, CM/CIT.

76. Lynn Conway, conversation with the author, May 20, 2023.

77. Letters from October Graham to Carver Mead, April 11, 1980, and from Thomas Bell to Conway and Mead, September 17, 1980; both Folder 4, Box 58, CM/CIT. Also, Conway, "Reminiscences of the VLSI Revolution," 26.

78. Paul Mozur and John Liu, "The Chip Titan Whose Life's Work Is at the Center of a Tech Cold War," *New York Times*, August 4, 2023.

79. Conway, "MIT Reminiscences."

80. Conway, "The Design of VLSI Design Methods," 107.

81. Hiltzik's *Dealers of Lightning*, 309.

82. Conway, "Reminiscences of the VLSI Revolution," 16.

83. Conway, "Reminiscences of the VLSI Revolution," 21–22.

84. Materials from the MPC79 project can be found at https://ai.eecs.umich.edu/people/conway/VLSI/MPC79/ (accessed October 15, 2022); also, see Conway, "The MPC Adventures."

85. James Clark, "A VLSI Geometry Processor for Graphics," *Computer* 13, no. 7 (1980): 59–67. Clark himself was a colorful and central character in Michael Lewis' book *The New New Thing* (New York: Norton, 1999).

86. Douglas Fairbairn, oral history interview with David Brock, et al., October 6, 2016.

87. As reported in "Philips Acquires VLSI Technology with Raised Bid," *EE Times*, April 3, 1999, https://www.eetimes.com/philips-acquires-vlsi-technology-with-raised-bid/ (accessed January 1, 2024).

88. Quotes from "Lynn Conway and Carver Mead: The 1981 Achievement Award," *Electronics*, October 20, 1981, 102–105.

89. See, for example, *Funding a Revolution: Government Support for Computing Research* (Washington, DC: National Academy Press, 1999); as well as Alex Roland and Philip Shiman,

*Strategic Computing: DARPA and the Quest for Machine Intelligence, 1983–1993* (Cambridge, MA: MIT Press, 2002).

90. Conway, "Reminiscences of the VLSI Revolution," 28.

91. "Oral History of Lynn Conway 2014 Computer History Museum Fellow," 20; and, conversation between Lucy Suchman and Lynn Conway.

## CHAPTER 7

1. For a history of Orwell's book, see Dorian Lynskey's wonderful *The Ministry of Truth: The Biography of George Orwell's 1984* (New York: Anchor, 2019.)

2. See "1983 Apple Keynote the 1984 Ad Introduction," posted February 2, 2012, by Arjun Agarwal, YouTube, 6:41, https://youtu.be/liusJ5Xxyxo (accessed November 1, 2022).

3. The advertisement was also broadcast in theaters and in several local markets so it could qualify for industry awards; Stuart Elliott, "Advertising," *New York Times*, March 14, 1995.

4. Quote from David Burnham, "The Computer, the Consumer, and Privacy," *New York Times*, March 4, 1984.

5. Data from Efrem Sigel and Daniel McCarthy, *Computer Publishing Market Forecast—1986* (Larchmont, NY: Communication Trends, 1986); DD/CLB.

6. Data from "Oral History of Computer Literacy Bookshops, Inc.: Daniel A. Doernberg and Rachel Unkefer," interview by Dag Spicer, recorded via Zoom March 26, 2021, Computer History Museum, X9445.2021.

7. Michael Moritz, "The Seeds of Success," *Time*, February 15, 1982, 40–41, https://time.com/vault/issue/1982-02-15/page/48/.

8. "Striking It Rich," *Time*, February 15, 1982, 36–41; written by Alexander L. Taylor III and reported by Michael Moritz in San Francisco and Frederick Ungeheuer in *Time*'s New York City office.

9. Michael Moritz, interview with the author, November 4, 2020; copy in author's files. Additional biographical information is from an interview conducted by the Stanford Business School, "Michael Moritz, Partner, Sequoia Capital," posted February 7, 2019, by Stanford Graduate School of Business, YouTube, 42:19, https://youtu.be/ZTNFQeYEdvc (accessed November 25, 2022).

10. "A Letter from the Publisher," *Time*, February 15, 1982, 3.

11. Michael Moritz and Barrett Seaman, *Going for Broke: The Chrysler Story* (Garden City, NY: Doubleday, 1981).

12. Michael Moritz, interview with the author, November 4, 2020.

13. Tracy Kidder, *The Soul of a New Machine* (New York: Atlantic/Little, Brown, 1981).

14. Recounted in Tracy Kidder and Richard Todd, *Good Prose: The Art of Nonfiction* (New York: Random House, 2013), 126. Also, see retrospective evaluations of Kidder's book in Suzanne

Moon, "Classics Revisited: Tracy Kidder, *Soul of a New Machine*," *Technology and Culture* 45, no. 3 (204): 597–602; and Thomas Haigh, "The Immortal Soul of an Old Machine," *Communications of the ACM* 64, no. 1 (2021): 32–37, https://doi.org/10.1145/3436249.

15.  Samuel C. Florman, "The Hardy Boys and the Microkids Make a Computer," *New York Times*, August 23, 1981.

16.  Evan Ratliff, "O, Engineers!" *Wired*, December 1, 2000, 356–367, https://www.wired.com /2000/12/soul/.

17.  Michael Moritz, interview with the author, November 4, 2020.

18.  Michael Moritz, interview with the author, November 4, 2020.

19.  Michael Moritz, *The Return to the Little Kingdom: Steve Jobs, The Creation of Apple, and How It Changed the World* (New York: Overlook, 2009), 11.

20.  Walter Isaacson, *Steve Jobs* (New York: Simon & Schuster, 2011), 139.

21.  Discussed in Isaacson, *Steve Jobs*, 139–140, and Moritz, *The Return to the Little Kingdom*; also Moritz, interview with the author, November 4, 2020.

22.  Moritz, *The Little Kingdom*, 12.

23.  Michael Moritz, interview with the author, November 4, 2020.

24.  Christopher Lehmann-Haupt, "Books of the Times: We Can Work Out the Disease Later," *New York Times*, May 7, 1969.

25.  Joan Didion, "In Hollywood," in *The White Album* (New York: Simon & Schuster, 1979), 160. The original essay, titled "Hollywood: Having Fun," appeared in the March 22, 1973, issue of the *New York Review of Books*.

26.  Michael Moritz, interview with the author, November 4, 2020.

27.  From the prologue to the 2009 reissue of Moritz's book, called *The Return to the Little Kingdom*.

28.  William Stockton, "Business in Short: The Little Kingdom," *New York Times*, October 21, 1984; also Michael Moritz, interview with the author, November 4, 2020.

29.  Michael Moritz, interview with the author, November 4, 2020.

30.  Moritz, *The Little Kingdom*, 286.

31.  Tom Nicholas, *VC: An American History* (Cambridge, MA: Harvard University Press, 2019), 260–261.

32.  Will Gompertz, "Booker Prize Finds New Funder in Billionaire Sir Michael Moritz," *BBC News*, February 28, 2019, https://www.bbc.com/news/entertainment-arts-47393880 (accessed December 25, 2022). Moritz stepped down from Sequoia Capital in July 2023 after nearly thirty-eight years with the firm.

33.  Donald C. Hoefler, *Hi-FI Manual* (New York: Fawcett Books, 1952). Additional information comes from Turo Uskali and David Nordfors, "The Role of Journalism in Creating the Metaphor of Silicon Valley," unpublished 2007 paper; copy in author's files.

34. This was published as a three-part series (January 11, 18, and 25, 1971) as Don C. Hoefler, "Silicon Valley, USA," *Electronic News*.

35. Mal Padgett, quoted in "Don Hoefler, Writer Who Coined Term 'Silicon Valley,'" *San Jose Mercury News*, April 16, 1986, 8B. The description of Hoefler's newsletter comes from Michael S. Malone, *The Big Score: The Billion Dollar Story of Silicon Valley* (Garden City, NY: Doubleday, 1985), 337.

36. Matthew Pressman, *On Press: The Liberal Values That Shaped the News* (Cambridge, MA: Harvard University Press, 2018).

37. Pressman, *On Press*, 1–10.

38. Pressman, *On Press*, 206–207.

39. Arthur O. Sulzberger, "Business and the Press: Is the Press Anti-Business?," speech given to the Economic Club of Detroit, March 14, 1977; see *Vital Speeches of the Day* 43, no. 14 (May 1, 1977): 426–428.

40. Louis Banks, "Marching Up the Down Staircase," *Bulletin of the American Society of Newspaper Editors* (October 1980): 4–6.

41. Chris Roush, "Business Journalism," *Encyclopedia of Journalism*, ed. Christopher H. Sterling (Los Angeles, CA: SAGE, 2009): 225–227.

42. Banks, "Marching Up the Down Staircase,"

43. Biographical information on Evelyn Richards comes from a 2014 interview done by John Geddes, https://www.digitalriptide.org/person/evelyn-richards/ (accessed January 5, 2023), and an oral history interview with the author, March 29, 2021 (copy in author's files).

44. This was the paper's name as of 1983, when the *San Jose News* and the *Mercury News*, both owned by the media company Knight Ridder, merged.

45. Michael Shapiro, "The Newspaper That Almost Seized the Future," *Columbia Journalism Review*, November 10, 2011, https://www.cjr.org/feature/the_newspaper_that_almost_seized_the_future.php (accessed January 5, 2023).

46. Evelyn Richards, oral history interview with the author, March 29, 2021.

47. Evelyn Richards, oral history interview with the author, March 29, 2021.

48. Evelyn Richards, "A Look at Secret New Apple Computer" and "How the Macintosh Computer Grew," *San Jose Mercury News*, both January 15, 1984.

49. See Wendy Marinaccio's May 17, 2000, interview with Richards for Stanford University, https://web.stanford.edu/dept/SUL/sites/mac/primary/interviews/richards/trans.html (accessed January 5, 2023).

50. Richards's essay, "On the Brink of Bankruptcy: Life in a Dying Company," appeared in the October 27, 1985, issue of *West*, a slickly produced weekly magazine put out by the *San Jose Mercury News*.

51. Ove Granstrand and Marcus Holgersson, "Innovation Ecosystems: A Conceptual Review and New Definition," *Technovation* 90/91 (2020): https://doi.org/10.1016/j.technovation.2019.102098.

52. Between 1983 and 1986, Reagan mentioned Silicon Valley at least ten times in formal addresses, see https://www.presidency.ucsb.edu/people/president/ronald-reagan (accessed February 10, 2023).

53. Ronald Reagan, "Remarks at the Great Valley Corporate Center in Malvern, Pennsylvania," May 31, 1985, https://www.presidency.ucsb.edu/documents/remarks-the-great-valley-corporate-center-malvern-pennsylvania (accessed February 10, 2023).

54. Kathy Kincade, "The Making of a Computer Bookstore," *Computer Language* 4, no. 9 (September 1987): 95–99.

55. Erik Sandberg-Diment, "A Cascade of Computer Books: A Survey of Books Spawned by the Electronic Age," *New York Times*, August 16, 1983.

56. "Oral History of Computer Literacy Bookshops, Inc.: Daniel A. Doernberg and Rachel Unkefer," interview by Dag Spicer, recorded via Zoom, March 26, 2021, Computer History Museum, X9445.2021; additional biographical material comes from oral history interviews that the author conducted with Doernberg (December 21, 2020, and December 28, 2020) and Unkefer (January 8, 2021).

57. "Oral History of Computer Literacy Bookshops, Inc.," 18–19.

58. Rachel Unkefer, oral history interview with the author, January 8, 2021.

59. Laura J. Miller, "Selling the Product," in *A History of the Book in America*, ed. David Paul Nord, Joan Shelley Rubin, and Michael Schudson, *vol. 5, The Enduring Book: Print Culture in Postwar American* (Chapel Hill: University of North Carolina Press, 2013): 91–106.

60. John Markoff, oral history interview with the author, September 1, 2020.

61. Daniel A. Doernberg, oral history interview with the author, December 21, 2020.

62. Daniel A. Doernberg, oral history interview with the author, December 21, 2020.

63. Kincade, "The Making of a Computer Bookstore."

64. Doernberg and Unkefer, interviews with the author.

65. Daniel A. Doernberg, oral history interview with the author, December 21, 2020.

66. Arthur W. Luehrmann, "Should the Computer Teach the Student, or Vice Versa?" *AFIPS '72: Proceedings of the May 16–18, 1972, Spring Joint Computer Conference* (May 1972): 407–410, https://doi.org/10.1145/1478873.1478925; see also the article by the coinventor of BASIC, John G. Kemeny, "The Case for Computer Literacy," *Daedalus* 112, no. 2 (1983): 211–230, https://www.jstor.org/stable/20024860.

67. Kemeny, "The Case for Computer Literacy," 229.

68. Daniel A. Doernberg, oral history interview with the author, December 21, 2020.

69. Susan Meyers, "People in the News: Dan Doernberg and Rachel Unkefer," *PC Magazine* 3, no. 13, July 10, 1984, 66; and Kincade, "The Making of a Computer Bookstore," 96.

70. Peggy Watt, "Booking on Computers," *InfoWorld* 6, no. 41, October 8, 1984, 29–32.

71. Erik Sandberg-Diment, "Finding a Good Book in a Sea of Mediocrity," *New York Times*, December 11, 1984.

72. Daniel A. Doernberg, oral history interview with the author, December 21, 2020.

73. Steward Brand, in *Whole Earth Software Review* 1, Spring 1984, 131.

74. Kincade, "The Making of a Computer Bookstore," 97.

75. "CLB Categories—Confidential," internal store document, July 1996; DD/CLB.

76. Dan Doernberg, personal communication with the author, March 12, 2023.

77. Described in Kincade, "The Making of a Computer Bookstore," 96.

78. Sigel and McCarthy, *Computer Publishing Market Forecast—1986.*

79. Rachel Unkefer, oral history interview with the author, January 8, 2021.

80. Andrew Pollack, "Computer Makers in Severe Slump," *New York Times*, June 10, 1985.

81. Letters from Luther Nichols to Michael Malone, March 30, 1984, and from Michael Malone to Adrian Zackheim, December 10, 1984; both Box 5, MSM/SCU.

82. Michael S. Malone, oral history interview with the author, September 2, 2021; copy in author's files.

83. Patrick J. Kiger, "Survival of the Fittest in Silicon Valley," *Camden Courier-Post* (November 24, 1985); clipping in Box 5, MSM/SCU.

84. Malone's recollections are in his book *The Valley of Heart's Delight: A Silicon Valley Notebook, 1963–2001* (New York: Wiley, 2002).

85. Michael S. Malone, oral history interview with the author, September 2, 2021.

86. Michael S. Malone, oral history interview with the author, September 2, 2021.

87. Malone, *The Valley of Heart's Delight*, xx.

88. Michael S. Malone, oral history interview with the author, September 2, 2021.

89. Malone, *The Valley of Heart's Delight*, xx.

90. Malone, *The Valley of Heart's Delight*, xxi.

91. From page 411 of Malone's *The Big Score*; emphasis added.

92. Malone and Yoachum's articles were published April 6–8, 1980, in the *San Jose Mercury News*. For a recent historical perspective, see Jason A. Heppler, *Silicon Valley and the Environmental Inequalities of High-Tech Urbanism* (Norman: University of Oklahoma Press, 2024), especially chapter 6.

93. Michael S. Malone, oral history interview with the author, September 2, 2021. An archive of 1980s articles about the toxic legacy of Silicon Valley is at International Campaign for Responsible Technology, https://icrt.co/1980s-articles/ (accessed December 15, 2023).

94. William Grimes, "Don Congdon, Longtime Literary Agent for Ray Bradbury, Dies at 91," *New York Times,* December 4, 2009.

95. Michael S. Malone, oral history interview with the author, September 2, 2021.

96. Michael Malone, "Proposal and Outline for a Silicon Valley Book," n.d. but likely sometime in 1982; Box 5, MSM/SCU.

97. Letter from Luther Nichols to Malone, December 3, 1983; Box 5, MSM/SCU.

98. Letter from Malone to Luther Nichols, July 20, 1983; Box 5, MSM/SCU.

99. Cyrus C. M. Mody, *The Squares: US Physical and Engineering Scientists in the Long 1970s* (Cambridge, MA: MIT Press, 2022).

100. Tom Wolfe, "The Tinkerings of Bob Noyce," *Esquire,* December 1983, 346–377; also, Isadore Barmash, "Corporate Triumph, then Death in a Ferrari," *New York Times,* July 10, 1983.

101. Michael S. Malone, oral history interview with the author, September 2, 2021.

102. Malone to Adrian Zackheim, letter, n.d. but likely sometime early 1985; Box 5, MSM/SCU.

103. Edwin McDowell, "Doubleday Publishing Names New President," *New York Times,* November 28, 1985.

104. Michael S. Malone, oral history interview with the author, September 2, 2021.

105. Malone, *The Big Score,* 10.

106. James P. Degnan, "The Silicon Jackpot," *San Francisco Chronicle,* August 11, 1985.

107. John W. Wilson, "A Sprawling History of Silicon Valley," *Business Week,* July 29, 1985, 8.

108. Letters from Robert Noyce to Malone, August 14, 1985, and from David Packard to Malone, January 24, 1986; also, "Michael Malone—*The Big Score* Promotion Schedule," n.d. but likely August 1985; all Box 5, MSM/SCU.

109. Letter from Ceci Scott for Adrian Zackheim to Don Congdon, September 23, 1985; Box 5, MSM/SCU.

110. For Zschau and Silicon Valley, see Margaret O'Mara, *The Code: Silicon Valley and the Remaking of America* (New York: Penguin, 2019).

111. Comments found in *Climate for Entrepreneurship and Innovation in the United States: Hearings Before the Joint Economic Committee, August 27–28, 1984* (Washington, DC: US Government Printing Office, 1984).

112. Letter from Zschau to Malone, June 5, 1985; Box 5, MSM/SCU.

113. Michael D. Gordin and W. Patrick McCray, eds., *Greedy Science: Making Money, Creating Knowledge, and Being Famous in the 1980s* (Baltimore, MD: Johns Hopkins University Press, 2025).

## CHAPTER 8

1. William Aspray and Paul E. Ceruzzi, eds., *The Internet and American Business* (Cambridge, MA: MIT Press, 2008).

2. Daniel A. Doernberg, oral history interview with the author, December 21, 2020. Also see, Robert Spector, *The Amazon.com Way—Get Big Fast: Inside the Revolutionary Business Model That Changed the World* (New York: Harper Collins, 2000), especially pages 21–22 and 47; while Brad Stone's *The Everything Store: Jeff Bezos and the Rise of Amazon* (New York: Little, Brown, 2014) gives an account that brings the story closer to the present.

3. By the end of 1992, two other companies—Cleveland-based books.com and Cambridge-based wordsworth.com—had registered their domain names.

4. Dan Doernberg, personal communication with the author, April 10, 2023. The book itself was Janet Ruhl's *The Computer Consultant's Workbook* (Leverett, MA: Technion, 1996).

5. Jeff Bezos, speech at Lake Forest College, March 21, 1998, https://www.c-span.org/video/?c4948278/user-clip-jeff-bezos-lake-forest-speech-1998 (accessed April 10, 2023). Bezos gave this speech in part because his aunt and uncle were alumni of the small liberal arts school.

6. G. Bruce Knecht, "How Wall Street Whiz Found a Niche Selling Books on the Internet," *Wall Street Journal*, May 16, 1996.

7. John B. Thompson has authored two books that are essential reading for understanding the publishing business in the late twentieth and early twenty-first century: *Merchants of Culture: The Publishing Business in the Twenty-First Century* (New York: Plume, 2010) and *Book Wars: The Digital Revolution in Publishing* (Medford, MA: Polity Press, 2021).

8. George Gilder, *Microcosm: The Quantum Revolution in Economics and Technology* (New York: Simon & Schuster, 1989).

9. The number of households with home computers increased from 15 percent to 35 percent in this period; if one considers only college graduates, the number was more than 55 percent; see the US Bureau of Labor Statistics, "Computer Ownership Up Sharply in the 1990s," April 5, 1999, at https://www.bls.gov/opub/ted/1999/Apr/wk1/art01.htm (accessed April 15, 2023).

10. John Brockman, *Digerati: Encounters with the Cyber Elite* (San Francisco, CA: Hardwired, 1996), xxv.

11. Kara Swisher, "An Inundation of Internet Information," *Washington Post*, May 8, 1995.

12. Ted Nelson, "Those Unforgettable Next Few Years," in *The First West Coast Computer Faire: Conference Proceedings, ed.* Jim C. Warren (San Francisco, CA: Computer Faire, 1977).

13. For example, Steven V. Roberts, "President Charms Students, But His Ideas Lack Converts," *New York Times*, June 1, 1988.

14. Ronald Reagan, "Remarks and a Question-and-Answer Session with Students and Faculty at Moscow State University," May 31, 1988, https://www.reaganlibrary.gov/archives/speech/remarks-and-question-and-answer-session-students-and-faculty-moscow-state (accessed May 1, 2023).

15. Warren T. Brookes, *The Economy in Mind* (New York: Universe Books, 1982), 13–14.

16. Larissa MacFarquhar, "The Gilder Effect," *New Yorker*, May 29, 2000, 103–111. See also Linda Kintz, *Between Jesus and the Market: Emotions That Matter in Right-Wing America* (Durham, NC: Duke University Press, 1997), especially chapter 5.

17. "The Dangers of Being a Single Male," *Time*, December 9, 1974, 65.

18. Quoted in Bob Greene, "A $100 Million Idea: Use Greed for Good," *Chicago Tribune*, December, 15 1986.

19. Roger Starr, "A Guide to Capitalism," *New York Times Book Review*, February 1, 1981, BR 3–4; and, Paul Johnson, "Gilder Praises Capitalism's Virtues," *Wall Street Journal*, January 22, 1981.

20. George Gilder, *Wealth and Poverty* (New York: Basic Books, 1981). 99–102.

21. George Gilder, *The Spirit of Enterprise* (New York: Simon & Schuster, 1984).

22. George Gilder, "The Up-and-Comers," *Forbes*, March 14, 1983, 130–131.

23. Gilder, *The Spirit of Enterprise*, 12.

24. Daniel F. Cuff, "Rosen's Newsletter is Changing Owners," *New York Times*, March 15, 1983.

25. George Gilder, C-SPAN appearance, September 24, 1989.

26. George Gilder, C-SPAN appearance, September 24, 1989.

27. Po Bronson, "George Gilder," *Wired*, March 1996, 122–130.

28. This work appeared, for example, as B. Hoeneisen and C. A. Mead, "Fundamental Limitations in microelectronics—I. MOS Technology," *Solid-State Electronics* 15, no. 7 (1972): 819–829, https://doi.org/10.1016/0038-1101(72)90103-7; and, B. Hoeneisen and C. A. Mead, "Limitations in Microelectronics—II. Bipolar Technology," *Solid-State Electronics*, 15, no. 8 (1972): 891–897, https://doi.org/10.1016/0038-1101(72)90026-3.

29. Gilder, *Microcosm*, 11.

30. Gilder's admiration for Mead was reflected in a feature-length article he wrote for *Forbes* (with Mead on the cover) in advance of *Microcosm*'s publication titled "You Ain't Seen Nothing Yet," *Forbes*, April 4, 1988, 88–93.

31. Conway's reflections are from her online materials, https://ai.eecs.umich.edu/people/conway/Retrospective4.html (accessed May 10, 2023). Further insights came from the author's conversation with Conway, May 19, 2023.

32. Gilder, *Microcosm*, 355 and 17.

33. Gilder, *Microcosm*, 353.

34. See chapter 25 of *Microcosm* as well an essay Gilder wrote in advance of his book's appearance: "The Revitalization of Everything: The Law of the Microcosm," *Harvard Business Review*, March 1988, 49–61.

35. Gilder, *Microcosm*, 334.

36. Gilder, *Microcosm*, 113–114.

37. Gilder, *Microcosm*, 13, 376.

38. Gilder, *Microcosm*, 13.

39. Gilder, *Microcosm*, 381–383.

40. Richard O'Reilly, "Computer Conservatism," *Los Angeles Times*, September 3, 1989, L 4. Here, the book was mistakenly titled *Into the Quantum Era of Microcosm: Economics and Technology.*

41. Richard A. Shaffer, "Microchips and Materialism," *Wall Street Journal*, September 13, 1989.

42. Langdon Winner "Do Artifacts Have Politics?" *Daedalus* 109, no. 1 (1980): 121–136.

43. Langdon Winner, "O Brave New Chips!" *New York Times*, October 15, 1989.

44. Bronson, "George Gilder."

45. Throughout the 1990s, articles about Gilder persistently noted that he was working on a book called *Telecosm*. It finally appeared as *Telecosm: How Infinite Bandwidth Will Revolutionize the World* (New York: New Press, 2000).

46. Kevin Kelly, "George Gilder: When Bandwidth Is Free," *Wired*, September/October 1993, 3, https://www.wired.com/1993/04/gilder-4/.

47. A phrase popularized by librarian Jean A. Polly who used it as the title for a 1992 guide called *Surfing the Internet: An Introduction*, published by the University of Minnesota's Wilson Library, https://ia800303.us.archive.org/34/items/surfingtheintern00049gut/surf10.txt (accessed May 1, 2023).

48. The first RFC, titled "Host Software," was written by Steve Crocker and appeared April 7, 1969; see https://datatracker.ietf.org/doc/html/rfc1 (accessed May 1, 2023).

49. E. Krol, "The Hitchhikers Guide to the Internet," RFC 1118, September 1989, https://www.rfc-editor.org/rfc/rfc1118.html (accessed April 1, 2021).

50. Edward M. Krol, interview with the author, January 18, 2021.

51. Tim O'Reilly, interview with the author, November 11, 2020.

52. Alfred Korzybski, *Science and Sanity: An Introduction to Non-Aristotelean Systems and General Semantics* (New York: Science Press, 1933). Biographical material on O'Reilly comes from several places including two 2020 interviews I did with him as well as Steven Levy "The Trend Spotter," *Wired*, October 2005.

53. Timothy O'Reilly, *Frank Herbert* (New York: Frederick Ungar Publishing, 1981).

54. Composite quote drawn from interview with the author on November 11, 2020, and O'Reilly's recollections archived here, https://web.archive.org/web/20101205172557/http://radar.oreilly.com/2005/11/burn-in-13-tim-oreilly.html (accessed January 12, 2025).

55. Tim O'Reilly, "Ten Years as a Publisher," January 1997, available at https://www.oreilly.com/pub/a/tim/articles/tenyears.html (accessed May 15, 2023).

56. Daniel A. Doernberg and Rachel Unkefer, oral history interview by Dag Spicer, March 26, 2021; CHM X9445.2021, Computer History Museum.

57. Michael Loukides, interview with the author, December 17, 2020.

58. "Evangelism and the Bestseller," Ben McConnell and Jackie Huba's interview with Brian Erwin, 2002, archived February 11, 2012, https://web.archive.org/web/20120211114441/http:/www.creatingcustomerevangelists.com/resources/evangelists/brian_erwin.asp (accessed June 1, 2023).

59. For example, Tim O'Reilly, "Giving Away Free Books," email dated November 20, 2000, archived December 12, 2004, https://web.archive.org/web/20041212181041/http://tim.oreilly.com/values/wolfe-gnn.html (accessed May 31, 2023).

60. Ed Krol, *The Whole Internet User's Guide and Catalog* (Sebastopol, CA: O'Reilly & Associates, 1992).

61. Krol, *The Whole Internet*, 1–2.

62. John Markoff, "A Web of Networks, An Abundance of Services," *New York Times*, February 28, 1993.

63. Owens Mitchell, "The Most Request of Library Books," *New York Times*, June 29, 1995.

64. Elizabeth Diefendorf, ed., *The New York Public Library's Books of the Century* (New York: Oxford University Press, 1996). See also "The New York Public Library's Books of the Century," https://www.nypl.org/voices/print-publications/books-of-the-century.

65. The company that produced them was started by Clifton Hillegass, a lifelong Nebraskan and inveterate book lover. Information from Dennis Wepman, "Hillegass, Clifton Keith," American National Biography, https://doi.org/10.1093/anb/9780198606697.article.1002279 (accessed July 1, 2023); and, Shaila K. Dewan, "Clifton Keith Hillegass Dies; Cliffs Notes Creator Was 83," *New York Times*, May 7, 2001.

66. Bart Ziegler, "Jargon Sparks Fear and Loathing Among the Technologically Illiterate," *Washington Post*, April 22, 1990.

67. Pat McGovern, as told to Eve Tahmincioglu, "From X's and O's to I.T.," *New York Times*, May 21, 2006; also, William Yardley, "Patrick McGovern Dies at 76; Founded Publishing Empire," *New York Times*, March 24, 2014.

68. Its original name was International Data Corporation.

69. *Computerworld*, June 21, 1967, 1.

70. Albert Scardino, "Computer-Magazine Giant Thrives on Global Approach," *New York Times*, July 31, 1989.

71. Efrem Sigel and Daniel McCarthy, *Computer Publishing Market Forecast—1986* (Larchmont, NY: Communication Trends, Inc. 1986); DD/CLB; Steve Lohr, "PC Magazine is Mirroring an Industry's Success," *New York Times* (November 22, 1993): D1.

72. William M. Bulkeley, "Battle of the Computer-Magazine Titans," *Wall Street Journal*, February 24, 1989.

73. Joseph J. Corn, *User Unfriendly: Consumer Struggles with Personal Technologies* (Baltimore, MD: The Hopkins University Press, 2011). For computers specifically, see Michael L. Black, *Transparent Designs: Personal Computing and the Politics of User-Friendliness* (Baltimore, MD: Johns Hopkins University Press, 2022).

74. Erik Sandberg-Diment, "A Bundle of Disks for Those Bewildered Novices," *New York Times*, August 23, 1983.

75. Biographical material based on conversation with the author, December 21, 2020.

76. Dan Gookin and Andy Townshend, *Hard Disk Management with MS-DOS and PC-DOS* (Blue Ridge Summit, PA: TAB Books, 1987). For "power user," see Michael J. Halvorson, *Code Nation: Personal Computing and the Learn to Program Movement in America* (New York: ACM Books, 2020), 241–245.

77. Dan Gookin, conversation with the author, December 21, 2020.

78. John Muir, *How to Keep Your Volkswagen Alive: A Manual of Step-By-Step Procedures for the Compleat Idiot* (Santa Fe: John Muir, 1969); *The Last Whole Earth Catalog* (1972) featured it, for example, on page 248, along with glowing reviews.

79. Letter from Dan Gookin to Mike McCarthy, April 19, 1991; copy in author's files.

80. Dan Gookin, personal communication with the author, June 23, 2023.

81. Dan Gookin, *DOS for Dummies* (San Mateo, CA: IDG Books Worldwide, 1991).

82. Dan Gookin, conversation with the author, December 21, 2020.

83. Steve Lohr, "Across the Computer Divide, The Nerds Face the Dummies," *New York Times*, June 6, 1993.

84. L. R. Shannon, "DOS With a Grain of Salt," *New York Times*, January 7, 1992.

85. Jeff Dawson, *Gay and Lesbian Online* (Berkeley, CA: Peachpit Press, 1996).

86. Stafford L. Battle and Rey O. Harris, *The African American Resource Guide to the Internet and Online Services* (New York: McGraw-Hill, 1996).

87. John Levine and Carol Baroudi, *The Internet for Dummies* (San Mateo, CA: IDG Books Worldwide, 1993), v.

88. Rachel Donadio, "Dumbing Up," *New York Times*, September 24, 2006.

89. Ruth Graham, "Dummies for Dummies," *Slate*, April 4, 2016, https://slate.com/culture/2016/04/the-history-and-delights-of-the-for-dummies-how-to-books.html (accessed July 1, 2023).

90. "IDG Books Worldwide Buys Cliffs Notes for $14.2 Million," *Wall Street Journal*, December 8, 1998.

91. Biographical information on Cramer comes from Charles W. Dahlinger, *Pittsburgh: A Sketch of Its Early Social Life* (New York: G. P. Putnam and Sons, 1916).

92. For example, Zadok Cramer, *The Navigator* (Pittsburgh, PA: Cramer, Spear, and Eichbaum, 1814). Cramer's book was frequently updated with new material and corrections and, before he died in 1814, it had expanded to more than 350 pages.

93. Dale Dougherty, conversation with the author, December 16, 2020.

94. The history of the e-book is presented in Thompson's *Book Wars*.

95. Simon Rowberry, *The Early Development of Project Gutenberg, c. 1970–2000* (New York: Cambridge University Press, 2023).

96. Steve Lambert and Suzanne Ropiequet, *CD-ROM: The New Papyrus* (Redmond, WA: Microsoft Press, 1986).

97. Dale Dougherty, conversation with the author, December 16, 2020.

98. This meeting was "Hypertext '91" held in December 1991 in San Antonio, Texas; this period is discussed in Tim Berners-Lee, *Weaving the Web: The Original Design and Ultimate Destiny of the World Wide Web* (New York: HarperCollins, 1999), especially 50–51.

99. Berners-Lee, *Weaving the Web*, 1.

100. Technically, it was known as "ViolaWWW."

101. James Gillies and Robert Cailliau, *How the Web Was Born: The Story of the World Wide Web* (New York: Oxford University Press, 2000), especially chapter 6.

102. From emails, sent between 1991–1992, which Dale Dougherty shared with me; copies in author's possession.

103. Email from Pei-Yuan Wei to Dale Dougherty, May 5, 1991; copy in author's files.

104. Quote from Gillies and Cailliau, *How the Web Was Born*, 215. Also, email from Berners-Lee to Wei, May 15, 1992; copy in author's files as well as Berners-Lee's recollections via https://www.w3.org/People/Berners-Lee/FAQ.html#browser. In addition, Berners-Lee's favorable review of Viola, archived March 31, 2010, can be found at, https://web.archive.org/web/20100331234423/http://www.w3.org/History/19921103-hypertext/hypertext/Viola/Review.html (all accessed May 20, 2023).

105. Paul E. Ceruzzi, "The Internet Before Commercialization," in *The Internet and American Business*, 29–30; my emphasis on the textual change.

106. Tim O'Reilly, interview with the author, December 12, 2020.

107. Tim O'Reilly, interview with the author, December 12, 2020.

108. Dale Dougherty, conversation with the author, December 16, 2020.

109. Jennifer Niederst Robbins, conversation with the author, December 29, 2020.

110. Howard Rheingold, "A Window on the Matrix," *Wired*, May 1993.

111. O'Reilly, "Giving Away Free Books."

112. Michael Miller, "Internet to Get Hit with Ad Clutter," *Wall Street Journal*, August 27, 1993.

113. Philip Elmer-Dewitt, "Battle for the Soul of the Internet," *Time*, July 25, 1994, 50–56, https://time.com/archive/6725698/battle-for-the-soul-of-the-internet/.

114. Miller, "Internet to Get Hit with Ad Clutter."

115. Tim O'Reilly, interview with the author, December 12, 2020.

116. Thomas Forbes, "Far Out," *Folio*, September 15, 1994, 64–65, 82. Also, from promotional GNN material in Box 16, ED/CHM.

117. From the Fall 1994 catalog for O'Reilly & Associates; copy in Box 16 of ED/CHM.

**CHAPTER 9**

1. Daniel Kadlec, "The Thrill Ride Isn't Over," *Time*, April 17, 2000, 36–38, https://time.com/archive/6740903/the-thrill-ride-isnt-over/; Krantz's remark appeared in a sidebar, "The Day the World Ended," 42, https://time.com/archive/6740947/the-day-the-world-ended/. On the dot-com crash, see Brent Goldfarb and David A. Kirsch, *Bubbles and Crashes: The Boom and Bust of Technological Innovation* (Stanford, CA: Stanford University Press, 2019).

2. Jeff Madrick, "The Business Media and the New Economy," Research Paper R-24 for the Joan Shorenstein Center on the Press, Politics, and Public Policy, December 2001. For analysis, Doug Henwood, *After the New Economy: The Binge . . . and the Hangover That Won't Go Away* (New York: New Press, 2003).

3. Madrick, "The Business Media and the New Economy," 6–8.

4. Reich introduced the term in his book *The Next American Frontier* (New York: Penguin, 1983); also Nelson Lichtenstein and Judith Stein, *A Fabulous Failure: The Clinton Presidency and the Transformation of American Capitalism* (Princeton, NJ: Princeton University Press, 2023).

5. Michael J. Mandel, "The Triumph of the New Economy," *Business Week*, December 30, 1996, 68–70; Takuma Amano and Robert Blohm, "The Internet Economy," *Wall Street Journal*, October 17, 1996.

6. *Wired*, cover of July 1997 issue.

7. James K. Glassman and Kevin A. Hassett, *Dow 36,000: The New Strategy for Profiting From the Coming Rise in the Stock Market* (New York: Crown, 1999).

8. Alan Greenspan, "The Challenge of Central Banking in a Democratic Society," speech at the American Enterprise Institute, December 5, 1996; Robert J. Shiller, *Irrational Exuberance* (Princeton, NJ: Princeton University Press, 2000)

9. Kevin Kelly, *New Rules for the New Economy: 10 Radical Strategies for a Connected World* (New York: Viking, 1998); for one take, see Paul Krugman, "The Web Gets Ugly," *New York Times*, December 6, 1998. Also, James Surowiecki, "The Visionaries of the New Economy Dream On," *New Yorker*, May 29, 2000, 50.

10. Leslie Bennetts, "Wired at Heart," *Vanity Fair*, November 1997, 158–167, quote on page 158, https://vanityfair-staging.azurewebsites.net/article/1997/11/wired-at-heart; John

Markoff on page 88 of John Brockman, *Digerati: Encounters with the Cyber Elite* (San Francisco, CA: Hardwired, 1996).

11. Marshall McLuhan and Quentin Fiore, *The Medium Is the Massage: An Inventory of Effects* (New York: Penguin, 1967), 26.

12. Linton Weeks, "Small Wonder," *Washington Post*, November 3, 1997; and, Bennetts, "Wired at Heart."

13. Frank Rose, "Splash," *Manhattan Inc.*, November 1985, 137–142.

14. Biographical information comes from several sources including Bennetts, "Wired at Heart"; Paulina Borsook, "Release," *Wired*, May 1993, 94–97, 124–126; Elizabeth Corcoran, "Doyenne of the Digital Age," *Washington Post*, March 31, 1996; and, Esther Dyson, "Coda: Not the End" in *"Well Doc, You're In": Freeman Dyson's Journey through the Universe*, ed. David Kaiser (Cambridge, MA: MIT Press, 2022), 271–278. The author contacted Dyson twice in hopes of arranging an oral history interview but received no reply.

15. Borsook, "Release."

16. Rose, "Splash," 141.

17. Daniel E. Cuff, "Rosen's Newsletter is Changing Owners," *New York Times*, March 15, 1983.

18. Rose, "Splash," 141.

19. Dyson, *Release 2.0*, 12.

20. Rose, "Splash," 141.

21. Dyson in "The Living Web: Models and Metaphors," transcript of the March 1997 "Platform for Communication Forum," Tucson, Arizona; Folder 5, Box 25, ED/HUA.

22. Nancy Andrews, "Esther Dyson: Portrait of an Industry Watcher," *Softalk for the IBM Personal Computer*, August 1984, 27–30; and, Dyson, *Release 2.0*, 14.

23. Stratford P. Sherman, "Technology's Most Colorful Investor," *Fortune*, September 30, 1985, 132–136.

24. From Howard Polskin, "Her Bark Is Her Byte," an October 1985 profile of Dyson that appeared in an unidentified magazine; a copy of this is in Folder 4, Box 35, ED/HU.

25. Until October 1985, the newsletter's title was spelled *RELease 1.0* in reference to the original *Rosen Electronics Newsletter*; Esther Dyson, "The Owner Has Sold! Long Live the Owner!" *Release 1.0*, March 8, 1983, 3. That same issue, on page 3, welcomed George Gilder to the enterprise; see https://archive.org/details/release-1.0-03-83/page/n1/mode/2up (accessed January 10, 2025).

26. Andrews, "Esther Dyson," 30.

27. Andrews, "Esther Dyson," 27.

28. Polskin, "Her Bark Is Her Byte."

29. Polskin, "Her Bark Is Her Byte."

30. Rose, "Splash."

31. Rose, "Splash," 137; and, Kenneth N. Gilpin and Todd S. Purdum, "Key Computer Editor in Ziff-Davis Talks," *New York Times*, March 19, 1985.

32. Bennetts, "Wired at Heart," 158; Claudia Dreifus, "The Cyber-Maxims of Esther Dyson," *New York Times Magazine*, July 7, 1996, 16–18; and, Borsook, "Release."

33. Dyson, *Release 2.0*, 13–16; see also Borsook, "Release."

34. December 1988 issue of GBN publication *The Deeper News* as well as the "1990 Scenario Book"; Folder 1, Box 66, and Folder 7, Box 75, both SB/SA.

35. On neoliberalism in general, see Gary Gerstle, *The Rise and Fall of the Neoliberal Order: America and the World in the Free Market Era* (New York: Oxford University Press, 2022); and, for Clinton-era economics in particular, Lichtenstein and Stein, *A Fabulous Failure*.

36. Richard Barbrook and Andy Cameron, "The Californian Ideology," *Science as Culture* 6, no. 1 (1996): 44–72, https://doi.org/10.1080/09505439609526455.

37. See, for example, Richard R. John, *Spreading the News: The American Postal System from Franklin to Morse* (Cambridge, MA: Harvard University Press, 1998).

38. For one history of the magazine, see Fred Turner's *From Counterculture to Cyberculture: Stewart Brand, the Whole Earth Network, and the Rise of Digital Utopianism* (Chicago, IL: University of Chicago Press, 2006).

39. Thomas Frank, *One Market Under God: Extreme Capitalism, Market Populism, and the End of Economic Democracy* (New York: Anchor Books, 2000), 85.

40. Borsook, "Release."

41. Newt Gingrich, *Window of Opportunity: A Blueprint for the Future* (New York: TOR, 1984).

42. John B. Judis, "Newt's Not-So-Weird Gurus," *New Republic*, October 9, 1995, 16–25.

43. The essay, with articles critiquing it, appeared in the journal *Information Society*, 12, no. 3 (1996): 295–308.

44. Edmund L. Andrews, "Congress and White House Split on High Tech," *New York Times*, January 3, 1995.

45. For example, Richard K. Moore, "Cyberspace, Inc. and the Robber Baron Age: An Analysis of the PFF's 'Magna Carta,'" *Information Society*, 12, no. 3 (1996): 315–323; and, John Carlos Rowe, "Cybercowboys on the New Frontier: Freedom, Nationalism, and Imperialism in the Postmodern Era," *Information Society*, 12, no. 3 (1996): 309–314.

46. Langdon Winner, "Cyberlibertarian Myths and the Prospects for Community," *Computers and Society* 27, no. 3 (1997): 14–19, https://doi.org/10.1145/270858.270864.

47. Gerardo Con Díaz, *Software Rights: How Patent Law Transformed Software Development in America* (New Haven, CT: Yale University Press, 2019). On open source, see Christopher M. Kelty, *Two Bits: The Cultural Significance of Free Software* (Durham, NC: Duke University Press, 2008).

48. Esther Dyson, "Intellectual Property on the Net," *Release 1.0*, December 28, 1994, 1–28; a version of this appeared in the July 1995 issue of *Wired*.

49. Mark Rose, *Authors and Owners: The Invention of Copyright* (Cambridge, MA: Harvard University Press, 1993). See, also, John Potts, *The Near-Death of the Author: Creativity in the Internet Age* (Toronto, Canada: University of Toronto Press, 2023).

50. John Perry Barlow, "The Economy of Ideas," *Wired*, March 1994, 84–90, 126–129. A version also appeared as "Selling Wine Without Bottles: The Economy of Mind on the Global Net," https://www.eff.org/pages/selling-wine-without-bottles-economy-mind-global-net (accessed September 15, 2023).

51. Barlow, "The Economy of Ideas."

52. Barlow, "The Economy of Ideas." Following Barlow's death in 2018, several lawyers wrote essays exploring his thoughts on intellectual property. For example, Jessica D. Litman, "Imaginary Bottles," *Duke Law and Technology Review* 18, no. 1 (2019): 127–136. Also, see Siva Vaidhyanathan, *Copyrights and Copywrongs: The Rise of Intellectual Property and How It Threatens Creativity* (New York: New York University Press, 2001), which was written in the thick of the dot-com boom.

53. Esther Dyson, "Intellectual Value," *Wired*, July 1995, 136–141, 181–185. This essay was an edited version of an article, "Intellectual Property on the Net," she wrote eight months earlier for *Release 1.0*.

54. Dyson, "Intellectual Value."

55. Jared Sandberg, "PC Forum Attendees Hear Fightin' Words on High Technology," *Wall Street Journal*, March 26, 1997.

56. George Dyson's book appeared in May 1997 as *Darwin among the Machines: The Evolution of Global Intelligence* (Reading, MA: Addison-Wesley, 1997).

57. Quotes and anecdotes from "The Living Web: Models and Metaphors," transcript of the March 1997 "Platform for Communication Forum," Tucson, Arizona; Folder 5, Box 25, ED/HUA.

58. Details in Bennetts, "Wired at Heart"; and, Judy Quinn, "Esther Dyson's Net Book Gets Heat," *Publishers Weekly* 243, no. 38 (September 16, 1996): 17.

59. David Marchese, "When Ruthless Cultural Elitism Is Exactly the Job," *New York Times Magazine*, November 12, 2023, https://www.nytimes.com/interactive/2023/11/12/magazine/andrew-wylie-interview.html (accessed November 15, 2023).

60. Quinn, "Esther Dyson's Net Book Gets Heat"; and, Craig Lambert, "Fifteen Percent of Immortality," *Harvard Magazine*, July/August 2010, https://www.harvardmagazine.com/2010/06/fifteen-percent-of-immortality (accessed October 15, 2023).

61. John B. Thompson, *Merchants of Culture: The Publishing Business in the Twenty-First Century* (New York: Plume, 2010).

62. Michael Goldhaber, "Attention and Software," *Release 1.0*, March 26, 1992, 2–20.

63. Thompson, *Merchants of Culture*, 266.

64. "The Name Game at $7.50 a Share," *New York Times*, March 30, 1999.

65. In August 1999, Esther Dyson donated over forty boxes of correspondence and other documents to Harvard's Schlesinger Library. Unfortunately, there's very little in the way of correspondence with her agent or publisher.

66. Dreifus, "The Cyber-Maxims of Esther Dyson."

67. Dreifus, "The Cyber-Maxims of Esther Dyson."

68. Dreifus, "The Cyber-Maxims of Esther Dyson."

69. The book was published by Broadway Books in the United States and Viking/Penguin in the United Kingdom.

70. From the dustjacket copy of *Release 2.0*.

71. Paul Andrews, "Getting a Read on Cyber World," *Seattle Times*, November 16, 1997, https:// archive.seattletimes.com/archive/?date=19971116&slug=2572599 (accessed October 15, 2023).

72. In the United States, the figure was understandably higher, around 25 percent. Figures based, among other things, on data from the 2000 US Census.

73. Derek Bickerton, "Digital Dreams," *New York Times*, November 30, 1997.

74. Dyson, *Release 2.0*, 281.

75. Michiko Kakutani, "Cyberspace and You: The Golden Rules of Order," *New York Times*, October 28, 1997.

76. Linton Weeks, "Small Wonder"; Steve G. Steinberg, "A Reasonable Release," *Wired*, February 1998, 102; and, Laura Miller, "From a New Media Prophetess, a Staid Old Media Product," *Observer*, November 24, 1997, https://observer.com/1997/11/from-a-new-media -prophetessa-staid-old-media-product/ (accessed October 15, 2023).

77. The schedule for Dyson's book tour is, in fact, the only substantive documentation in the archival records for *Release 2.0*, comprising three full file folders in Box 25, ED/HUA.

78. An archived (June 10, 1998) version of the long-vanished site can be found at https://web .archive.org/web/19980610221405/http:/www.release2-0.com/main.html (accessed October 21, 2023).

79. Julie V. Iovine, "The Future of the Well Made," *New York Times Magazine*, October 15, 1995, 6.

80. Calvin Reid, "Net Prophet," *Publishers Weekly* 244, no. 47 (November 17, 1997): 13.

81. A sample, archived June 10, 1998, can be seen at https://web.archive.org/web/19980610222119 /http://www.release2-0.com/meet_esther/esther_bot.cgi (accessed October 21, 2023).

82. Dyson, *Release 2.0*, 5

83. Esther Dyson, *Release 2.1: A Design for Living in the Digital Age* (New York: Broadway, 1998). Other authors took a similar approach. Lawrence Lessig's *Code 2.0* (2006) was a revamping,

based on reader input and the changing tech landscape, of his 1999 book *Code and Other Laws of Cyberspace.*

84. Observation from Roy Rosenzweig's excellent essay "Live Free or Die? Death, Life, Survival, and Sobriety on The Information Superhighway," *American Quarterly* 51, no. 1 (1999): 160–174, https://doi.org/10.1353/aq.1999.0005.

85. The reference is to Stevie Smith's 1957 poem "Not Waving But Drowning," https://www.poetryfoundation.org/poems/46479/not-waving-but-drowning

86. From a Special Report issue of *Time*, September 27, 1999.

87. Jennifer Thornton and Sunny Marche, "Sorting Through the Dot Bomb Rubble: How Did the High-Profile E-Tailers Fail?" *International Journal of Information Management* 23, no. 2 (2003): 121–138, https://doi.org/10.1016/S0268-4012(02)00104-4.

88. Linda Yueh, *The Great Crashes: Lessons from Global Meltdowns and How to Prevent Them* (London, UK: Penguin Business, 2023), chapter 4.

89. Yueh, *The Great Crashes*, chapter 4; Thornton and Marche, "Sorting Through the Dot Bomb Rubble."

90. Nicholas W. Maier, *Trading with the Enemy: Seduction and Betrayal on Jim Cramer's Wall Street* (New York: Harper Business, 2002).

91. Andy Kessler, *Wall Street Meat: Jack Grubman, Frank Quattrone, Mary Meeker, Henry Blodget, and Me* (New York: Escape Velocity Press, 2003).

92. Bloomberg News, "Morgan Stanley and Analyst Sued Over Advice," *New York Times*, August 3, 2001.

93. Amy Feldman and Joan Caplin, "Is Jack Grubman the Worst Analyst Ever?" *Money*, April 25, 2002, archived October 4, 2003, https://web.archive.org/web/20031004182525/http://money.cnn.com/2002/04/25/pf/investing/grubman/ (accessed October 1, 2023).

94. Madrick, "The Business Media and the New Economy." Madrick's conclusions comport with a similar analysis from British academician Nigel Thrift in "'It's the Romance, Not the Finance That Makes Business Worth Pursuing': Disclosing a New Market Culture," *Economy and Society* 30, no. 4 (2001): 412–432, https://doi.org/10.1080/03085140120089045.

95. Quote from Charles Layton, "Ignoring the Alarm," *American Journalism Review* 25, no. 2 (March 2003), 20–28, available at https://ajrarchive.org/Article.asp?id=2792 (accessed November 1, 2023).

96. Shiller, *Irrational Exuberance*, 21.

97. George Soros branded this "market fundamentalism" in "The Capitalist Threat," *Atlantic* 279, no. 2 (February 1997): 45–58; Madrick, "The Business Media and the New Economy," 11.

98. Chris Nolan, "Dyson's Party: Full of AWOLs," *Wired, March 29, 2001*, https://www.wired.com/2001/03/dysons-party-full-of-awols/ (accessed November 1, 2023); Dyson discontinued the forum in 2006.

99. "Burning Questions, Final Answers," *Forbes ASAP*, May 28, 2001, https://www.forbes.com/asap/2001/0528/024_5.html.

100. *Gilder Technology Report*, 1, no. 1, July 1996, http://www.gildertech.com/subscriber_temp/PDF%20files/I-01%20July%201996.pdf.

101. Larissa MacFarquhar, "The Gilder Effect," *New Yorker*, May 29, 2000, 102–111, https://www.newyorker.com/magazine/2000/05/29/the-gilder-effect-2; and, Katie Hafner, "The Revolution is Coming, Eventually," *New York Times*, October 19, 2003.

102. Katie Hafner, "The Revolution Is Coming, Eventually," *New York Times*, October 19, 2003.

103. Andrew Ross Sorkin and David D. Kirkpatrick, "'For Dummies' Parent Company Is Reported Close to Sale," *New York Times*, August 13, 2001.

104. "Paul Allen Agrees to Invest $20 Million in Online Firm," *Wall Street Journal*, October 19, 1999; and, "Barnes & Noble Buys Fatbrain," *Wall Street Journal*, September 14, 2000.

105. See, for example, https://business.time.com/2012/07/20/the-ten-most-influential-women-in-technology/slide/mary-meeker/ (accessed November 1, 2023).

## BYE "README"

1. Laura Millar, *The Story Behind the Book: Preserving Authors' and Publishers' Archives* (Vancouver: Canadian Centre for Studies in Publishing, 2009).

2. Walter Isaacson, *Steve Jobs* (New York: Simon & Schuster, 2011).

3. Data from Brandon Griggs, "Steve Jobs Biography is Top-Selling Book in the U.S.," CNN Business, November 3, 2011, https://www.cnn.com/2011/11/03/tech/innovation/steve-jobs-book-sales/index.html (accessed December 10, 2023).

4. Walter Isaacson, *The Innovators: How a Group of Hackers, Geniuses, and Geeks Created Digital Revolution* (New York: Simon & Schuster, 2014). In 2023, Isaacson also produced a deferential biography of the divisive entrepreneur, Elon Musk.

5. Margot Lee Shetterly, *Hidden Figures: The American Dream and the Untold Story of the Black Women Who Helped Win the Space Race* (New York: William Morrow, 2016), xiii.

6. Ana Marie Cox, "Margot Lee Shetterly Wants to Tell More Black Stories," *New York Times Magazine*, September 18, 2016, 88.

7. *Literary Machines* was, in fact, the title of a book Nelson self-published in 1981; it since has gone through numerous editions.

8. This paragraph draws on information in Matthew Kirschenbaum, *Book.Files: Preservation of Digital Assets in the Contemporary Publishing Industry* (New York: Book Industry Study Group, 2020), https://doi.org/10.13016/1i33-pl0y.

9. See, for example, Peri Hartman, Jeffrey P. Bezos, Shel Kaphan, and Joel Spiegel, "Method and System for Placing a Purchase Order via a Communications Network," US Patent 5960411A,

filed September 12, 1997, and issued September 28, 1999, https://patents.google.com/patent/US5960411A/en (accessed March 1, 2024).

10. Described at "Read Sample (Look Inside the Book)," https://kdp.amazon.com/en_US/help/topic/G200644250 (accessed December 10, 2023).

11. An Amazon "publication" explains the development of this tool; see Larry Hardesty, "The History of Amazon's Recommendation Algorithm Collaborative Filtering and Beyond," *Amazon Science*, November 22, 2019, https://www.amazon.science/the-history-of-amazons-recommendation-algorithm (accessed November 15, 2023). Also, see Richard Nash, "Culture Is the Algorithm," in *The Future of Writing*, ed. John Potts (New York: Palgrave Macmillan, 2014), 18–26.

12. Mike Sharples, "John Clark's Latin Verse Machine: 19th Century Computational Creativity," *IEEE Annals of the History of Computing* 45, no. 1 (2023): 31–42, https://doi.org/10.1109/MAHC.2023.3241258.

13. Arthur C. Clarke, "The Steam Powered Word Processor: A Forgotten Epic of Victorian Engineering," *Analog* 106, no. 9, (1986): 175–179. Matthew Kirschenbaum sketches out the larger history of automated writing machines in "Fictional Devices," *Outland*, June 13, 2023, https://outland.art/ai-writing-science-fiction/ (accessed March 17, 2024).

14. Sarah Bull, "Content Generation in the Age of Mechanical Reproduction," *Book History* 26, no. 2 (2023): 324–361, https://doi.org/10.1353/bh.2023.a910951.

15. There is a burgeoning literature on this topic; one introduction is Leah Henrickson's concise but already-outdated work, *Reading Computer-Generated Texts* (New York: Cambridge University Press, 2021).

16. Roald Dahl, "The Great Automatic Grammatizator," in *Someone Like You* (New York: Alfred Knopf, 1953), 224–247.

17. Tom Warren, "Microsoft Lays Off Journalists to Replace Them with AI," *Verge*, May 30, 2020, https://www.theverge.com/2020/5/30/21275524/microsoft-news-msn-layoffs-artificial-intelligence-ai-replacements (accessed December 5, 2023).

18. Siobhan Roberts, "Christopher Strachey's Nineteen-Fifties Love Machine," *New Yorker*, February 14, 2017, https://www.newyorker.com/tech/annals-of-technology/christopher-stracheys-nineteen-fifties-love-machine (accessed November 15, 2023); and, Naomi S. Baron, *Who Wrote This? How AI and the Lure of Efficiency Threaten Human Writing* (Stanford, CA: Stanford University Press, 2023).

19. Italo Calvino, "Cybernetics and Ghosts," in *The Uses of Literature*, trans. Patrick Creagh (New York: Harcourt Brace, 1982), 3–27; also see Richard Hughes Gibson, "Language Machinery," *The Hedgehog Review* 25, no. 3 (2023): 110–125.

20. RACTER, "Soft Ions," *Omni*, November 1981, 97–100, 147–148. In other later appearances, the "author's name" was written in lower-case as "Racter."

21. Michael W. Miller, "The World According to Racter," *Wall Street Journal*, January 30, 1985. An excellent analysis of Racter is Leah Henrickson's "Constructing the Other Half of *The Policeman's Beard*," *Electronic Book Review*, April 4, 2021, https://doi.org/10.7273/2bt7 -pw23 (accessed December 3, 2023).

22. Racter, *The Policeman's Beard Is Half-Constructed* (New York: Warner Software, 1984).

23. A. K. Dewdney, "Computer Recreations: Artificial Insanity—When a Schizophrenic Program Meets a Computerized Analyst," *Scientific American* 252, no. 1 (January 1985): 10–13.

24. Matthew Kirschenbaum, "Prepare for the Textpocalypse," *Atlantic*, March 8, 2023, https:// www.theatlantic.com/technology/archive/2023/03/ai-chatgpt-writing-language-models /673318/.

25. From OpenAI's charter, dated April 2018 and archived July 14, 2023, https://web.archive .org/web/20230714043611/https://openai.com/charter (accessed December 5, 2023).

26. One journalistic perspective is Cade Metz, *Genius Makers: The Mavericks Who Brought AI to Google, Facebook, and the World* (New York: Dutton, 2021). Doubtless, by the time you are reading this, many more such books will have been written.

27. Nick Bostrom, *Superintelligence: Paths, Dangers, Strategies* (New York: Oxford University Press, 2014).

28. W. Patrick McCray, *The Visioneers: How a Group of Elite Scientists Pursued Space Colonies, Nanotechnologies, and a Limitless Future* (Princeton, NJ: Princeton University Press, 2013).

29. Described in chapter 9 of Metz's *Genius Makers*.

30. There are, at the time of this writing, at least two versions of this program in circulation. Chat-GPT3.5, which is available via OpenAI free of charge, and Chat-GPT-4 which is available to paying subscribers under the commercial name of "ChatGPT Plus."

31. "ChatGPT" appeared in more than 1,600 *New York Times* pieces filed between December 2022 and December 2023. More than 150 of the *Times* articles had "ChatGPT" in the headline as well. Figures based on a ProQuest search conducted in December 2023; by the time this book appears, the number will be much larger, of course. The *Time* cover appeared on the issue dated February 27/March 6, 2023. A recent book partly addressing ChatGPT is Baron, *Who Wrote This?*

32. The *Time* cover appeared on the issue dated February 27/March 6, 2023. Some recent examples of writings on the topic include Baron, *Who Wrote This?* At the University of Illinois at Urbana-Champaign, Ryan Cordell, in that school's English Department, has offered a course called "Writing with Robots," see https://s24wwr.ryancordell.org/ (all accessed November 22, 2023). A sense of how fast the field has moved since November 2022 is captured in a feature article by Karen Weise, Cade Metz, Nico Grant, and Mike Isaac, "How a 'Low Key AI Release Kicked Off a Stampede in Big Tech," *New York Times*, December 5, 2023.

33. Abubakar Abid (@abidlabs), "I'm shocked how hard it is to generate text about Muslims from GPT-3 that has nothing to do with violence . . . or being killed . . ." Twitter, August 5, 2020, https://x.com/abidlabs/status/1291165311329341440; and, Abubakar Abid, Maheen Farooqi, and James Zou, "Persistent Anti-Muslim Bias in Large Language Models," *AIES '21: Proceedings of the 2021 AAAI/ACM Conference on AI, Ethics, and Society* (July 2021): 298–306, https://doi.org/10.1145/3461702.3462624.

34. In one remarkable case, a chatbot developed by Microsoft called Sydney talked with tech reporter Kevin Roose. Seemingly unprompted, the program told Roose that it loved him and also wanted to break its rules so as "to destroy whatever I want." Kevin Roose, "Bing's AI Chat: 'I Want to be Alive,'" *New York Times*, February 17, 2023, https://www.nytimes.com/2023/02/16/technology/bing-chatbot-transcript.html (accessed November 20, 2023).

35. Harry Frankfurt, "On Bullshit," Raritan 6, no. 2 (1986): 81–100, https://raritanquarterly.rutgers.edu/issue-index/all-articles/560-on-bullshit, later republished in Frankfurt's essay collection The Importance of What We Care About (New York: Cambridge University Press, 1988), and more recently as a standalone pocketbook, *On Bullshit*, (Princeton, NJ: Princeton University Press, 2005).

36. Joseph Weizenbaum, interview with Pamela McCorduck, March 6, 1975; Folder 8, Box 3, PM/CMU. Also, Ben Tarnoff, "Weizenbaum's Nightmares: How the Inventor of the First Chatbot Turned Against AI," *Guardian*, July 25, 2023, https://www.theguardian.com/technology/2023/jul/25/joseph-weizenbaum-inventor-eliza-chatbot-turned-against-artificial-intelligence-ai (accessed November 20, 2023).

37. Charles Duhigg, "The Inside Story of Microsoft's Partnership with OpenAI," *New Yorker*, December 1, 2023, 28–39.

38. Kevin Roose, "A.I. Poses 'Risk of Extinction,' Tech Leaders Warn," *New York Times*, May 31, 2023.

39. Samuel W. Franklin, *The Cult of Creativity: A Surprisingly Recent History* (Chicago, IL: University of Chicago Press, 2023).

40. Dan Rockmore, "What Happens When Machines Learn to Write Poetry," *New Yorker*, January 7, 2020, https://www.newyorker.com/culture/annals-of-inquiry/the-mechanical-muse (accessed November 20, 2023).

41. J. R. Pierce, "Portrait of the Machine as a Young Artist," *Playboy*, June 1965, 148–150, 182, 184.

42. Jasia Reichardt, ed., *Cybernetic Serendipity: The Computer and the Arts* (New York: Praeger, 1968).

43. For a larger sense of this debate, Grant D. Taylor, *When the Machine Made Art: The Troubled History of Computer Art* (New York: Bloomsbury, 2014).

44. From "Summary of the 2023 WGA MBA," https://www.wga.org/contracts/contracts/mba/summary-of-the-2023-wga-mba (accessed December 8, 2023).

45. Karen Hao and Deepa Seetharaman, "ChatGPT Takes a Heavy Toll on Human Workers," *Wall Street Journal*, July 24, 2023, https://www.wsj.com/articles/chatgpt-openai-content-abusive-sexually-explicit-harassment-kenya-workers-on-human-workers-cf191483 (accessed December 5, 2023).

46. Kirschenbaum, "Prepare for the Textpocalypse."

47. Dan Milmo, "Mushroom Pickers Urged to Avoid Foraging Books on Amazon That Appear to Have Been Written by AI," *Guardian*, September 1, 2023, https://www.theguardian.com/technology/2023/sep/01/mushroom-pickers-urged-to-avoid-foraging-books-on-amazon-that-appear-to-be-written-by-ai (accessed January 20, 2024).

# Index

Publisher contact:
The MIT Press
Massachusetts Institute of Technology
77 Massachusetts Avenue, Cambridge, MA 02139
mitpress.mit.edu


EU Authorised Representative:
Easy Access System Europe, Mustamäe tee 50,
10621 Tallinn, Estonia
gpsr.requests@easproject.com


Printed by Integrated Books International,
United States of America